CLINICAL MECHANICS OF THE HAND

SECOND EDITION

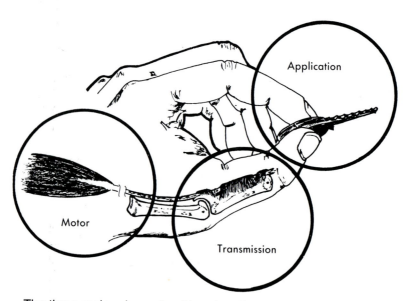

The three major elements of hand motion: motor, transmission, and application.

Clinical Mechanics of the Hand

Second Edition

Paul W. Brand, M.B., B.S., F.R.C.S.
Clinical Professor Emeritus
Department of Orthopaedics
University of Washington
Seattle, Washington

Anne Hollister, M.D.
Assistant Professor
Department of Orthopaedic Surgery
University of California at San Francisco
San Francisco, California

*To Michael Freeman
c̄ highest regard

anne Hollister*

Mosby Year Book

St. Louis   Baltimore   Boston   Chicago   London   Philadelphia   Sydney   Toronto

**Mosby
Year Book**

Dedicated to Publishing Excellence

Sponsoring Editor: James D. Ryan
Associate Managing Editor, Manuscript Services: Deborah Thorp
Production Supervisor: Karen Halm
Proofroom Manager: Barbara Kelly

1 2 3 4 5 6 7 8 9 0   CL/MV 97 96 95 94 93

Library of Congress Cataloging-in-Publication Data
Brand, Paul W.
    Clinical mechanics of the hand / Paul W. Brand, Anne Hollister.—
2nd ed.
        p.   cm.
    Includes bibliographical references and index.
    ISBN 0-8016-6978-2
    1. Hand—Mechanical properties.   2. Hand—Surgery.   I. Hollister,
Anne.   II. Title.
    [DNLM: 1. Biomechanics.   2. Hand—physiology.   3. Hand—surgery.
WE 830 B817c]
RD559.B73   1992
DNLM/DLC   92-48783
for Library of Congress   CIP

Acknowledgments

Almost all that is new in this book has been worked out in the Biomechanics Laboratories of the Rehabilitation Research Department of the National Hansen's Diseases Center, United States Public Health Service at Carville, Louisiana. Some of the research has been funded by what is now the National Institutes of Handicapped Research and most of the rest by the United States Public Health Service.

Much of the inspiration and most of the hard work have come from the staff of our Rehabilitation Research Laboratory, and from our collaborating engineers from Louisiana State University and also from a succession of fellows, paid and unpaid, who have been part of the team from time to time.

The cover of the first edition carried only the name of Dr. Paul Brand, because he had done all the writing. The present edition includes other authors, and one of them, Dr. Anne Hollister, has now joined Dr. Brand as co-editor.

ENGINEERS

David E. Thompson, Ph.D., Professor of Mechanical Engineering, Louisiana State University, has been our chief bioengineering consultant for many years. He has been working with us under contract with a variety of other L.S.U. staff members and students. He has authored and co-authored Chapters 6 and 14.

William L. Buford, Ph.D., Biomedical Engineer, has worked full time on our own staff and has been responsible for the conceptual and design work of the computer applications and biomedical devices that we use.

David Giurintano M.S., M.E., is a full-time member of our staff and has worked on the mechanical aspects of joint function including the modeling of joint mechanics and kinematics.

SURGEONS

Many have come on short-term fellowships or on a project basis and have contributed new ideas of their own.

Dr. John Agee, worked in our research laboratory at his own expense for many months, during which time he designed and built much of the apparatus we now use for moment arm studies on fresh cadaver arms. He remains a valuable consultant from his own biomechanics laboratory in Sacramento and has contributed a new chapter on the wrist.

Dr. Ronnie Matthews, Associate Professor of Orthopaedic Surgery, Louisiana State University, is the hand surgeon at Carville and the Chief of Hand Surgery at Earl K. Long Hospital, Baton Rouge.

Dr. Anne Hollister, while still an orthopaedic resident at Tulane University, New Orleans (by arrangement with the late Dr. R.J. Haddad, Professor and Chairman Department of Orthopaedic Surgery, Tulane) has contributed greatly of her own time and energy and remarkable anatomical mechanical insights in the hand. Then, as Chief of

Hand Surgery at Harbor UCLA Medical Center, she continued to maintain close liaison with the staff at Carville, and to actually initiate and lead aspects of the research by frequent visits from across the continent. A product of this fruitful collaboration has been the new chapter, "How Joints Move."

Dr. Alvin Freehafer is the only contributor who has not worked with us at Carville. He has independently developed similar ideas about the mechanics of tendon transfers and has integrated them into his practice of hand surgery. His article, written with Dr. Peckham, is now Chapter 13.

Dr. Lynn Ketchum, after working with us an an unpaid associate for some months, returned to continue similar studies in Kansas City. His special interest was in the analysis of muscle strength in the intact living hand. His contribution, with Dr. Thompson, is now Chapter 14.

We also appreciate the advice and help of **Dr. Dan Riordan,** the late **Dr. Richard Smith,** and **Dr. George Omer,** who have read parts or all of the manuscript and given valuable advice.

THERAPISTS

Our own staff therapists have contributed a very great deal.

Robert Beach, P.T., former head of the research department, has detailed knowledge of the mechanics of the muscles of the forearm and hand that must exceed that of most hand surgeons.

Judy Bell, O.T., Chief of Hand Therapy and Occupational Therapy, works in both research and clinical care. Judy is best known for her leadership in the theory and practice of sensory testing.

Wim Brandsma, consultant and physical therapist from Holland and Ethiopia, has added his experience and ingenuity to our studies and now has been most helpful in the revision of the previous edition.

SUPPORT STAFF

Carol Langlois, Research Health Technician, has the longest tenure in the department and has competence with its rabbits, rats, machines, and staff that is matched only by her ability to collect bibliography and create order out of chaos, especially in the writing of books and articles.

Jerry Simmons, at first on deputation from the Mechanical Engineering Department at Louisiana State University, and now on our own staff, has proved to be not only our computer programmer in residence but also an excellent diagram artist. Most of the illustrations in this book have come from his pen.

The artisitic drawings of muscles in Chapter 12 are by **Don Alvarado,** previously Associate Professor of Medical Illustrations, Louisiana State University Medical School, New Orleans, and now resident of West Branch, Iowa.

Preface to the Second Edition

The primary purpose of this book has not changed. There has been a temptation to become more sophisticated in an engineering sense, but our lack of sophistication in the first edition was not for lack of knowledge, but for the sake of the readers that we wanted to attract. We were writing, and we still are writing, for the surgeons and therapists who want to understand the mechanical basis for the working of the hand, but have never mastered the terminology and symbols of engineering.

The first edition had only one author. The second has several, all of whom contributed to the ideas and work behind the first edition and who now have written or revised sections under their own names. Two of the new authors are real engineers. Dr. David Thompson, who was active in the earliest stages of our program, and who has now revised and extended the old chapter we called "Drag", bringing its terminology more into line with acceptable mechanical usage, while still avoiding the formulae and jargon of his natural sophistication. David Giurintano is the engineer who has worked with Dr. Anne Hollister, providing the mathematics and interactive three-dimensional computer imaging that have helped to establish Dr. Hollister's new insights into the way joints move and how they are moved by muscles and tendons.

Dr. Agee, who worked with us at Carville for many months, and whose ideas have infiltrated the philosophy of all our work, has now contributed a new chapter on some of the mechanical features of the control of the wrist.

The biggest change is that Dr. Anne Hollister, who has worked frequently in the research laboratories at Carville, and who attributes much of her early interest in biomechanics to what she learned there, has now become co-editor of the book. Since Dr. Brand's retirement, Dr. Hollister, from her position as Chief of Hand Surgery at Harbor UCLA Medical Center, has maintained active liaison with Carville. She has been responsible for much of the overall revision of the book, as well as for the writing of the new, outstanding chapter, "How Joints Move."

I have now retired from active surgery, although not from involvement in teaching. I have moved to Seattle where I serve as Clinical Professor, Emeritus, in the Department of Orthopaedics at the University of Washington. I have maintained overall guidance over the direction of the book, and have done some of the work of revision, extending and updating sections I wrote in the first edition. As a grandfather in a biological sense, so also in the field of ideas, I observe with great pride that the next generation has not been content just to maintain a tradition. They have enhanced it!

Paul W. Brand, M.B., B.S., F.R.C.S.

Preface to the First Edition

This book is written by a surgeon for surgeons and for therapists. It has a well-defined object. It is not to serve as a primer of mechanics or bioengineering. It is to help surgeons and therapists to use sound mechanical principles as they operate on the hand and as they plan its conservative management.

If this book is to fulfill its primary objective, it must succeed in a preliminary objective. It must be read by the surgeons and therapists for whom it is written. These are not the ones who ordinarily read articles on bioengineering. My personal experience in teaching has convinced me that the majority of surgeons who operate on the hand simply pass over the excellent articles on biomechanics that appear from time to time in journals of hand surgery. These surgeons are turned off by complex formulae and by terminology and symbols they do not understand. It seems to them that they would have to study calculus and trigonometry and engineering to make sense of the material that is presented.

Therefore I will not use engineering or mathematical terminology or symbols that are not familiar to everybody. I will try to convince my colleagues that the practical mechanical principles that are so important in hand surgery and therapy are as simple and down to earth as those involved in changing a tire, sailing a boat, or building a crane with an erector set. I will use the language of the home hobbyist and of the backyard mechanic.

I am afraid that this will offend my friends in bioengineering, but to them I will say that my hope is that at least some of my readers will become so interested in the mechanics of the hand that they will then gladly learn the language and discipline that will open up to them the real literature of this fascinating subject. Think of this book as bait. If I can catch them, then you can train them.

Missing from this book are those aspects of biomechanics that have had most of the attention in other writings. There is nothing here on bones or on the shape or configuration of joints. The need to design joint replacements has attracted some of the best minds in bioengineering. Hard tissue mechanics has been better documented than the mechanics of soft tissues.

Also missing is a full account of the extensor apparatus of the fingers. This has been well covered conceptually by Landsmeer and others, and our own attempts to put numbers to the various elements of the whole interacting tube of tendon fibers around the fingers is still incomplete and must await another publication.

What we have tried to do is to bring together all of the basic elements of the forces that result in movement of the hand and those that limit or hinder movement. We have tried to find ways to quantify them in a practical way, so that surgeons and therapists can judge their own measure of success and failure as they try to restore to damaged hands that flexibility, strength, sensibility, and beauty that are normal to the human hand.

Paul W. Brand, M.S., B.S., F.R.C.S.

Contents

CHAPTER 1

Terminology

Although we shall try not to introduce unfamiliar terms, we do want to be precise about the terms we use. Sometimes a word that is loosely used in conversation may have a precise meaning in mechanics.

AXIS

When two bones move around each other at a joint there is usually one line that does not move in relation to either bone (Fig 1–1). This is the axis. In the hand it often lies at about the center of the convex head of the more proximal bone. If a joint is capable of more than one kind of movement, there will be an axis for each. For example, a transverse axis for flexion-extension, a vertical axis for abduction-adduction, and maybe a longitudinal axis for rotation. (See also Chapter 4, How Joints Move.)

STRESS

Stress is a force per unit area. If the force is normal (perpendicular) to a surface, it is called a *normal stress* or alternatively, a *pressure*. If a force is applied parallel to a surface, it is termed a *shear stress*.

STRAIN

The nondimensional deformation of an object is defined as the ratio of a change in some physical length to the original length. If one stretches a rubber band, its strain is calculated as the change in its length divided by the unstretched length of the rubber band.

TENSION STRESS

Tension stress occurs typically in a tendon. If a tendon is fixed at one end and is hanging with a weight of 1 kg on its free end (Fig 1–2), and if this is at sea level, then the tendon is subjected to a tensile stress of 1 kg force. When we speak loosely of a tension of 1 kg, we mean kilogram force as defined above.

COMPRESSION STRESS AND PRESSURE

Compression stress and pressure may be defined as force per unit area. When 1 kg force presses on a surface at right angles to that surface, through an area of 1 cm^2, the pressure is 1 kg/cm^2. (Fig 1–3) The word *normal* in relation to force means "perpendicular to." So pressure is caused by a force that is normal to a surface.

SHEAR STRESS

Shear stress is caused by forces that are parallel to a surface. It occurs in tissue that

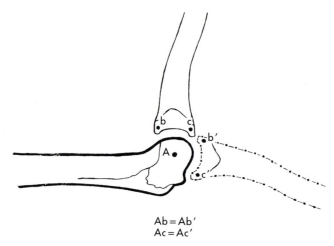

$$Ab = Ab'$$
$$Ac = Ac'$$

FIG 1–1.
A is the axis, which does not move in relation to either bone, no matter what the position of the joint. The axis, seen here in cross section as a point, is actually a line at right angles to the plane of movement of the bones.

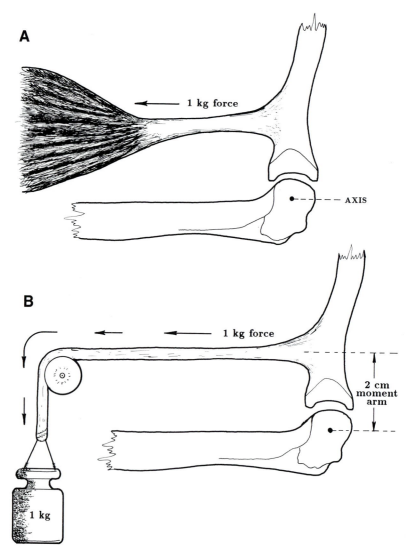

FIG 1–2.
A, 1 kg of weight subject to gravity at sea level results in 1 kg force on the rope or tendon that supports it. In this book, when we use the term *1 kg of tension,* we mean 1 kg force as defined above. **B,** in this diagram, the joint is subject to a moment or torque of 1 kg × 2 cm = 2 kg · cm of moment.

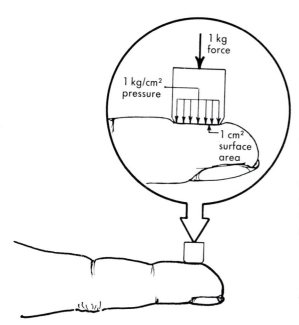

FIG 1–3.
The same force will cause lower pressure if it is applied through a larger area.

is subjected to two opposing forces that are not exactly in line. The two blades of scissors cut paper by shear stress. A carpet is ruckled under moving chairs by shear stress between the chair and the floor (Fig 1–4). Shear is difficult to measure but may be inferred either by the position and direction of

known forces or by the effect on tissues. Friction and shear tend to be more damaging to tissues than normal stress, as when a loose shoe rubs a blister on a heel.

FORCE

Force is defined as that which will cause acceleration. However, for our purposes we shall think mainly of the effect of force as it causes stress in bones, joints, and soft tissues.

FRICTION

The resisting force parallel to and resulting from direct contact between two surfaces.

STATIC FRICTION

A frictional force produced when the two surfaces do not move relative to one another. This force is usually proportional to that component of the force between the

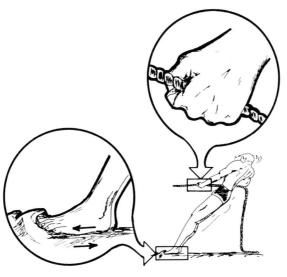

FIG 1–4.
Friction and shear stress. If the rope is held loosely, it will slip and cause friction effects (perhaps a burn). If it is held tightly (using more pressure), the friction between hand and rope will increase and prevent slippage, but this will cause shear stress within the tissues of the hand, between skin and bone.

two surfaces that is perpendicular to the two surfaces at the point of contact.

MOVING FRICTION (COULOMB FRICTION)

A frictional force resulting between two surfaces which move relative to one another. This is often also termed *dry friction* because of the absence of fluids between the two surfaces which would enable them to glide past each other.

RADIAN

We include this term only because it has one very useful application in hand surgery.

A radian is a unit of angular measurement and happens to be about 57.29 degrees. The reason it is useful is its definition. On the circumference of a circle, if you measure and mark a distance equal to the radius of the circle, and then join the two ends of that line to the center, you will have an angle of 1 radian between the 2 radii (Fig 1–5,A). To put it another way, if a radius rotates through a radian, it will have moved a distance of 1 radius along the circumference in doing so. Moreover, every point on any lever that rotates through a radian will have moved exactly the distance along its arc that it is from the axis (Fig 1–5,B). So, if a person sits on a seesaw 1.5 m from the axis and it tilts up and down through 1 radian, he will move 1.5 m up and down. If he sits 2 m from the axis, then he will move 2 m up and down for the same angle. This is important

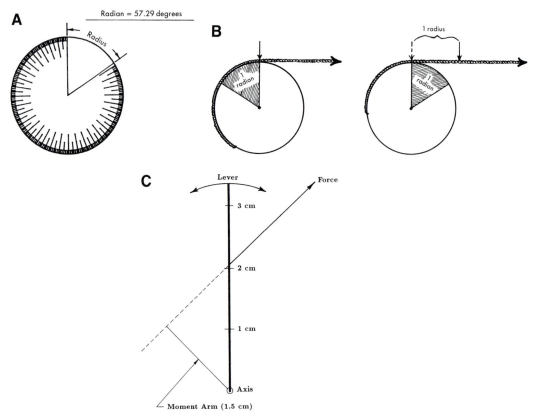

FIG 1–5.
A, one radius, laid along the circumference of the circle subtends a *radian* at the center. **B,** the circle may represent a pulley wheel and a rope, or it may represent the convex articular surface of a joint with a tendon attached to the other bone. When the wheel (or joint) turns through a radian, the rope or tendon *excursion* is equal to its own perpendicular distance from the axis at the wheel (or joint). **C,** tension is applied to the lever at 2 cm from the axis but the *line* of action of the tension is not perpendicular at that point, it is oblique. The *dashed line* projected from that force crosses closest to the axis at 1.5 cm from it. That is the *moment arm*.

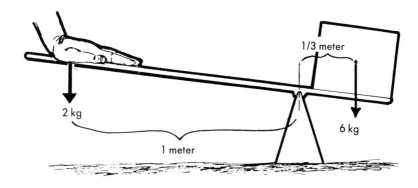

Mechanical advantage of arm = 3:1
Moment of hand = 2 kg M
Moment of weight = 2 kg M

FIG 1–6.
Because the hand is three times as far from the axis as the weight, it has a 3:1 mechanical advantage. It can therefore use one-third less force to balance the weight, but will move three times as far as the weight does.

to hand surgery and splinting because we usually cannot actually see and identify the axis of any joint, but we can measure angular movement and we can measure excursions, so we infer the position of the axis and the relative position of a tendon by measuring the excursion of the tendon while the joint moves through 1 radian.

However, many forces that move joints are not perpendicular to the bone (or lever). Engineers may use trigonometric formulas to calculate the moment. We suggest it is usually simpler to project the line of the force until it crosses closest to the axis. Then the perpendicular distance may be measured directly (Fig 1–5,C).

LEVERAGE, TORQUE, AND MOMENTS

When a bone moves around at a joint, we may think of it as a lever moving around an axis. Every point on the lever describes a segment of a circle with the axis as center.

Any force that acts on the lever tends to make it move one way or the other. The effect of a force on the circular or rotational movement of a lever or bone is called the *moment* of that force. It is also called *torque*.

Moment is measured by multiplying the force by its perpendicular distance from the axis of the lever. So 1 kg force at 10 cm from the axis will have the same moment (the same torque) as 0.5 kg force at 20 cm if both are perpendicular to the lever.

The perpendicular distance from the axis may be called the lever arm or the *moment arm* so

Force × moment arm = moment
2 kg force × 2 cm = 4 kg force · cm

MECHANICAL ADVANTAGE

When one force, acting on a lever, has double the moment arm of an opposing force (or load), it is said to have a mechanical advantage of 2:1 (Fig 1–6). In general, it may be said that mechanical advantage is a ratio of

$$\frac{\text{Moment arm of force}}{\text{Moment arm of load}}$$

which is the same as

$$\frac{\text{Distance moved by force}}{\text{Distance moved by load (or object)}}$$

ELASTICITY

The property of a material that resists any change in its shape and tends to restore its shape following the removal of all forces.

Perfectly elastic materials will always return to their exact original shape following the removal of all loads. Elasticity is in effect whether the loads are tensile, as in a stretched rubber band, or compressive, as in a rubber ball. Thus, any tissue which hinders the movement of a joint or tendon and which tends to return the tissue toward its initial state has some measure of elasticity.

materials continuously deform when loaded and have no restoring force. Thus viscous effects do not prevent motion, they merely retard it. One might imagine the retarding force of moving one's hand through water or syrup. The latter material would offer the greatest resistance and would have the greater viscosity.

VISCOSITY

The property of a material which causes it to resist motion in an amount proportional to the rate of deformation. Purely viscous

VISCOELASTICITY

A combination of both viscous and elastic material properties. One important quality of a viscoelastic material is a variation between the loading and unloading load-de-

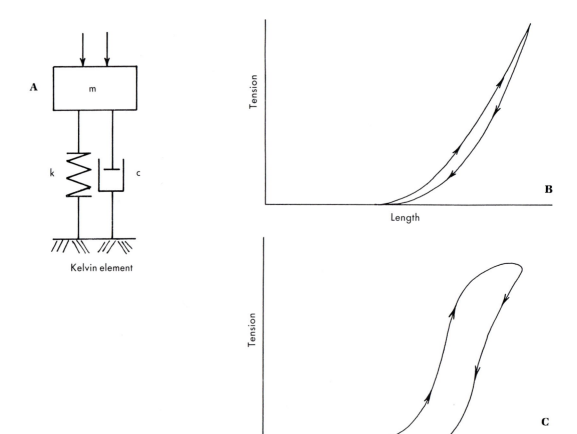

FIG 1–7.
A, *k* is a spring, *c* is a shock absorber, and *m* is a mass. The two elements, in parallel, give an automobile a smooth ride and also give firm compliance to the palm of the hand. **B,** when tension is slowly applied to this material and then slowly relaxed, the length-tension curve does not follow the same path on the way back. This is a *hysteresis loop.* **C,** if excessive tension is applied to a material, a permanent deformation, or *plastic change,* may occur at the highest tension. The relaxation curve will then follow a different path and will never return to its original zero.

flection curve. This difference is termed *hysteresis* and its presence denotes both elastic and viscous components. The model of a viscoelastic material is the Kelvin model with a spring and a dashpot standing side by side (Fig 1–7,A).

HYSTERESIS

An elastic material, when it has been stretched or compressed, will spring back to its original shape when the stress is removed. If it returns only slowly, perhaps because of viscosity, the length-tension diagram will be different on the way back. This is hysteresis, and the two curves—the way out and the way back—form a loop (hysteresis loop) (Fig 1–7,B).

WORK

The energy required to move a force through a distance. Only that component of the force in the direction of the motion produces or requires work. In purely elastic systems, all of the work input into the system to achieve a displacement is returned when the displacement reverses. In purely viscous systems, all of the energy input is dissipated or lost. Viscoelastic systems, demonstrating a combination of the properties of the two effects, allow some energy recovery following displacement but also suffer viscous losses. These effects are discussed at more length in Chapter 6, Mechanical Resistance.

POWER

The time rate of doing work.

PLASTIC CHANGE

If a material with elastic qualities is stretched beyond its elastic limit, it may be permanently deformed. It will never return fully to its original shape. This is called *plastic change* (Fig 1–7,C).

CREEP

There are certain biologic materials with elastic qualities that, if stretched and held under moderate tension for a long time, will gradually change their structure and remain in a lengthened state after the tension is released. This is called *creep* (see Chapter 6).

VECTOR

A vector diagram is a means of analyzing a force or a movement into convenient component parts. It may also be used to find the resultant of two or more forces or movements. For resolving two forces, it is done by completing a rectangle or a parallelogram of which the forces form two sides. The diagonal is then the resultant.

If an object is being pulled north by a force of 2 kg and pulled east by another force of 2 kg, a vector diagram will show that in effect it is being pulled northeast by a force that can be measured on the diagram (Fig 1–8).

In the same way, if a point moves 10 cm north and then 20 cm east, a diagram will show the final resultant position of the point and the direction it lies from its starting point.

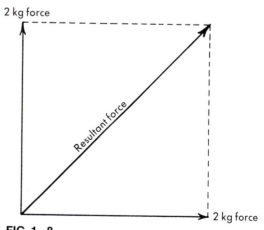

FIG 1–8.
Vector diagram to resolve the resultant effect of two known forces.

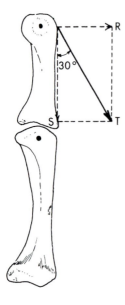

FIG 1−9.
A tendon at 30 degrees to a bone may exert a known tension *(T)*. This may be resolved into two parts by a vector diagram that shows that it is in part a rotator *(R)* of the joint and in greater part a stabilizer *(S)*, by its vector along the length of the bone tending to impact the two bones at the joint.

In application to joints and tendons, a tendon at a known tension may approach a bone at an angle of 30 degrees. If a vector diagram is drawn with the known force as a diagonal of a rectangle with one side along the bone and one side at right angles to the bone, it is easy to see (and measure) how much force is available for moving the bone transversely (or turning it around a joint) and how much remains for stabilizing the joint, by thrusting along the bone (Fig 1−9).

DRAG

The forces or shear stresses resulting from the action of a fluid on a surface. When two moving surfaces are separated by a thin layer of fluid, the resulting forces are not considered to be frictional but rather effects resulting from surface-fluid shear. In severe cases of arthritis, however, even the presence of synovial fluid in joints does not prevent the surfaces from direct contact owing to the size of the defects on the articular surfaces which interdigitate and approach or exceed coulomb friction potential.

YOKE OR BRIDLE

A device to link two structures together, so that they may share a common source of strength or a common effect from one or more sources, is called a *yoke* or *bridle* (Fig 1−10).

LENGTH-TENSION CURVES (STRESS-STRAIN CURVES)

Many of the graphs in this book relate to the relationship of tension to length. In all such cases the tension is on the vertical scale (ordinate) of the diagram and the length is on the horizontal scale (abscissa). In the case of active contraction of a muscle the tension

FIG 1−10.
A yoke is still used in many parts of the world to permit one force to be used to affect two loads. In hand surgery, the term *yoke* is sometimes used to speak of one muscle that is made to affect two different joints or two sides of one joint.

may be produced by the muscle. The change of length is active. In the case of passive tissues the tension is imposed on the tissue and the change of length is passive. This is true also of muscle, which may be passively stretched while not actively contracting.

TORQUE-ANGLE CURVES

When it is not possible to measure the change in length of a muscle or other tissue, because it is part of an intact limb, we make a comparable evaluation by applying torque to a joint and measuring the change of angle produced by the torque. Every tissue that crosses the joint is affected by that change, and a record of changing torque angles gives a good idea of changing tissue length inside the limb. A full account of torque-angle curves is given in Chapter 6.

VOLAR

In this book volar and palmar are used interchangeably.

CHAPTER 2

Overview of Mechanics of the Hand

The hand is more than a machine. In writing a book about the mechanics of the hand, we have to debase a thing of exquisite beauty and sensitivity to compare it with the crude and callous things that we call machines.

A number of mechanical hands are used in industry. They are used to handle radioactive fuels, to apply very large forces to gross materials, and sometimes to undertake repetitive tasks that would be boring to a human. We may admire the ingenuity that designs an artificial hand, but to the amputee who tries to use a prosthetic hand even a single finger of a real hand would be preferred.

Let us imagine that it is 1976, and we have put an engineer in the front row of a concert by Arthur Rubinstein. The ten fingers of that 90-year-old man are flying over the keys of the piano. Each digit is doing a different job, each using different and precise pressure and timing, each feeding back to the brain a moment-by-moment account of its environmental pressures and joint positions. We remind the engineer that none of the bearings or parts of those hands have been serviced for 90 years. The lubrication system that was built 90 years ago is still functional and those same fingers can control a screwdriver or a wrench and can work under water, in the dark, in snowstorms, and in tropical heat. They can heal their own injuries, can detect, by touch, etched lines on glass 0.0001 in. deep, can handle rough rocks and stroke a baby to sleep. The engineer looks at his own hands and knows that he will never see a machine that can compare with them.

But hands do get injured and diseased beyond their capacity to repair themselves, so we have to try, in our crude way, to serve the injured hand and assist it to recover some of its lost ability. Even in this task we defer to the wisdom of the healing cells and seek only to give them the best environment in which to exercise their own skill.

In this task of helping an injured hand to repair itself, we have to understand the mechanical principles that made it work in the first place, and the modified mechanics of any new joint or tendon placement that we can offer. So we still have to think of the hand as a machine; now we shall find some marked differences and a few limitations that we have to get used to.

We shall be considering three major aspects of the hand machine: (1) Muscles—the motor; (2) the transmission—tendons, bones, and joints; and (3) the application through skin and pulp tissues to the objects the hand has to control. Along with these, we shall consider the hindrances to movement—friction and other forms of internal resistance. The three major aspects may be seen in Figure 2–1, which brings out some factors we discuss later.

THE MOTOR

An industrial machine is based on wheels. Wheels can go round and round independently and thus can pull on cables that keep the same tension over long excursions. There is no wheel in the human hand.

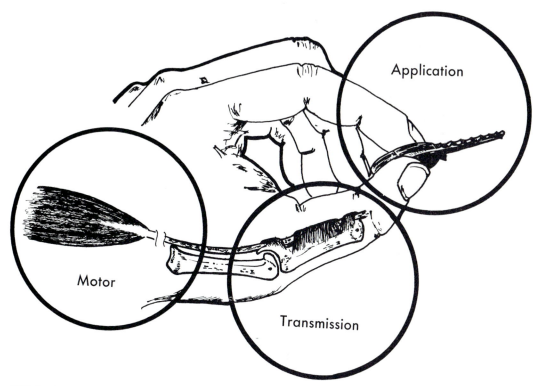

FIG 2–1.
The three major elements of hand motion, controlled and moderated by motor nerves and by sensation.

There are sections of wheels in joints, but a free wheel that goes full circle is not possible if nourishment of all parts of the hand by the bloodstream is to be maintained. Muscles are the motors of the hand. They are like the piston and cylinder units of a two-stroke gasoline motor in that the power stroke of one piston (muscle) provides the energy for a return of the other one of the pair. In the engine the power is on the push cycle while the return is passive. In a pair of muscles the power is in the pull stroke (or tension) while the lengthening is passive. The superiority of the muscle system is seen in its fine control, based on sensory feedback; its fuel regulation; its long life; and its matchless cooling system and emissions control.

Muscles are limited in the range through

FIG 2–2.
Caricature of old-fashioned cable-operated trench digger.

which they can pull. Even in that limited range the tension is variable. Thus the whole design of the hand (and the redesign by the surgeon) is based on matching the power and placement of available muscles to the need of the fingers that do the work.

THE TRANSMISSION

In the figure of an old-fashioned cable-operated trench digger we have a pretty good analogy to the hand (Fig 2–2). The cables are like tendons and they work in just the same way. Note that the cable that operates the distal joint crosses the middle joint very close to the axis. It goes through a kind of carpal tunnel that prevents it from bowstringing and exerting too much leverage at the proximal joint.

We shall explore some of the special features of the cables in the hand, but anyone who understands pulley systems and gear ratios will readily understand the special adaptations in the hand. They will also understand the absolute need to have a well-defined axis for every joint, forming the basis for the operation of the levers that control movement.

THE APPLICATION

Here is a great contrast. The metal teeth of the earth-moving equipment have no real counterpart in the surface tissues of the hand. We use our hands for everything and the key words are not hardness and sharpness but sensitivity, compliance, and self-renewal. Problems occur in the hand when sensory feedback is lost after nerve injury or neuropathy, and a patient may treat his hand as if it were a steel claw and be surprised when it breaks down. The discussion of external forces on skin and interface tissue (Chapter 7) will be of special significance to therapists and those who design and fit splints and orthoses to the hand.

FRICTION AND RESISTANCE

We seldom give a thought to the mechanics of hand mobility until a twinge of arthritis reminds us that joints have feelings. The stiff hand is a disaster that may result from disease, from injury, or even from surgery or therapy.

We may think of it as a failure of gliding of joint surfaces or of tendons in sheaths, but it is more complex than that. Unlike the naked arm of our caricature, our own limbs are clothed with skin, fat, and fascia and intertwined with blood vessels and nerves. All of these have to bend and stretch with every movement. We can refer to the need to move and bend all this passive mass of tissue as a passive resistance to all movement. The extraordinary thing is that in health it amounts to so little. We need to understand how this resistance is ordinarily minimized, why it sometimes increases and dominates all effort to move, and how we may try to overcome it when this happens. At least let us learn not to make it worse.

CHAPTER 3

Muscles: The Motors of the Hand

The output of a muscle is tension. A muscle is important to the body because it is uniquely able to shorten to order, and to do it with enough force to enable the body to do work. As soon as we begin to study the way in which muscles exert tension, we realize that the tension comes from two different mechanisms—active contraction and passive elastic recoil after being stretched. These two usually act together but need to be understood separately because they can sometimes augment each other and sometimes work against each other.

We shall study both by means of length-tension curves in which the tension is recorded on the vertical scale and the elongation or shortening on the horizontal scale. In considering the passive elastic properties of muscle, the curve will always be of the general shape shown in Figure 3–1,A, with tension rising as the muscle is passively lengthened. The curve of active contraction will be a different shape, and will show that there is an optimal length for the muscle fiber at which it can exert its maximum *active* contractile force and that this ability to contract falls off quickly when it becomes shorter and also when it becomes longer (Fig 3–1,B).

Later we shall explore the tension output of a muscle when we combine the two curves together, as they are in nature. After that, we shall see how tension is affected when opposing muscles work together in the intact hand.

ACTIVE CONTRACTION

The active unit in all striated muscle is the sarcomere. This beautiful and intricate biologic device is identical in the muscles of man, mouse, and elephant. It is identical both in size and in mechanical qualities. Big muscles and small; strong muscles and weak, differ chiefly in the number of sarcomeres they have and in the way they are arranged.

Each sarcomere is composed of two Z plates with many filaments of actin arising from each and interdigitating with filaments of myosin that project between them (Fig 3–2,A). This is rather like two hairbrushes facing each other with their bristles overlapping or as the fingers of one hand interdigitating with the fingers of the other hand (Fig 3–2,B).

When the unit is activated by a motor nerve impulse, the actin and the myosin exercise a strong attraction on each other, which pulls the filaments together into a deeper interdigitation. Muscle fibers are made up of long chains of sarcomeres. So when the sarcomeres shorten, the whole fiber becomes shorter. This is the basis of all contraction of striated muscle.

We should look first at the mechanics of one sarcomere. Since the attraction of actin and myosin for each other is the basis for contraction, the maximum contractile force

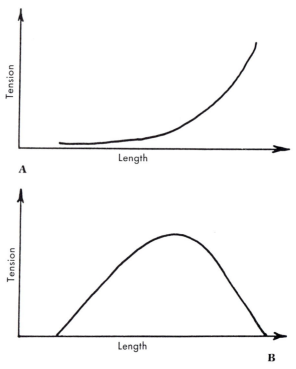

FIG 3–1.
A, the general shape of the length-tension curve of a muscle when it is passively stretched. It may require only a little tension to result in lengthening near the resting length of the muscle. It will require more tension to produce less lengthening as it gets near its elastic limit. **B,** the general shape of active contraction of a muscle fiber. It can produce its highest tension at its resting length. It is able to produce less active tension when it gets shorter or longer.

is obtained when there is a maximum overlap between the interdigitating fibers. This force is diminished if the plates are pulled apart, as there is then less overlap between the actin and myosin. Under the microscope, the Z plates can be seen as dark lines transverse to the length of the fiber. The lines become closer together in contraction, and the dark zone of overlapping actin and myosin fibers widens as the overlap occurs. The force is also diminished if they are pulled in so close that the fibrils bunch up together.

Elftman[9] has drawn a diagram of the active length-tension relationship of a sarcomere (Fig 3–3). Since a muscle fibril is just a chain of sarcomeres, the diagram will be the same shape for the length-tension relationship of a whole fiber.

There are three important features to note in this diagram. The first is the concept of the *resting length*. This is the length that a sarcomere (or a muscle fiber) assumes when the whole limb is in its resting and balanced condition. For example, when the wrist is in its neutral position, between flexion and extension, both flexor and extensor muscles are at their resting length.

If one tendon is cut, its muscle will shorten by elastic recoil even though it is not stimulated, but that new length is not called its resting length.

The second important lesson from this diagram is that the highest tension that the fiber (or sarcomere) can produce is at the resting length. The peak strength of active contraction is not when the muscle is stretched (and has high elastic tension) but when it is in about the middle of its total range from maximum stretch to full contraction. It is in the middle of their normal range of excursion that muscles are poised for their best active contraction.

The third feature to note in this length-tension diagram is the relationship between

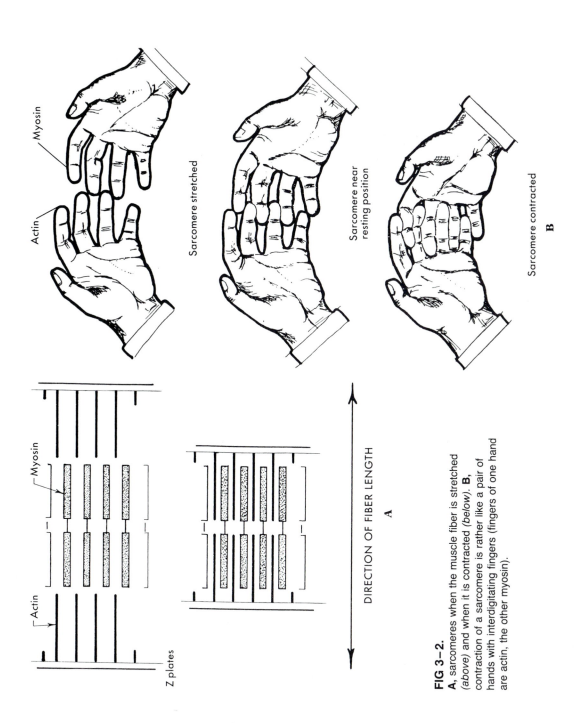

FIG 3–2.
A, sarcomeres when the muscle fiber is stretched (*above*) and when it is contracted (*below*). **B,** contraction of a sarcomere is rather like a pair of hands with interdigitating fingers (fingers of one hand are actin, the other myosin).

Myosin

Actin

Sarcomere stretched

Sarcomere near resting position

Sarcomere contracted

B

Myosin

Actin

Z plates

DIRECTION OF FIBER LENGTH

A

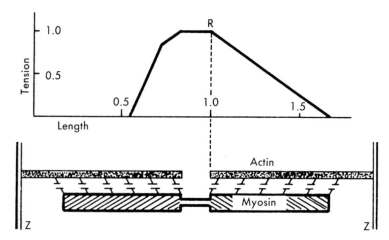

FIG 3–3.
Above: length-tension diagram of a single sarcomere unit. R = physiologic resting length. *Below:* diagram of a sarcomere showing a pair of actin and myosin filaments between a pair of Z plates. *(Redrawn from Elftman, H.: J. Bone Joint Surg. 48A:363–377, 1966.)*

the resting length of the sarcomere and the distance between the stretched length and the fully contracted length (the potential excursion of the sarcomere). We have drawn a copy of the Elftman diagram on which we have inserted three lines (Fig 3–4).[6] The first *(A)* is from zero to the resting length, the second *(B)* is from maximum stretch to maximum contraction, and the third *(C)* marks the longest excursion that can be

achieved without the active tension falling below half of its maximum.

Now it may be observed that the first two lines are approximately equal in length. This means that, in theory, a sarcomere (and therefore a fiber) has the possibility to contract actively from maximum stretch to maximum contraction through a distance about equal to its resting length. We have called this its *potential excursion.* So a fiber that is

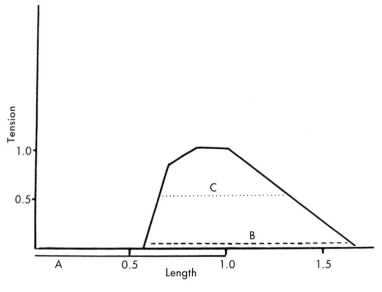

FIG 3–4.
A = resting length; *B* = potential excursion from maximum active length to maximum contraction; *C* = potential excursion using a minimum of 50% of maximum tension.

5 cm, measured at its resting length, could have an excursion of about 5 cm, but at each end of the excursion its active tension capability would be almost zero.

The third line indicates the shorter distance (excursion) of the same fiber contracting within its *effective* range of at least 50% of its maximum tension.

THE ELASTIC BEHAVIOR OF MUSCLE FIBERS

When a muscle fiber at its resting length is detached from its insertion, it shortens. When it is stretched to its maximum, using external force, it is in a state of tension and will spring back as soon as it is released. This is all caused by the elastic behavior of the muscle and is independent of nerve stimulus.

A.V. Hill[15, 16] was one of the first to identify two types of elastic elements in muscle, the *series* elasticity and the *parallel* elasticity (Fig 3–5).

Series elasticity refers to the elastic flexibility in line with the contractile element. For example, if the tendon were to stretch when the muscle contracted, the joint might not move at all, because the tendon would lengthen while the muscle shortened. This does actually happen, but to such a small extent that we may neglect it. The series elasticity is a small adaptability that allows sarcomeres to shorten just a little during isometric contraction of muscle.

Parallel elasticity refers to the whole muscle. In addition to the contractile ele-

ments, there is collagen and elastic tissue that need to stretch when the muscle elongates and that will recoil and shorten when the muscle shortens. This is a very important factor in muscle behavior and profoundly modifies the action of muscle and therefore should modify the way we handle a transferred muscle. From here on, when we refer to muscle elasticity, it is the parallel element that we are talking about.

The elastic tissues in and around muscle, when fully stretched, will limit the extent to which muscle fibers may be further lengthened by opposing muscles. It has been shown that if the actin and myosin fibrils are pulled apart too far, they have almost no ability to exert active tension and may experience some delay getting back into normal interdigitation again.[6]

Thus the elastic elements of a muscle may be useful in preventing the passive overstretch of the contractile elements when the latter are relaxed. Once the muscle has been stretched and is ready for active contraction, the elastic elements are a help, because they add their elastic recoil to the true contraction and make the muscle as a whole stronger and more effective.

ACTIVE MUSCLE CONTRACTION COMBINED WITH PASSIVE RECOIL

The diagram in Figure 3–6, often known as the Blix curve, after Magnus Blix,[3–5] is a useful summation of two different tension curves. The first is the length-tension diagram of a muscle fiber, plotting active contraction

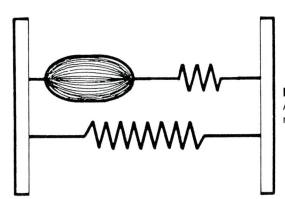

FIG 3–5.
A.V. Hill's concept of elastic element in series with muscle and elastic element in parallel with muscle.

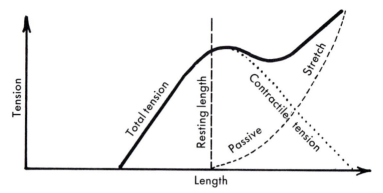

FIG 3–6.
Basic concept of Blix curve.

vs. length, and the second is the length-tension diagram of elastic stretch and recoil in the same fiber. Together and summated they give us a picture of the tension in a muscle that contracts from a stretched position.

However, this curve is deceptive if we fail to think about the two elements separately. The first is under the control of the motor nerve. The second is dependent only on the passive stretch of the muscle. In cases of very severe paralysis of the hand, in which a single active muscle may remain opposed only by flaccid paralyzed muscles, the active muscle may appear surprisingly limited in range and power until we remember that to function in its normal range it needs an antagonist to lengthen it and to "wind up" its elastic spring.

MUSCLE SYSTEMS IN SITU

There is thus another element that should be added to any composite diagram of muscle contraction and tension. This is the effect of the opposing muscles. The trouble with scientific specialists is that they like to isolate the thing that they are studying so that it can be accurately observed. In real life things are not usually isolated. In biology they are never isolated. No skeletal muscle ever works alone. When we study muscles in pairs we recognize that a large part of a muscle's duty is to control the elasticity of the opposing muscle. This takes work.

When a muscle contracts, the opposing muscle on the other side of the limb relaxes and offers no *active* resistance to elongation. However, the muscle offers increasing passive resistance because its elastic elements are being stretched. Now, while this is happening, the elastic elements of the opposing muscle are storing potential energy and are ready to spring back and release that energy as muscle shortening (not "contraction"). This elastic recoil is not under the control of its own nerve-muscle unit. It is controlled by the opposing muscle. If a person is performing a continuing reciprocal action like walking or hammering, it may be that as soon as a muscle *A* relaxes, its antagonist, *B*, will contract. In such a case the elastic stretch that the contraction of *A* has put into *B* will be used by *B* to augment the strength of its own (*B*'s) contraction. Otherwise, all muscle tone and elastic recoil will tend only to bring muscles back to their neutral or resting position (Fig 3–7). We have never seen a diagram that takes note of the fact that every time a muscle contracts it is using much of its energy stretching the elastic element of its opposing muscles.

We now append our own diagram of a modified Blix curve (Fig 3–8,B) showing the length-tension curve of a muscle fiber (or sarcomere) in situ where it is opposed by a muscle of equal amplitude and strength. This new curve assumes that the antagonists are equal in size. It would be a little different if they were unequal.

Below the zero line is the length-tension diagram of the elastic element only of the opposing muscle. It matches the length-tension diagram of the muscle under study but

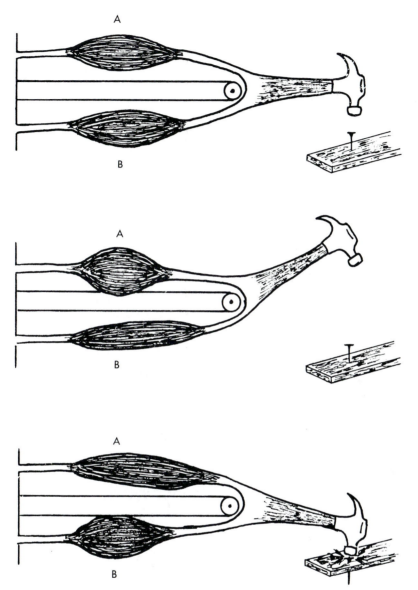

FIG 3–7.
When the hammer strikes, it is using both the active contraction of *B* plus the elastic recoil in *B* that has been put into it by *A*.

is a negative quantity and increases as the other decreases.

Observe that the final composite diagram of the muscle in situ (see Fig 3–8,B) shows not only the tension of muscle *A* enhanced by its own elastic element when the muscle is stretched[3–5] but also the sharp cutoff of muscle *A*'s ability to contract and shorten at the other end as a result of the elastic element of the opposing muscle.[6]

The net result of combining the length-tension diagrams of the active contraction of a muscle with the diagrams of its own *and its opposing muscle's* elastic behavior is that we have a more nearly square curve (Fig 3–8,C). The total range of potential excursion is somewhat diminished and the peak tension is more nearly level for the greater part of the range.

ISOMETRIC CONTRACTION

When a muscle is stimulated, it creates tension. This tension tends to shorten the

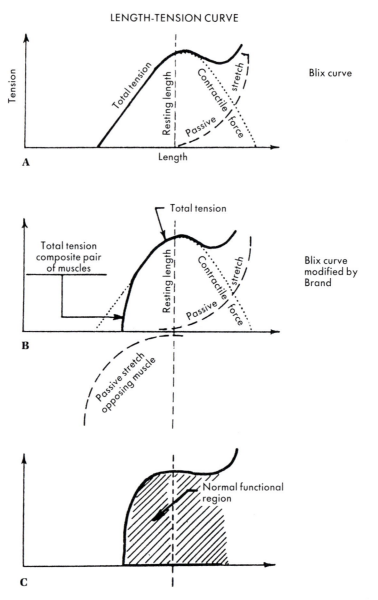

FIG 3–8.
A, Blix curve, integrating the active contraction and elastic recoil. **B,** Brand curve integrating the above with the elastic curve of the opposing muscle subtracted from the active output of the primary muscle. **C,** final approximate shape of muscle curve in situ in intact limb. (*Note:* The curve is more nearly square in shape compared with the isolated sarcomere in Fig 3–3. Also note the shorter excursion than the potential excursion in Fig 3–4,B.)

muscle, but it may not actually result in shortening. The tension in an opposing muscle, or an opposing external force, may balance the tension of the muscle under study so there is no movement or shortening in it (isometric contraction). Or it may be that a muscle may actually be lengthened while it is "contracting," because the opposing force may be greater than the muscle tension.

This should not be regarded as a defeat for the muscle. It may serve to moderate the opposing force or to stabilize a joint while the opposing muscle does its work. Thus when an athlete lands on his feet after a jump, his knees will bend while he absorbs the shock, and his quadriceps muscle will "contract" quite strongly while it is being lengthened by knee flexion. The athlete has created tension in the quadriceps tendon to slow and smooth his deceleration.

Thus the term *contraction* in a muscle may mean only the creation of active tension.

MUSCLE TONE

Even in the totally relaxed individual, during sleep, all muscles have a certain tone. There is a tendency to shorten or to resist lengthening even without any apparent nerve stimulus. This tone is caused by a variable mixture of two components. The first is the elasticity of muscle and of the connective tissue that invests it. The term *viscoelasticity* is better than simple *elasticity* here, because a muscle behaves differently when it is subjected to fast movement than when it is subjected to slow movement. This shows that there is a damping influence like a shock absorber, and this is called a viscous quality. At its resting length, all the tissues of the composite muscle are somewhat stretched and will passively shorten when the tendon is divided. The other component of tone is the involuntary contraction of muscle fibers that respond by some active contraction to any sudden or unexpected passive stretch or lengthening. This reaction is familiar in the knee-jerk and other tendon reflexes. This active resistance to passive lengthening is inhibited during normal voluntary coordinated movement and is increased when there is interruption of normal voluntary muscular control as in upper motor neuron disease or spinal injury (spasticity).

We have made no special studies of the mechanics of spastic muscles and will not explore the subject here except to comment that the laws of muscle mechanics discussed in this chapter do not apply to spastic muscles. They are (literally) a law unto themselves. Attempts to passively stretch contracted spastic muscles may be necessary in order to keep joints mobile and tissues healthy, but such stretching may only increase the tendency of the muscle itself to persist in its independence and contracture.

MUSCLE SYSTEMS AND EXTERNAL FORCE

Muscles should be studied not only in pairs but also with the effects of external force. In the diagram in Figure 3–9, we see a pair of muscles that oppose each other so that as one shortens the other lengthens or they can both contract together and produce a stable equilibrium with no movement. We can use this model while we summarize the mechanical nature and effects of muscle tension.

Consider three possible positions of the joint that is controlled by muscles A and B. Both muscles are about the same size. There is a neutral position at which they balance each other at rest. Each muscle fiber will now be at what is called its *resting length* (X). At this length, if it is activated, it can exert its maximum active contraction and develop its maximum tension. It is also under its best control.

Now consider a position in which muscle A has contracted and shortened by 40% and muscle B has relaxed and been stretched 40% (Y). If both muscles are now relaxed completely (if there are no more impulses from either of their motor nerves), they will not stay where they are, they will return to their balanced resting positions. The force that returns them is not active muscle contraction, it is the passive elastic recoil of muscle B. Thus when muscle A contracted 40%, it not only had to move the distal part of the limb, it also had to stretch the elastic tissues in muscle B.

If muscle A contracts further from its 40% shortened position (muscle B is relaxed), the net effect on the joint will be very small indeed. This is because (1) the power of a muscle to contract becomes very small when it approaches the end of its potential range and (2) it cannot contract without having to stretch the elastic component of muscle B. Even though B is totally relaxed as a nerve-muscle unit, it is unable to cancel its elastic element, which still works against muscle A. In fact, the elastic resistance to muscle A has increased as muscle B is further stretched.

Using the same model (see Fig 3–9), consider the situation when the joint is moved a little further with the help of external force, so that the origin and insertion of A move still closer together. A is now shortened by 60% and is only 40% of its original resting length, and B is stretched 60% longer than its resting length. In this position A will be loose and useless even when it is fully stimulated. B will be overstretched and may begin to suffer damage (Z).

We will not study this extreme situation further at this time because in a normal

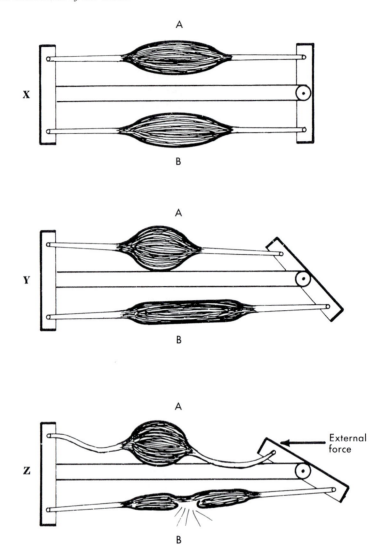

FIG 3–9.
X, opposing pair of muscles, resting. **Y,** *A* contracts 40% of length, *B* stretches 40% of length. **Z,** external force causing overstretch of *B*.

hand it occurs only rarely. In a normal hand muscle fibers are long enough to allow normal range of motion of joints they control. Relative fiber shortness may occur in muscles that cross several joints that ordinarily do not all flex at the same time. The fibers of flexor digitorum profundus (FDP) and extensor digitorum are rarely required to accommodate to a flexed wrist with fully flexed fingers. If a wrist is forced to flex by external force while the fingers are flexed, the flexor digitorum becomes loose and ineffective, the extensor digitorum becomes stretched to the point of pain, and the hand lets go of whatever it was holding.

The stretching of a muscle fiber beyond the range it can tolerate does sometimes occur postoperatively in hands in which a short-fibered muscle has been transferred to do a long-fibered job. The management of such a case is discussed later.

CHARACTERISTICS OF INDIVIDUAL MUSCLES

It should now be clear that there are two variables that characterize every muscle. These are (1) the tension it can develop and

(2) the length or distance through which it can contract.

It is of the utmost importance that the surgeon know something about the capabilities of the muscles that he or she may want to transfer to different situations. A certain muscle may be in a convenient position for transfer, but is it strong enough or perhaps too strong, and will it be able to stretch and contract far enough in its new situation?

At this time we have no final figures for any of these factors. Our studies and those of our colleagues have resulted in tentative lists of relative tensions and potential excursions for all muscles below the elbow. These figures are useful approximations. They are being modified as others repeat our experiments and add to our experience (An et al., 1991; Kaufman et al., 1989).

In order that the figures we use for muscle tension and excursion may be understood, we will describe how they were obtained and what we think they represent.

Since the sarcomere is the basic unit of muscle contraction, the variables that distinguish muscle from muscle must relate to (1) the number of sarcomeres in series (i.e., the length of muscle fibers) and (2) the number of sarcomeres in parallel (i.e., the cross-sectional area of all the fibers of a muscle).

The length of fibers will be proportional to the potential for excursion and the cross section of all fibers will be proportional to total maximum tension.

Since the sarcomere units are all mechanically identical, it would seem obvious that there should be a very simple relationship between fiber length and tendon excursion. Not so. One has to read the German literature of the last 100 years to uncover the passionate debate on the relationship of fiber length to tendon excursion. Weber and Weber[30] and the Ficks,[10, 11] in making some of the earliest attempts to record muscle fiber length-function relationships, developed what came to be known as the *Weber-Fick law;* which states that, in general, muscle fibers are twice as long as the distance through which their tendons move at the joint (i.e., fiber length = 2 × excursion). This view was hotly disputed by others, most effectually by Murk Jansen[19] who stud-

ied many individual muscles in the course of his investigations into the nature and management of spastic muscles in children. His measurements show a wide variety of fiber length-to-excursion ratios, from a low of 0.76:1 for the sartorius to 2.55:1 for the tibialis posterior and peroneus brevis. Most of his readings were from leg muscles and unfortunately neither he nor Weber nor the Ficks defined the conditions under which either the fiber length or the tendon excursion at the joints was measured.

Our colleagues and we have now made similar measurements of muscle fiber length for all muscles of the forearm and hand and have matched them with what we call the *required excursion* of their tendons. We have made an attempt to standardize the conditions for measurement. We hold the elbow, wrist, and digits at their neutral or resting position, which we have assumed to be (1) elbow—60 degrees flexed from the fully extended position, (2) forearm—midprone, (3) wrist—10 degrees extended with the third metacarpal aligned with the radius, (4) all digits flexed 45 degrees at all joints, (5) the thumb pulp beside the middle segment of the index finger, and (6) the thumbnail at right angles to the plane of the palm (Fig 3–10). We agree with Jansen that there is wide variation in the ratio of fiber length to tendon movement needed for full joint motion.

The FDP has about a 1:1 ratio whereas the extensor carpi radialis brevis (ECRB) is more like 2:1. In general, the muscles that cross two or more joints are less able to accommodate to the positions of full flexion and full extension of both joints than single-joint muscles are to the requirements of their one joint.

This is probably because in almost all multiple joint systems there is some degree of synergistic reciprocal joint action whereby a proximal joint will extend while distal joints flex. Thus a tendon that crosses the wrist and the distal joints will ordinarily only have to contract by the amount required by the digital joints *less* an amount conserved by the wrist. If we thus calculate the excursion ordinarily required by the normal interaction of the joint system, we may find that

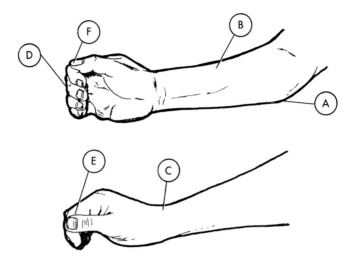

FIG 3–10.
Resting position of hand for evaluation of fiber length. *A,* elbow 60 degrees flexed from the fully extended position; *B,* forearm midprone; *C,* wrist 10 degrees extended; *D,* all finger joints 45 degrees flexed; *E,* thumb tip opposite the side of the middle segment of the index finger; *F,* thumbnail 90 degrees to the plane of the palm.

there is a more constant relationship between fiber length and required excursion.

Until we are able to come to final conclusions about all this, we have chosen simply to record the resting length of all muscle fibers and state that this probably is directly proportional to the *potential excursion* from the fully stretched to the fully contracted muscle fiber.

There is yet one more factor to modify the ratio of fiber length to range of motion. This is the factor of *ordinary* range of motion. In a cadaver limb we measure total range, but as one watches normal people at work and play it is very apparent that they have no ambition to use extremes of most movements. We predict that when we finally map out the ordinary ranges of commonly used movements and factor in the reciprocal relaxation of some joints to facilitate contraction at others, we shall find that they require just about two thirds of the resting length of muscle fibers as excursion. This may be true of all muscles.

MEASURING MUSCLE FIBER LENGTH

Since fiber length is so importantly linked to the ability of a muscle to lengthen and shorten enough for joint movement, we need to be able to measure the fiber length of muscles. The hand surgeon does not need

to expose muscles to check fiber lengths at surgery, because we now have tables that give us average values. However, as surgeons repeat our work we shall eventually have better figures. More important, all hand surgeons should dissect a fresh (or a fresh-frozen) human forearm and hand to convince themselves how easily we get misconceptions about the qualities of the muscles we all work with. Only by direct observation and measurement will some of these errors be dispelled. The major error is to confuse muscle length with muscle fiber length.

Only a few very weak muscles can afford the luxury of parallel muscle fibers that all start at about the same point and end at the same point. The lumbricals are rather like this, so that the length of the muscle is only a little longer than the length of its average fiber. If a muscle is to be strong, it will have thousands of fibers of the same length as one another. If they were all to start at the same point and end at the same point, there would be a tremendous local bulge between those points. Instead, such muscles have long tapering tendons (Fig 3–11). The tendon of insertion may traverse three quarters of the total muscle length and will receive fibers all along that length. These fibers will take origin in sequence along a length of bone or tendon of origin. The flexor carpi ul-

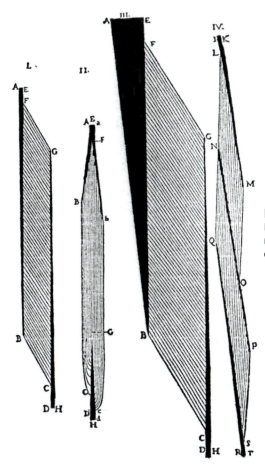

FIG 3–11.
Diagram by Steno (1667)[26] showing uniformity of muscle fiber lengths and the progressive thickening of the tendon as it accepts more and more fibers.

naris (FCU), for example, may have a muscle belly that is 25 cm long, but the tendon of insertion traverses about 21 cm of the muscle, receiving 4-cm fibers all along the way. So we have a very strong muscle with very short fibers and thus a very limited excursion. It would be a mistake to transfer the FCU to replace a paralyzed flexor digitorum, which has long fibers. The ECRB has much longer fibers than the FCU although it is a shorter muscle and a weaker one. The ECRB or the extensor carpi radialis longus (ECRL) would make a better substitute for a paralyzed finger flexor.

The best way to estimate the fiber length of a muscle is to look for the most distal fiber and the most proximal fiber and trace them directly from origin to insertion on bone or tendon and measure them in situ. Another way is to cut the tendon in the cadaver hand and attach hemostats all along the tapering tendon that runs along the muscle. Then the hemostats are pulled laterally while a blade is used lightly to free the fibers from the light areolar tissue that holds them together side to side. In this way, the fibers will all come to lie parallel to one another and be at an angle to the tendons (Figs 3–12 and 3–13).[6]

It is now easy to measure the length of many of the fibers and obtain a mean. The tendon should then be returned under the same tension to its resting position and the distal joints held in their neutral position. Now if the two cut ends of the tendon overlap, it shows that the muscle fibers have been overstretched during measurement. We measure the amount of overlap and subtract it from the measured mean fiber length to obtain a fiber length corrected for the resting joint position. This is how the figures in Chapter 12 were obtained.

Variability in Length of Fibers.—In most muscles, all of the fibers lie in one segment of the limb between the joint above and the

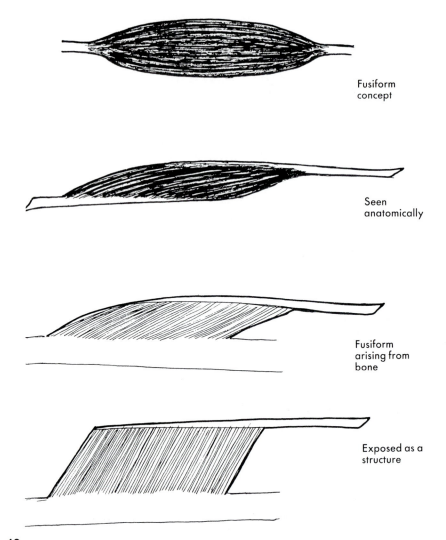

Fusiform
concept

Seen
anatomically

Fusiform
arising from
bone

Exposed as a
structure

FIG 3–12.
Method of lifting tendon of insertion away from origin in order to demonstrate fiber pattern.

joint below. It is only the tendon that crosses the joints. In such cases, all the fibers of the muscle will be about the same length.

In some muscles, however, the muscle fibers themselves cross a joint, and in such cases some fibers will lie closer to the axis of the joint than other fibers, and thus individual fibers will require more excursion than others and will thus be longer than others. In such cases, we have simply tried to estimate a mean fiber length and must state frankly that we have little confidence in the accuracy of our estimate. Thus in such muscles as the supinator and pronators and first dorsal interosseous, we may be wide of the mark in a true mean fiber length and cross-sectional area.

MUSCLE TENSION CAPABILITY

Just as muscle fiber length is proportional to excursion, so the physiologic cross-sectional area of muscle fibers is proportional to the maximum tension they can produce.

There is so much variation between the strength of one person compared to another, and even in the same person from one time to another, that there is not much to be gained by calculating the exact strength of any muscle. What is very important to the hand surgeon, however, is the *relative* strength of one muscle to another in any given hand. It is muscle balance that is important. Fortunately, although actual strength varies so much, relative strength of muscle to muscle varies

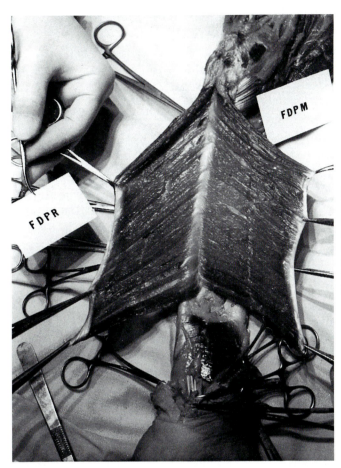

FIG 3–13.
Flexor digitorum profundus to middle finger *(FDPM)* and ring finger *(FDPR)* demonstrates a parallelogram of fibers. All fibers are nearly equal in length. *(From Brand, P.W., et al.: J. Hand Surg. 6:209–219, 1981. Used with permission.)*

much less. Therefore we have set out to produce a list of all the muscles below the elbow, graded by their strength relative to one another.

Most of the early work on muscle strength was done by German anatomists such as Weber and Weber[30] and the Ficks,[10, 11] who identified the significance of the cross-sectional area of all the muscle fibers. Many measurements and estimates have been made of the relationship between area and actual strength. The figure of 3.6 kg/cm^2 was used by von Recklinghausen[21] in 1920 and was accepted by Steindler[24, 25] and Arkin.[2] The actual maximum tension varies with the speed of contraction and whether it is isotonic or isometric or whether the contraction occurs during a lengthening of the muscle by opposing forces.[14] All of this is outside the scope of this book; we will be content with the figure of 3.6 kg/cm^2 and will be concerned mainly with relative tension, which we record as a percentage of the total tension of all muscles below the elbow.

We obtained our estimates by dissection of a number of fresh-frozen cadaver arms. Having measured the mean fiber length of each muscle, we removed the whole muscle by careful dissection and weighed it. Having weighed all the muscles below the elbow, we totaled their weights and worked out the percentage of the total for each muscle. We called that the *mass fraction* for that muscle. Mass is directly proportional to volume, and we used the volume of the muscle divided by its mean fiber length to obtain a figure proportional to the physiologic cross-sectional area of the muscle.

Thus we have three measurable variables for each muscle: (1) mass, (2) fiber length, and (3) physiologic cross-sectional area (Table 3–1). By these three, we may have a good idea of the performance of each muscle relative to all other muscles that affect the hand. Functionally these represent (1) rela-

TABLE 3–1.

Comparison of Extensor Carpi Radialis Brevis (ECRB), Extensor Carpi Radialis Longus (ECRL), and Flexor Carpi Ulnaris (FCU)

	ECRB	ECRL	FCU
Mass fraction*	5.1	6.5	5.6
Mean fiber length (cm)	6.1	9.3 (range, 6.3–12.3)	4.2
Tension fraction*	4.2	3.5	6.7

*In this and other tables we use no units of mass or tension. The numbers are simply percentages of all muscles below the elbow. See Tables 11–1 and 11–2 for additional data.

tive work capacity, (2) excursion potential, and (3) relative tension potential.

WORK CAPACITY

Having already explained the relation between tension and excursion, we now need to define *work*.

Work is measured by force multiplied by distance. If 1 kg is lifted against gravity through 1 m, 1 kg · m of work has been done. If 0.5 kg is lifted through 2 m, the same 1 kg · m of work is done.

Now look at a hypothetical muscle composed of only four parallel fibers of equal cross-sectional area and equal length. We will assume that when they contract, each fiber can pull with 1 g tension through a dis-

tance of 1 cm. Arranged side by side they can pull 4 g through 1 cm (Fig 3–14,A), so the *work* they do will be 4 g · cm. Arranged end to end in pairs they will be able to pull 2 g through 2 cm (Fig 3–14,B). Arranged with all four end to end they will only be able to pull 1 g but through 4 cm (Fig 3–14,C).

The work they can do is thus 4 g · cm however they are arranged because 4 × 1 = 2 × 2 = 1 × 4.

Thus the work capability of a muscle depends on its bulk; on how much muscle there is. The range or excursion depends on fiber length. The tension depends on cross section. In Chapter 12 we have listed all three qualities for all muscles, so that the figures will be available for reference.

DEFINITIONS FOR EXCURSION

In the interest of clear understanding, we must define three different terms for excursion of a muscle-tendon unit.

POTENTIAL EXCURSION

Since the fiber length of the muscle is its most constant and unchanging factor, and

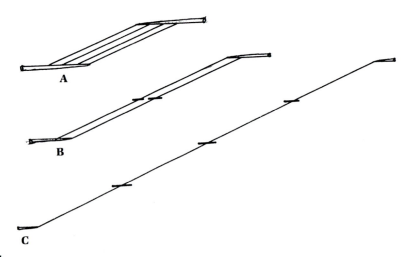

FIG 3–14.
A, four muscle fibers in parallel. *B,* two muscle fibers in parallel, each twice the length of the fibers in *A. C,* one muscle fiber, four times as long as the fibers in *A.* System *A* can exert four times the tension of *C.* System *C* will have four times the excursion of *A.* Systems *A, B,* and *C* are all capable of the same amount of work as one another. Work = tension × excursion.

since it is directly related to the number of sarcomeres in series and to its capability to produce tension at various degrees of stretch and shortening, we propose that all muscles should ordinarily be characterized by their resting fiber length without reference to the state of connective tissue restraints. This should be called the *potential excursion.*

REQUIRED EXCURSION

Since a muscle is never required to lengthen or shorten through a larger excursion than it takes to move the joints that it crosses through their full passive range, we suggest that the term *required excursion* be used to define the maximum excursion that might be required of a muscle in situ. When a muscle is transferred, its new required excursion may be different and may be beyond the capability of the muscle to deliver.

AVAILABLE EXCURSION

Available excursion is a term used by Freehafer et al.,[12] and is a measure of the maximum excursion a muscle can deliver through its tendon when that has been cut and freed from its insertion. This available excursion may be measured at operation by electrical stimulation or by asking the conscious patient to relax and then contract the muscle. Omer et al.,[20] made careful studies of the physiologic length and tone of muscles at operation but did not attempt to develop standard tables of normal excursions. The figure for available excursion is often about the same as the required excursion but it is responsive to the pattern of usage. It is probably a measure of the limitation imposed by the surrounding connective tissue, paratenon, paramysium, and intramuscular collagen. This excursion can be changed by exercise. It may be lengthened by stretching over a period of time and it may be severely shortened by adhesions following injury or operations for tendon transfer. All of these factors, which are responsible for the difference between potential and available excursions, are discussed in Chapter 6.

MUSCLE CHANGE FOLLOWING TRANSFER

When a group of muscles is irreversibly paralyzed, the surgeon may have to redistribute the remaining muscles to balance the hand and allow it to perform its most important functions. In doing this, we try to match the strength and excursion capabilities of the transferred "donor" muscle to the needs of the paralyzed "recipient" muscle-tendon unit with due allowance for the overall balance of the hand. We can never get a perfect match, but many of us must wonder how much latitude we can allow ourselves. If we transfer a weak muscle to do a heavy job, will the transferred muscle become stronger? If we transfer a short-fibered muscle to do a long-fibered job, will its fibers grow longer?

WEAK MUSCLE—STRONG REQUIREMENT

There is little doubt that muscles respond to increased demand by progressive increase in cross-sectional area and increase in strength. This is routine in the build-up of muscle strength for athletes and weightlifters. Thus there is no reason to doubt that it can happen following tendon transfer. However, the stimulus for hypertrophy must be the frequent strong contraction of the muscle and this will happen only if the subject is actually using the muscle to its full capacity. This will usually happen only if it is contracting in phase with its new requirements. The stimulus to greater accomplishment is *some* accomplishment. This is as true of muscles as it is of human character.

If a transferred muscle is a total failure in its new situation, it will not become stronger. This failure may be because of tendon fixation, as from immovable adhesions, or it may be because of total inadequacy in the muscle itself. If a transferred muscle is so weak that it does not move the hand, it will be ignored by the motor neuron complex and will become weaker. We have seen a spastic hand in which the flexors of the wrist were in such constant overwhelming contraction that the ECRB and ECRL could never actually extend the wrist. On careful testing, even with the wrist supported in a neutral position, those muscles proved to be capable of only a twitch. Once the wrist was balanced by transferring half the tension of the FCU to the dorsum of the wrist by a bridle or yoke type of transfer,[7] the extensor muscles gradually developed strength and worked quite effectively. They had always been under fair voluntary control but had atrophied because their strength had never been able to overcome the opposition of the spastic FCU.

This need for a transferred muscle to be somewhat successful before it can really succeed is one reason why a therapist should be involved in the preoperative and postoperative education of a tendon transfer. The duty of the therapist is to identify the transferred muscle and help the patient to identify it in his own central nervous system and make sure that it is working in phase and that it is appreciated. This responsibility is doubly important if the muscle is rather weak and marginally adequate for its new job. The therapist need not always continue to work with it until it is strong; only until it is working in phase with other muscles and is contributing significantly to the use of the hand.

Although we know that a muscle does sometimes become stronger after transfer, we do not know how much. Certainly if a muscle has been weakened by disease or partial paralysis it is unwise to expect a better performance after transfer. There is an old rule about tendon transfers that says that a muscle may be expected to drop one grade of strength after transfer. This refers to grading of partially paralyzed muscles 0 to 5, where 5 is normal muscle.[22] We question that rule, or at least suspect that the loss of effectiveness of muscles experienced after transfer is brought about more by adhesions and drag and imperfect reeducation than by actual loss of muscle tension capability.

Even under the best circumstances, we doubt that one should expect a muscle to do better than double its strength, relative to other muscles, after transfer. It would be poor planning to select a muscle for transfer that was more than 50% stronger or 50%

weaker than the one it was to replace, assuming that one is dealing with a localized paralysis. The situation may be different in gross paralysis when only a few muscles remain. Here one muscle may have to do duty for several, and here also the total expectations from the hand will be very limited. In such a case, a single muscle may serve as wrist extensor and a single other muscle may share the responsibility for flexing all the digits. Each will be effective only to the extent that it is balanced by the other. Balance is always more important than strength.

SHORT FIBER–LONG EXCURSION REQUIREMENT

If a short-fibered muscle is transferred to do a long-fibered job, will it develop longer fibers? This is a question that we cannot answer with any assurance. Much less work has been done on excursion and fiber length than on the physiology of strength.

The French workers Tarbary et al.[27, 28] and the workers in the United Kingdom, Goldspink[13] and Crawford,[8] have all done beautiful animal experiments in which joints have been immobilized in flexion or extension and then after a few weeks the flexor and extensor muscles have been fixed in situ by glutaraldehyde and then removed for comparison with muscles on the control limb.

The results in all such cases show that muscle fibers adapt rapidly to the changed condition and that the change in all cases is for the fiber length to accommodate to its new resting position by lengthening or shortening so that at rest the fiber is neither loose nor tight and all sarcomeres rest at their physiologic resting length. The change is accomplished by adding or subtracting sarcomere units, probably at the ends of the fibers where they join the tendons. Thus, if we look at any muscle fiber under a microscope, we can tell whether that fiber is in a condition of stretch or contraction simply by measuring the length of any sarcomere against a grid.[1]

We are starting to do this in our fresh cadaver specimens as well as after experimental tendon transfers on animals. In the human cadavers, we do it to check the resting position and resting length of the fibers. In animals, we do it to see whether the muscle fibers have changed the true length of their fibers after transfer or have just "stretched." For example, if a resting muscle fiber was 15 mm long before transfer and was 20 mm long 6 weeks after transfer, this could mean either growth or stretch. If the sarcomeres are seen under the microscope to be at the standard resting length, then we know that true growth has occurred and sarcomeres have been added to compensate for the stretch that must have occurred at operation.

The stimulus for the adding and subtraction of sarcomeres is apparently the length and tension status of the sarcomeres at rest, not contraction or dynamic activity (so far as is known). For example, de la Tour[29] tried the effect of constant tetanic stimulation of muscles in cats, but this produced no change in resting fiber length or sarcomere number.

Thus, if a surgeon wants to change a short-fibered muscle to make it do a long-fibered job, the only thing he can do is to pull on the tendon at operation and hold the muscle fibers in a lengthened position while the attachment is made to the new tendon insertion. The hand must then be held in a position that keeps the muscle lengthened. If transferred muscles behave like muscles in experimental animals, then the fibers will grow by the addition of sarcomeres.

Some muscles share common fibers and have limited independent movement. If an individual flexor profundus tendon is fixed at a higher tension at operation, it may be permanently tighter than the other profundi because of cross-connections between adjacent units.

We have noted that when patients have tendon transfers attached at higher tension than normal, they feel a certain unease in the postoperative phase and they tend to contract the transferred muscle more readily than if it had been at neutral tension. Even at rest, the transferred muscle may show some electromyographic discharge in a biofeedback device. Both the sense of unease and the electromyographic discharge are likely to disappear in a few days or weeks after exercises have started. This may represent the stage at which the fibers have adjusted to their new length by adding more sarcomeres.

Much of this is still speculative, and we have no firm figures for changed fiber lengths for human hands. We have talked with many experienced hand surgeons and most of them state that they frequently suture tendon transfers at a slightly greater tension than what they regard as neutral. They claim simply that they think they get better results this way. Some say that they do it to get a stronger action (more tension). This concept of more strength is almost certainly not valid, but more excursion may be achieved. It is also true that a longer fiber may give better tension at extremes of required excursion than one which is only just long enough (see Fig 3–3).

However, if surgeons are to pull tendons tight before suturing, let them be aware of the increased tendency to avulsion at the attachment because of involuntary contraction of a stretched muscle postoperatively.

We have no evidence at this time that a short-fibered muscle will develop longer fibers simply because of demands on it for greater excursion in a new situation. That would have to involve the growth of the fleshy muscle fibers into the tendon, where sarcomeres would have to replace tendon. This probably does not happen.

Therefore in the intact adult limb the resting fiber length of a muscle probably does not change. It does not become shorter by disuse or longer by increasing the range of motion of joints. However, there is no doubt that the active *usable* range of excursion of a muscle does change with use and disuse. Those who newly take up jogging or ballet find out how limited their muscles are. They also find that after regular stretching exercises their calf muscles become more limber and allow a larger range of excursion that permits a wider range of motion at the ankle. This paradox can only be resolved if we assume that it is the connective tissue around and within the muscle and around the tendon that limits the range of excursion of most muscles. This connective tissue is constantly responsive to the demands on it. Its fibers become gradually longer in response to exercise, and they become gradually absorbed and shorter in persons who never use the full amplitude of their muscles and joints.[17, 23] There is a very considerable tissue interface between the millions of filaments of actin and myosin that must rub against each other as they interdigitate. The physics of that smooth movement and of its potential for friction and for improvement by exercise and stretching is beyond the scope of this book, but is a fascinating field for study.[23]

INHIBITION OF STRENGTH

When a patient complains of "weakness" in the use of the hand, he or she may not be saying that the muscles are weak or unable to contract. It is up to the physician or therapist to find out why the patient feels that way. A common problem is *inhibition* at the level of the motor neuron. This is not a diagnosis, it is simply a challenge to find out why the motor neuron is being inhibited.

Pain is the most obvious cause. If a person with active arthritis in a carpal joint begins to grasp an object strongly, the pressure or movement in the wrist causes sudden pain and forces the patient to relax his or her grip. This is a clear and direct connection. However, the subconscious memory of that pain may persist after the acute episode had passed, and an inhibition may remain that still makes the patient "weak" because of an unrealized fear of pain.

A second cause is structural weakness. Most skeletal structures have a sensory feedback of sorts that results in the limitation of the stresses that a patient might otherwise place on them. A ligament will accept tension when a joint reaches the end of its proper range of motion, and then, when its elastic limit is reached, will send back an impulse, short of pain, that will inhibit further muscular contraction. If that ligament is partially torn, it will inhibit muscular contraction at a lower level of stress.

An interesting interplay of inhibition occurs at the metacarpophalangeal (MCP) joints when the interosseous muscles of the hand are paralyzed. One of the major responsibilities of the interossei is to provide a force close to the axis of the MCP joint that forces the base of the phalanx and the head of the metacarpal together during strong flexion of the fingers. This allows the long flexor tendons to apply their uninhibited force across the joint and to have the whole of this force used for the flexion of the joint. The force vector that would tend to dislocate the joint (see Chapter 5) is balanced by muscles that stabilize the firm lodgment of the head in the shallow cup of the phalanx. If the interossei are paralyzed, the stability of the joint has to depend only on the ligaments and sagittal bands. If these become lax or stretched in disease, the shallow rim of the cup is inadequate to hold the head.

Patients with rheumatoid arthritis affecting the MCP joints experience an inhibition of their muscular power when the disease causes the weakening of ligaments and other joint structures. If excessive pain medication suppresses that inhibition, it may lead to freer use of strength and to earlier joint dislocation. Physicians who prescribe heavy pain medication need to warn their patients to be extra careful about the use of their full strength while their joints are flexed, in case the penalty of pain relief is earlier subluxation. Hand therapists need to be careful to respect a patient's own pain barriers and inhibitions. Passive motion must never be used to override the patient's pain, nor should one encourage patients to be too brave about the use of their hands.

In leprosy, patients often lose cutaneous sensation in the hand at an earlier stage than their loss of motor power. They often use their strength in a totally uninhibited way, resulting in damage to their fingers; yet they may complain that they feel weak. This sense of weakness may be a result, in part, of the loss of sensory feedback from the skin, which they have previously used to monitor the strength of their grasp.

Cases of nerve injury, after repair, are different from leprosy and other neuropathies in that paresthesias occur from the cut end of the nerve where a neuroma forms. The pins-and-needles sensation that is felt in the limb during activity is an unpleasant feedback and serves to inhibit use of the limb.

We have experimented with pressure-sensing gloves on patients who have partial paralysis. We find that patients with sensory neuropathy tend to use the insensitive fingers with excessive force, whereas patients with nerve injury, after repair, tend to use mainly the unaffected fingers and leave the partially insensitive fingers unused, "sticking out like a sore thumb."

The whole picture of muscle weakness vs. inhibition is complicated by the fact that a muscle that is normal but inhibited by joint pain or disease may later become wasted so that it finally becomes really weak. A patient of ours who had been an internationally acclaimed concert pianist gave up her career following a fall that caused pain in the base of the thumb. Later she said her thumb was too weak to strike a piano note. However, it proved to be quite strong when tested. It turned out that she was constantly afraid of the sudden pain that she occasionally had when moving the thumb into abduction to strike a note. Her inhibition at the mental level interfered with her music and persisted after her thumb had become pain-free through the whole of its range of motion. Instead of thinking about Rachmaninoff she was wondering if her carpometacarpal (CMC) joint was staying in its pain-free range.

The systems that interact to preserve the "hand machine" and keep it running are so complex and have so many automatic compensating mechanisms that when they go wrong it takes all our art and science and engineering to sort them out.

That is the challenge of hand surgery and rehabilitation.

MUSCLE SYNERGISM AND REEDUCATION

One cannot leave a general discussion of muscles without mention of synergism, even though it is not an easily defined or mea-

sured mechanical quality. Muscles work in groups and in patterns that are controlled at subconscious and even at lower reflex and spinal levels.

The conscious voluntary decision to flex a finger involves an involuntary contraction of various stabilizing muscles such as wrist extensors. It makes it easier to get a muscle moving after tendon transfer if its previous action had been a part of the same synergistic complex. A wrist extensor makes an easily reeducated finger flexor. However, this concept should not dominate the planning of tendon transfers. It is most important in older patients and may be almost neglected in children. Almost any muscle in the upper extremity may be ultimately reeducated to any other function, but it is not right to impose difficult reeducation on an older person if an easily reeducated synergistic muscle of similar mechanical output is available.

REFERENCES

1. Adams, J.P.: Personal communication, 1984.
2. Arkin, A.M.: Absolute muscle power: the internal kinesiology of muscle research, Research Seminar Notes, Dpt. Orth. Surg., State Univ. of Iowa, 12D:123, 1938.
3. Blix, M.: Die Länge und die Spannung des Muskels, Skand. Arch. Physiol. 3:295–318, 1891.
4. Blix, M.: Die Länge und die Spannung des Muskels, Skand. Arch. Physiol. 4:399–409, 1893.
5. Blix, M.: Die Länge und die Spannung des Muskels, Skand. Arch. Physiol. 5:149–206, 1894.
6. Brand, P.W., et al.: Relative tension and potential excursion of muscles in the forearm and hand, J. Hand Surg. 6:209–219, 1981.
7. Cranor, K.C.: Personal communication.
8. Crawford, G.N.C.: An experimental study of muscle growth in the rabbit, J. Bone Joint Surg. 36B:294–303, 1954.
9. Elftman, H.: Biomechanics of muscle, J. Bone Joint Surg. 48A: 363–377, 1966.
10. Fick, A.: Statische Betrachtung der Muskulature des Oberschenkels, Z. Rationelle Med. 9:94–106, 1850.
11. Fick, R.: Handbuch der Anatomie und Mechanik der Gelenke unter Berücksichtigung der bewegenden Muskeln, 1904–11,

Vol. 3, Spezielle Gelenk-und Muskel-mechanik, Jena, 1911, Gustav Fischer.

12. Freehafer, A.A., et al.: Determination of muscle-tendon unit properties during tendon transfer, J. Hand Surg. 4:331–339, 1979.

13. Goldspink, G.: The adaptation of muscle to a new functional length. In Anderson, D.J., and Matthew, B., editors: Mastication, Bristol, England, 1976, John Wright & Sons, Ltd. pp. 90–99.

14. Gordon, A.M., et al.: Tension development in highly stretched vertebrate muscle fibres, J. Physiol. 184:143–169, 1966.

15. Hill, A.V.: The series elastic component of muscle, Proc. R. Soc. Lond. (Biol.) 137:273–280, 1950.

16. Hill, A.V.: The mechanics of active muscles, Proc. R. Soc. Lond. (Biol.) 141:104–117, 1953.

17. Holland, G.J.: The physiology of flexibility: a review of the literature, Kinesiology Review 1968, p. 49.

18. Huxley, A.F. and Peachey, L.D.: The maximum length for contraction in vertebrate striated muscle, J. Physiol. (London) 156:150–165, 1961.

19. Jansen, M.: Ueber die Länge der Muskelbündel und ihre Bedeutung für die Entstehung der spastischen Kontrakturen, Z. Orthop. Chir. 36:1–57, 1917.

20. Omer, G.E., et al.: Determination of physiological length of a reconstructed muscle-tendon unit through muscle stimulation, J. Bone Joint Surg. 47A:304–312, 1965.

21. Recklinghausen, H. von: Gliedermechanik und Lähmungsprothesen, Berlin, 1920, J. Springer.

22. Seddon, H.J.: War injuries of peripheral nerves, Br. J. Surg. (War Supplement) 2:329, 1949.

23. Stanish, W.D.: Neurophysiology of stretching. In D'Ambrosia, R., and Drez, D., Jr., editors: Prevention and treatment of running injuries, Thorofare, N.J., 1982, C.B. Slack, Inc., Publishers. pp. 135–145.

24. Steindler, A.: Postgraduate lectures in orthopedics: Diagnosis, and indications, Springfield, Ill., 1950, Charles C Thomas, Publisher.

25. Steindler, A.: Kinesiology of the human body, Springfield, Ill., 1955, Charles C Thomas, p. 47.

26. Steno, N.: Elementorum myologiae specimen s. musculi descriptio geometrica, 1667. In Maar, V., editor: Opera Philosophico, Copenhagen, 1910, vol. 2, p. 108. Quoted in Bastholm, E.: The history of muscle physiology, Copenhagen, 1950, Ejnar Munksgaard.

27. Tarbary, J.C., et al.: Physiological and structural changes in the cat's soleus muscle due to immobilization of different lengths by plaster casts, J. Physiol. 224:231–244, 1972.

28. Tarbary, J.C., et al.: Functional adaptation of sarcomere number at normal cat muscle, J. Physiol. Paris, 72:277–291, 1976.

29. de la Tour, E.H., et al.: Decrease of muscle extensibility and reduction of sarcomere number in soleus muscle following a local injection of tetanus toxin, J. Neurol. Sci. 40:123–131, 1979.

30. Weber, W., Weber, E.: Mechanik der menschlichen Gehwerkzeuge, Göttingen, 1836, Dieterich.

ADDITIONAL READINGS

An, K.N., Linscheid, R.L., Brand, P.W.: Correlation of physiological cross-section areas of muscle and tendon, J. Hand Surg. 16:66–67, 1991.

Kaufman, K.R., An, K.N., Chao, E.Y.: Incorporation of muscle architecture into the muscle length tension relationship, J. Biomech. 22:943–948, 1989.

CHAPTER 4

How Joints Move

Anne Hollister, M.D.,
with David Giurintano, M.S.M.E.

Bones are responsible for rigidity within a segment of a limb, joints provide freedom of movement, and muscles serve to move rigid segments on each other. A great deal of mechanical design is necessary in making a joint, if it is to allow freedom of motion without instability or reduction of strength.

The simplest unplanned joint is a fracture. It provides a point at which the two segments of bone can angulate and move relative to each other. It is often surrounded on all sides by muscles, which, by their selective contractions, might control the directions in which the fracture might bend. We have known pain-free persons, such as leprosy patients, to use a fractured limb and explore their new range of freedom. It soon becomes obvious that an unplanned joint is not a source of freedom, but of weakness and progressive deformity. It rapidly develops either a gross shortening from overriding of the segments from the compression of the muscle pull or an irreversible angulation, because there is no control over the movements or the muscles' moment arms at the level of the fracture. As soon as angulation begins at an unplanned joint, the muscle on the angulating side begins to bowstring across the angle, multiplying its mechanical advantage until it is unopposable.

Starting at the point of free motion between two bone segments, it soon becomes clear that real freedom must depend on discipline and control. It is essential to limit the potential movements of the bones to just those specific angulations that can be controlled by specific muscles, which, in their turn, may have to be limited by sheaths or slings to moment arms that match the moments of those that oppose them. Stability and freedom of motion must occur in the face of the large compressive loads from the muscles.

It is not easy to analyze the freedoms and limitations of some of the synovial joints we have to work with. There is usually just one pair of opposing joint surfaces at which the bones meet. We tend to look at the shape of the opposing cartilage and try to figure out what it means in terms of motion. A better way is to analyze the motion itself in an intact limb, and define an axis of motion for each separate component of joint motion. Then we realize that the cartilaginous surfaces have taken their shape from the need to provide gliding for movements around one or more axes of rotation, which can be proximal or distal to the interface and at unexpected angles to it.

The *axis of rotation* between two bones may be defined as a line which does not move in relationship to either bone while they move around each other. In many axes, the line actually passes through one of the bones, and is thus clearly immobile in relation to that bone.

A simple mechanical system which demonstrates the three-dimensional position of the axis directly is desirable. Inman[7] used a mechanical method to demonstrate directly the three-dimensional position of the axes of rotation of the ankle and subtalar joints. He found that the axes were constant and were not perpendicular to the limb segments. Another simple mechanical method, the axis finder, can be used to find the axes of rotation of the joints in intact and living hands. The principle of the method can be understood if one considers a stick attached to and projecting out from a wheel. If a stick is put in a wheel anywhere but along the axis of rotation the stick will move in an arc when the wheel turns. The only place pure rotation of the stick can occur without describing an arc is along the axis of rotation. The "axis finder" (Fig 4–1) is a freely movable rod that can be attached to a bone which moves about a joint. The joint is then moved and the rod's position is adjusted until only rotation of the rod is observed with joint motion. This method, first described by Wright and Johns[11] has been used to locate the axes of rotation of the thumb[5] and index[1] finger metacarpophalangeal (MCP) joints, the thumb carpometacarpal[4] (CMC) and inter-phalangeal (IP) joints,[5] the forearm, the ankle,[10] the jaw, and the knee.[6] These studies have shown the location of the axes of rotation or hinges around which the bones move.

This concept, where apparently universal motion results from the combination of simple hinges, each of which represents a true axis of motion, is basic to the understanding of the mechanics of joint motion. Therefore, we have to simplify the concept by using the graphic symbol of hinges to represent the actual freedoms and limitations that exist in the joints of the hand.

Synovial joint movements occur about axes of rotation. The orientation of the axes determines the kind of motion which occurs. We have been taught to think of motion in terms of the anatomic planes. Flexion and extension are movements which occur in the sagittal plane, abduction and adduction in the coronal plane, and internal and external rotation in the transverse plane (Fig 4–2). Hand surgeons and therapists tend to think

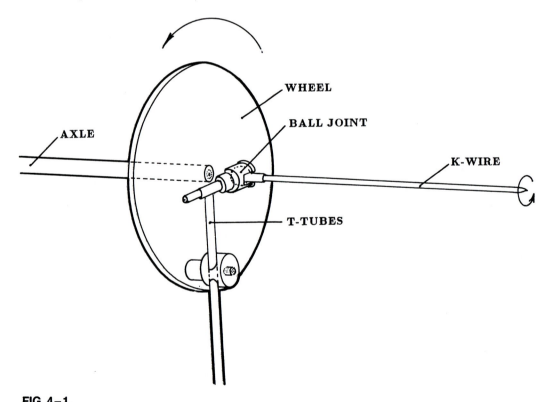

FIG 4–1.
The axis finder allows the rod to be moved about until it is in line with the axis of rotation of a joint. When it is so aligned, the rod only rotates when the wheel is moved. If it were in any other position, it would describe an arc when the wheel moved.

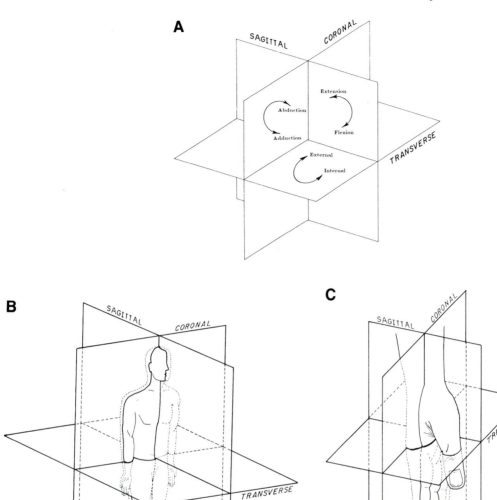

FIG 4–2.
A, the rotations of limb segments defined relative to the anatomic planes. Flexion and extension occur in the sagittal plane, abduction and adduction in the coronal plane, and internal and external rotation in the transverse plane. **B,** the anatomic planes are shown in relation to the body in the standard position. Note that the hands are not aligned with the anatomic planes. **C,** the plane of the palm is the coronal plane, a plane perpendicular to the coronal plane running the length of the hand is the sagittal plane, and the third plane, which is perpendicular to both the coronal and sagittal planes, is the transverse plane. In resting or neutral position, the thumb sticks out from the palm and is not oriented with the anatomic axes.

of flexion as being the closing motion, extension the opening of a joint, abduction the moving away, and adduction the moving toward the center of the hand. These terms are useful since they describe the position of the hand to the various members of the

treating team, but they are not anatomically accurate or precise.

For a joint to exhibit pure flexion and extension, its axis of rotation must be perpendicular to the sagittal plane (Fig 4–3,A). If the hinge is tilted in the transverse plane

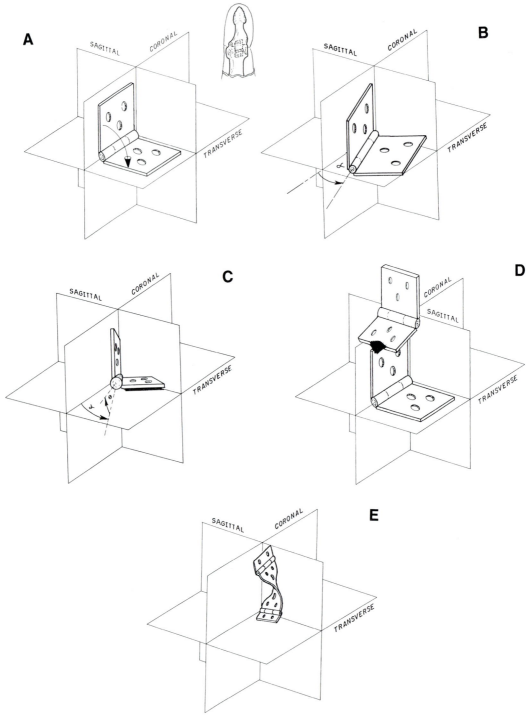

FIG 4–3.
A, the hinge is perpendicular to the sagittal plane, so the motion about the hinge's axis is pure flexion and extension. **B,** the hinge is now tipped in the transverse plane. Now motion includes varus and valgus as well as flexion and extension. The amount of varus and valgus is determined by the amount of offset, and is constant throughout the range of motion. **C,** this hinge is tipped in two planes, and now motion includes varus and valgus and internal and external rotation, in addition to flexion and extension. Again, the amount of these conjoint rotations is determined by the amount of offset of the hinge's axis from the anatomic planes. **D,** these two hinges are welded together. The first hinge is perpendicular to the sagittal plane, and allows pure flexion and extension. The second

(Fig 4–3,B), pure flexion and extension will no longer occur. There will be a new plane of motion which is perpendicular to the hinge's axis, but the motion will include varus and valgus when viewed relative to the anatomic planes. This situation is similar to a door moving on its hinges. If the hinges are mounted perfectly perpendicular to the floor, the movements of the door will be in the plane parallel to the floor and the door will open and close freely. If the door is hung on hinges which are not aligned perpendicular to the floor, the plane of motion will not be parallel to the floor, and the door will run into the floor.

If the hinge is tipped in two planes, as in Figure 4–3,C, motion will include internal and external rotation as well as flexion and extension and varus and valgus. The amount of varus-valgus and internal-external rotation which occur with flexion is dependent on the amount of offset of the hinge and is constant throughout the range of motion. These associated movements are called conjunct rotations and occur whenever an axis of rotation is offset from the anatomic planes.

Most joints of the body have axes of rotation that are offset from the anatomic planes, and therefore have conjunct rotations. This is easily seen in the IP joint of the thumb. If you look at your thumb, the proximal and distal phalanges lie in nearly a straight line when the IP joint is extended. As you bend your IP joint, the tip pronates and adducts so that it can better interact with the index finger. If you look closely, you can see that there is a constant degree of varus and pronation with flexion. The tip of the thumb is actually tracing a shallow cone because the axis of rotation is not perpendicular to the proximal phalanx. In fact, the axis runs between the tips of the flexion creases on the lateral sides of the thumb and is parallel to the volar and dorsal flexion creases. Note

that these lines are not perpendicular to the proximal phalanx, and that you can roughly determine the offset of your IP joint axis from these lines. The varus and internal rotation with flexion are the conjunct rotations which occur because the IP joint axis is offset from each of the anatomic planes.

The effect of offset hinges can easily be demonstrated with a piece of paper. If the paper is folded along a line perpendicular to its length, the folded ends overlay each other. If the crease is at an angle to the edges, folding results in the paper bending unevenly. Now, in addition to flexion and extension, abduction occurs, i.e., the distal end of the paper moves away from the center as it folds.

This offset also affects the way tendons work on the joints. Recall that the effect of a tendon or any other force on a joint is determined by the distance from the axis of rotation as well as the angle of pull to the axis of rotation. If we place a tendon across the hinge shown in Figure 4–4,A, the tendon will cause pure flexion of the joint. If the hinge and tendon are tilted in the transverse plane, the tendon will cause flexion and varus (Fig 4–4,B). The mechanical advantage of this tendon for this joint is the same because the distance and the angle of pull relative to the axis are the same, but the effect of the tendon is quite different from the previous example. In Figure 4–4,C, the same tendon and hinge are now tipped in two planes and when the tendon moves the joint, flexion, with a small amount of varus and internal rotation, will occur. Again, we have not changed the mechanical advantage of the tendon for moving the joint, but have changed the direction of movement and the application of force to complement the orientation of the axis.

This situation is very similar to the tendons which cross the IP joint of the thumb.

---

hinge is perpendicular to the first hinge and to the coronal plane, allowing pure abduction and adduction. The two hinges together allow combinations of flexion and extension and abduction and adduction. No internal or external rotation can occur. **E,** now the two hinges are both tipped relative to the anatomic planes and to each other. Like the single hinge in **C,** movement about each hinge will include movements in all three anatomic planes, but the movements will be linked. There are two basic movements with these hinges, one about axis 1, and the other about axis 2, but these movements include flexion and extension, abduction and adduction, and internal and external rotation.

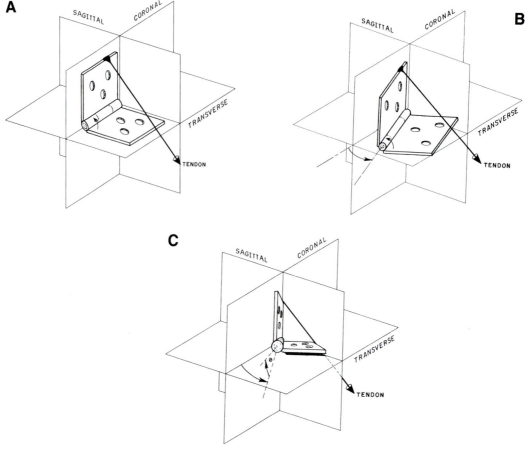

FIG 4–4.
A, the mechanical advantage of a tendon is determined by its distance from and angle of pull to the axis of rotation. The movement which results depends on the orientation of the hinge in space. This hinge is perpendicular to the sagittal plane, so pulling on the tendon will result in pure flexion of the joint. **B,** now the same hinge and tendon have been tipped in the transverse plane. The same pull on the tendon will result in flexion and adduction of the hinge, but the absolute force generated will be the same. **C,** when the hinge and tendon are tipped in two planes, pull on the tendon will cause flexion, adduction, and internal rotation. The magnitude of the force will be the same as in **A,** but the direction will be different.

Both the flexor pollicis longus (FPL) and the extensor expansion (EE) cross the offset axis of the IP joint. The FPL produces flexion, varus, and internal rotation when it moves the joint. Pulling on the EE causes extension, valgus, and external rotation because the axis of rotation of the IP joint is not perpendicular to the proximal phalanx. This arrangement of offset hinges is very efficient. It takes two motors to control a hinge, with one on each side of the hinge. If three different axes existed to produce the flexion, varus, and internal rotation at the tip of the thumb, one would need at least six motors to cross the joint. This would make the thumb a rather cumbersome object. The

varus and internal rotation needed to take a straight thumb in extension to a position which can interact with the index finger in flexion are provided by offsetting a single hinge. Now only two motors are needed for the distal joint, and the thumb can be delicately and precisely controlled, yet powerful.

Many joints have two axes of rotation which allow much greater freedom of movement. These include the wrist, CMC, and MCP joints. The movements of the two axis joints are similar to the classic drive shaft in automobiles and trucks by which the power of the engine is transmitted to the wheels through a universal joint. This joint permits the shaft from the engine to make an angle with the shaft that carries the power onto the rear wheels. It also permits the constant changes of angle between the two segments of the shaft that take place as the vehicle bumps its way over rough ground. The term *universal joint* suggests that universal motion should be possible, such as at a ball-and-socket joint. However, such a joint would be useless, since the function of the drive shaft is to transmit rotary forces. A glance at the joint (see Fig 4–10, A) itself makes it obvious that the so-called universal joint is actually two separate hinge joints, one of which allows only, as it were, flexion and extension, and the other only abduction and adduction. Since both function simultaneously, any angulation is possible at the composite joint of the two hinges, but absolutely no longitudinal rotation can occur. There is no third axis.

Two-axis joints can be thought of as two hinges which are welded together (Fig 4–3,D). The movements of a two-axis joint are combinations of motions about each axis. If the two hinges are perpendicular to the anatomic planes, as in Figure 4–3,D, the motion about the first hinge will be flexion and extension. The second hinge will allow abduction and adduction. The position of the joint can be described in terms of the amount of flexion (motion about hinge 1) and the amount of abduction (hinge 2). There is no internal or external rotation allowed with this axis arrangement.

If the axes of rotation are offset from the anatomic planes, conjunct rotations will occur about each axis. Because of the way the first hinge in Figure 4–3,E is offset from the anatomic planes, its motion is mainly flexion and extension with some varus and valgus, and internal and external rotation. The movements of the second hinge are mainly abduction and adduction, but flexion and extension and internal and external rotation also occur because the axis is not perpendicular to the anatomic planes. Only two kinds of movements occur in this joint—one about the first axis and the other about the second axis—but all three anatomic rotations are present. This is known as a two-degree-of-freedom joint because it has only

two axes of rotation. An interesting characteristic of this sort of joint is that if you know two positions, you know the third. That is to say, if you know the position of flexion and the position of abduction, there is one and only one possible position of internal rotation. Or if you know the position of rotation and abduction, there is only one possible position of flexion.

This model of offset hinges is actually a good model for the CMC joint of the thumb. Figure 4–5 shows the orientation of the thumb's CMC joint relative to the anatomic planes. Close examination of metacarpal movements (as opposed to the whole thumb which includes MCP and IP joint motion), reveal that extension is movement in a plane from the hypothenar eminence out of the palm. Flexion is in this same plane and is from the extended position toward the hypothenar eminence. Abduction and adduction occur about a cone whose apex is volar and ulnar to the joint. Take your metacarpal and place it in 15 degrees of flexion and then in 5 degrees of abduction. You will note that there is some pronation of the metacarpal as you move it from 15 degrees flexion, neutral abduction, to 15 degrees of flexion, 5 degrees of abduction. But once in that position, you could not change the amount of pronation without moving the thumb about either the flexion or abduction axis (Fig 4–5,B). The CMC joint has only two degrees of freedom and two axes of rotation, but the axes are not in the anatomic planes. Again, offset of the hinges allows for motion in three planes with only two axes to control. The thumb needs to pronate with flexion to allow the tip to contact the little finger and to place the thumb tip in proper position to grasp things like hammers. But it needs to be in neutral rotation when extended for wide grasp, so the pulp is facing the surface of the volleyball or coffee cup. If an independent axis of longitudinal rotation were used to achieve this pronation, at least two more motors would be needed at the base of the thumb and the palm would be so full of thumb muscles that grasp would be extremely awkward if not impossible.

The action of tendons that cross two-axis joints is a little more complicated than those

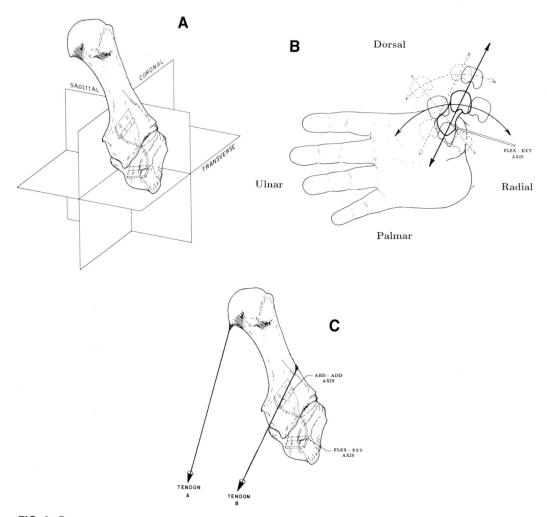

FIG 4–5.
A, these hinges demonstrate the axes of rotation of the thumb carpometacarpal (CMC)-joint. The flexion-extension axis is in the trapezium and the abduction-adduction axis is in the metacarpal. The axes are not perpendicular to the bones or to each other. The flexion-extension axis allows movement of the metacarpal into and out of the palm. Movement about the abduction-adduction axis is about a shallow cone. Neither of these movements occurs in the anatomic planes. **B,** the movements of flexion and extension occur in planes which are directed from the thumb toward the hypothenar eminence (shown by the *thick lines*). The abduction-adduction movements occur about a cone whose apex is volar and ulnar to the CMC joint (shown by the *broken arrows*). Note how pronation and supination occur with movement about both axes, especially the abduction-adduction axis. **C,** these two tendons cross both axes of the CMC joint. *Tendon A* passes palmar to the flexion-extension axis and on the ulnar side of the abduction-adduction axis. Pulling on this tendon will cause the metacarpal to flex and adduct. *Tendon B* is also palmar to the flexion-extension axis and is radial to the abduction-adduction axis. Tension on tendon B will cause the metacarpal to flex and abduct. If the adduction moment for tendon A and the abduction moment for tendon B were equal, pulling on both tendons at once would result in pure CMC flexion.

that cross only one axis. A tendon will have an effect on every hinge it crosses, so any tendon crossing a two-axis joint will have an effect on both axes. If we look at two tendons which cross the CMC joint (Fig 4–5,C), we can see that both tendons cross on the flexor side of the flexion-extension axis, so they both flex the joint. *Tendon A* crosses on the adduction side of the abduction-adduction axis, so it adducts the CMC joint as well as flexes it. *Tendon B* crosses on the abduction side of the abduction-adduction axis, so it acts as an abductor as well as a flexor. It we pull on tendon A, the CMC joint will flex and adduct. If we pull on tendon B, the CMC joint will flex and abduct.

Action of the two tendons produces two very different paths for the metacarpal because they have different effects on each of the joint's axes of rotation.

If we could load the two tendons in such a manner the the adductor moment for *A* was exactly equal to the abductor moment for *B*, only flexion of the CMC joint would occur. The forces about the abduction-adduction axis would be balanced. When the thumb is used for pinch and grip, both the abductors and adductors are often fired together in perfect balance to stabilize the joint and give powerful CMC flexion for pinch.

The wrist is also a two-axis joint. If we draw a transverse section of the wrist we can see the relationship of the extensor carpi radialis longus (ECRL) to the two axes of motion. The perpendicular distance of the ECRL from the abduction-adduction axis is more than the perpendicular distance from the flexion-extension axis. We sometimes forget that there is no way a muscle can exert its moment about one axis without exerting it about the other. Thus when a person

puts tension into his ECRL tendon with the intention of extending the wrist, he is *unavoidably* applying more moment for radial deviation than for extension. If extension without deviation is required, it is essential that another muscle be activated that will produce an equal ulnar-deviating moment to the ECRL's radial-deviating moment. The extensor carpi ulnaris (ECU) is such a muscle. However, in radial nerve palsy, if the only tendon transfer for wrist extension is put into the ECRL, then the patient who wants to avoid constant radial deviation has no wrist extensor that has an ulnar-deviating moment. He may have to use the flexor carpi ulnaris (FCU) to provide the ulnar-deviating moment. Again, however, he cannot use the FCU as an ulnar deviator without activating its moment for wrist flexion as well. This opposes the extensor effect of the extensor carpi radialis and thus weakens the grip. We will return to this dilemma later. It is mentioned here to emphasize the importance of considering both axes every time a muscle balance operation is planned.

There are joints with three axes of rota-

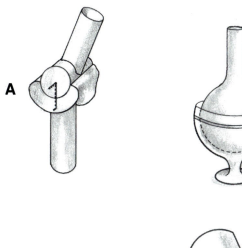

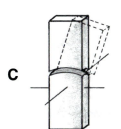

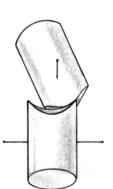

FIG 4–6.
These are pictures from R. Fick of models he made to show how joints of different shapes move. **A** is a model of a single-axis joint and is in the form of a simple cylinder. **B** is of a joint with three intersecting perpendicular axes and is a ball-and-socket joint. **C** and **D** are both two-axis joints. When both axes are on the same side of the joint, that side of the joint is made of two convex surfaces, like the metacarpal head, with the other side having two concavities. If one axis is in each bone, one concave and one convex curve form each of the joint surfaces. (From Fick, R.: Handbuch der Anatomie und Mechanik der Gelenke under Berücksichtisung der bewegenden Muskeln, 1908, Vol. 2, Spezielle Gelenk-und Muskelmechanik, Jena, 1911, Gustav Fisher.)

tion such as the hip and the glenohumeral joint. If the three axes are perpendicular to each other and intersect, a ball-and-socket joint is formed (Figs 4–6,B and 4–10,B). In these joints, there is independent movement about each of the axes. Consider the hip. The femur can be moved to a position of flexion, then to a position of abduction or adduction, and still have a range of internal and external rotation motion available. There are no conjunct motions. This principle is harder to demonstrate for the glenohumeral joint because the scapula and the humerus move together and pure motion of the glenohumeral joint is hard to achieve. These three-axis joints require many motors for control and stability, and are completely surrounded by layers of muscles. They are placed in the proximal part of the limb so that these heavy muscles are near the center

of gravity and need not be moved around very much when the limb is used. Stabilizing these joints after paralysis is quite difficult because of the many degrees of freedom that are present.

SHAPES OF JOINTS

The theory that joints move about discrete fixed axes of rotation was introduced by Adolf Fick[2] in 1854. Adolf Fick is better remembered for his law of gas diffusion and his equations for conjugate gaze, but he was also the father of orthopaedic mechanics. He was a brilliant mathematician and physicist in addition to being a professor of physiology. He studied the shapes of the joints and noted that they corresponded to mathematic

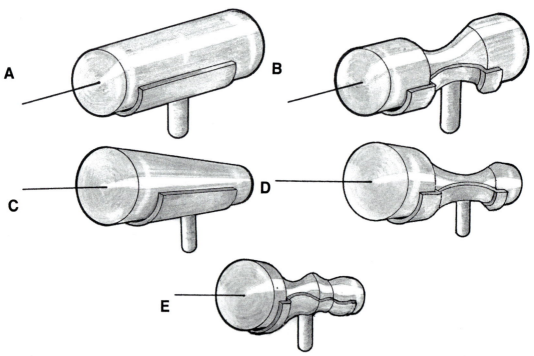

FIG 4–7.
A, joint with a single axis in which the axis is parallel to the joint surface forms a cylinder. **B,** we can turn the cylinder on a lathe and make a trochlea by varying the radii of the joint surface. The joint still moves about a single axis. **C,** when the axis is not perpendicular to the joint surface, a cone shape is formed. **D,** again, we can make a trochlea in the cone. The trochlea is perpendicular to the axis, not to the joint surface. This is similar to the IP joints of the thumb and fingers. **E,** the cone can be made any number of shapes by varying the sizes of the radii along the axis. Here we have created the shape of the distal humerus. All of the grooves are perpendicular to the axis and not the joint surface. (Adapted from Fick, R.: Handbuch der Anatomie und Mechanik der Gelenke unter Berücksichtigung der bewegenden Muskeln, 1908, Vol. 2, Spezielle Gelenk-und Muskelmechanik, Jena, 1911, Gustav Fischer.

curves consistent with two bodies moving about one or more fixed axes of rotation (Fig 4–6).

Single-axis joint movements generate shapes of modified cylinders or cones. In Figure 4–7,A, we see the shape of a simple cylinder with a single axis. If we vary the radius of the cylinder, we can produce a trochlea as in Figure 4–7,B, which is similar to the long finger proximal interphalangeal (PIP) joint. When the axis is not perpendicular to the joint surface, the joint will generate a conical shape, as seen in Figure 4–7,C, which is similar to the ankle joint. Varying the sizes of the radii of the joint can produce a joint surface with an offset trochlea like that of the IP joint of the thumb (Fig 4–7,D). A double trochlea shape like the elbow can also be formed (Fig 4–7,E).

The shapes generated by two-axis joints are more varied. Adolf Fick noticed that joints such as the CMC have saddle-shaped surfaces. This form is like the inner surface of a doughnut. If one small circle moves about another, larger circle, a doughnut shape is formed (Fig 4–8,A and B). The surfaces of two-axis joints are bits of the surface of the doughnut. If both axes are on the same side of the joint, the surface consists of two convex curves (Fig 4–8,C). When one axis is on each side of the joint, a saddle shape is formed (Fig 4–8,D).

Human two-axis joints do not have the symmetric surfaces of a doughnut. Again, if we vary the radius of the small circle, we can make different surface shapes which are still consistent with two fixed axes of rotation for the joint. If we gradually increase the radius of the outer circle, we get the deep saddle seen in Figure 4–9,A, which is similar to the CMC joints. Decreasing the radius of the outer circle gives a football shape to the joint surface, seen in Figure 4–9,B. If one incorporates the fact that the axes of rotation are not perpendicular to each other or parallel to the joint surface, and a changing radius for the circles, the asymmetric surfaces of human joints emerge.

The position of the axes of rotation determines the position the finger or thumb can take. Understanding this concept helps one to predict the effect of surgery, such as a transferred tendon, or an osteotomy, on the final range of positions available for the finger. A single-axis joint is pretty simple. We can figure out the amount of conjunct rotation that will occur with motion by noting the offset of the axis from the anatomic planes, as in Figure 4–3,C.

For a simple two-axis joint like a universal joint where the two axes are perpendicular and intersect, the space the body could occupy is the surface of a sphere (Fig 4–10,A). The finger can either abduct and adduct, or flex and extend, or have a combination of both movements. No pronation or supination can occur.

If the joint has three intersecting, perpendicular axes, the space the limb can occupy is still on the surface of a ball, but the limb can turn internally and externally as well (Fig 4–10,B).

The shape of motion for a two-axis joint in which the axes are perpendicular but do not intersect is a doughnut (Fig 4–10,C). Again, the finger can flex and extend and abduct and adduct, but no pronation or supination occurs. The shape of motion is much broader than that of a sphere in the flexion-extension range and much narrower in the abduction-adduction range, a shape that is very useful in the hand and in the extremity as a whole.

If the two axes are not perpendicular and do not intersect, a skewed doughnut results (Fig 4–10,D). The track of the finger is no longer limited to the anatomic planes, and conjoint rotations occur as the finger moves about this crooked doughnut. This shape of motion is very useful when one is designing fingers because it lets there be a wide range of flexion and extension to open and close the hand. The fingers can be abducted in extension and tightly clustered in flexion, and have a narrow range of motion about the abduction-adduction axis. Only four motors are needed to control the MCP joint, and yet provide this uniquely adapted range of positions for the proximal phalanx.

These concepts are important not only in the original design of the hand, but to those of us who would dare to attempt to construct a replacement arthroplasty. It is important that the mechanics, including the number

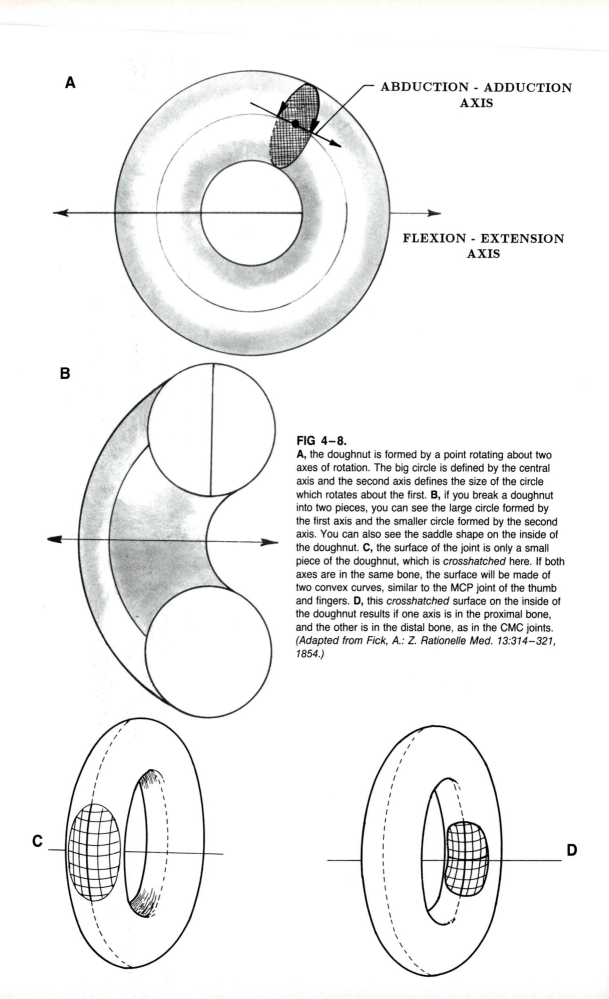

A

ABDUCTION - ADDUCTION
AXIS

FLEXION - EXTENSION
AXIS

B

FIG 4-8.
A, the doughnut is formed by a point rotating about two
axes of rotation. The big circle is defined by the central
axis and the second axis defines the size of the circle
which rotates about the first. **B,** if you break a doughnut
into two pieces, you can see the large circle formed by
the first axis and the smaller circle formed by the second
axis. You can also see the saddle shape on the inside of
the doughnut. **C,** the surface of the joint is only a small
piece of the doughnut, which is *crosshatched* here. If both
axes are in the same bone, the surface will be made of
two convex curves, similar to the MCP joint of the thumb
and fingers. **D,** this *crosshatched* surface on the inside of
the doughnut results if one axis is in the proximal bone,
and the other is in the distal bone, as in the CMC joints.
*(Adapted from Fick, A.: Z. Rationelle Med. 13:314–321,
1854.)*

C

D

A

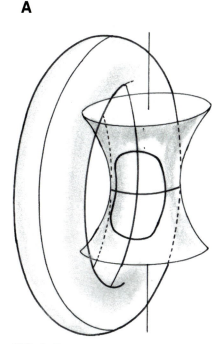

B

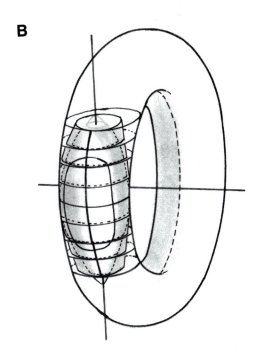

FIG 4–9.
A, if we vary the size of the radii for a two-axis joint, we can create a number of surfaces. Here we have decreased the radius for the smaller circle as we move about the bigger one, creating a sort of football shape. A body moving on this surface would still move about the same two axes as those on the surface of the original doughnut. **B,** we can change the shape of saddle in a similar fashion, and produce an asymmetric surface for two fixed axes if we wish. *(Adapted from Fick, A.: Z. Rationelle Med. 13:314–321, 1854.)*

and location of the naturally occurring axes, be reproduced in the artificial joint if normal function is to be achieved. For one, the moment arms for the tendons crossing the joints will be different if the axes' location is changed. The positions that the digit can assume will also be very different. If one replaces the CMC joint, which has two axes and a motion shape similar to that shown in Figure 4–10,C, with a ball-and-socket joint, the resulting envelope of motion would be a section of a sphere (see Fig 4–10,B), which is very different. The muscles and tendons would have to control three instead of only two axes of motion, and the distances of the muscles from the new axes would also be quite different.

The movements of the joints are also related to the shapes of the soft tissues. R. Fick[3] noted that only one- and two-axis joints have collateral ligaments. Inman[7] found that the origins of the ligaments of the ankle were the axes of rotation of the joints. The collateral ligaments of the joints of the

fingers and the thumb arise from the epicondyles, and the flexion-extension axes of the joints pass just volar to the epicondyles. This is also true for the elbow joint, where the flexion-extension axis is just distal and anterior to the epicondyles[9] which are the origins of the ligaments. In the MCP joints, the abduction-adduction axis passes through the point of attachment of the true collateral ligaments. This arrangement in the two-axis joints where the ligaments connect the two axes allows the ligaments to be nearly isometric throughout the normal range of motion of the joint.

The relationship of the ligament to the axis of rotation is beautifully demonstrated in the forearm, where the fibers of the interosseus membrane take their origin from the axis of rotation, first along the ulna, then along the radius. Where the axis runs inside the bones, there is no ligament, hence the holes in the membrane's distal and proximal ends.

The elbow is also a splendid example of

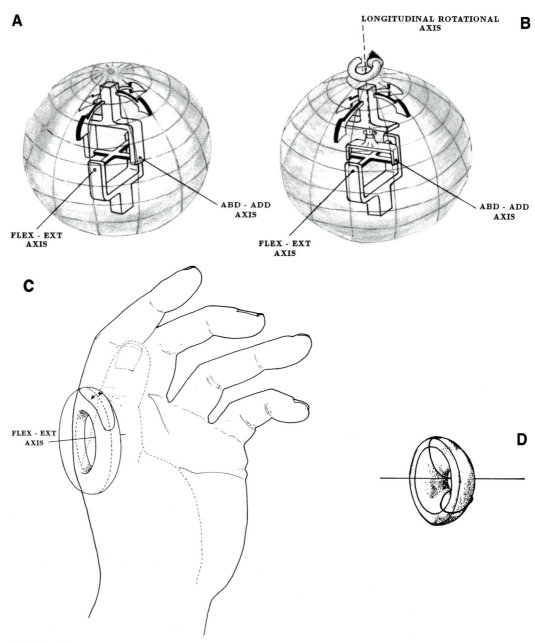

FIG 4–10.
A, the surface of motion for a universal joint which has two perpendicular intersecting axes is a sphere. The finger can move about the sphere in the east-west direction (abduction-adduction) or in the north-south direction (flexion-extension), but cannot turn around. **B,** for a ball-and-socket joint like the hip, the surface of motion is also a sphere, but the femur can turn around (internally and externally rotate) and face in any direction. **C,** if the joint has two perpendicular axes which do not intersect, the shape of motion is a part of a doughnut. Again, the finger can flex and extend, or abduct and adduct, but not pronate or supinate. This doughnut favors a wide arc of flexion and extension, and a narrow arc of abduction and adduction. It is no accident that the shape of motion of the joint and the shape of the joint surface are similar. The joint shape is actually much more complex (see Figs 4–7 and 4–9) than the shape of motion of a point on one of the bones. **D,** this doughnut results if the two axes do not intersect and are offset 45 degrees from each other. It is a skewed doughnut. The path of abduction-adduction in this doughnut is illustrated by the *line.* The finger appears to pronate and then supinate as it moves around the abduction-adduction axis. (Courtesy of A. Lupichuk.)

this principle. On the medial side, the elbow joint has only one axis, the flexion-extension axis and the medial collateral ligament is a fan-shaped structure arising from the epicondyle. The lateral side of the elbow has both the elbow flexion-extension axis,[9] under the lateral epicondyle, and the pronation-supination axis, which passes through the radial head. The only isometric point would be from the epicondyle to the center of the radial head. No ligament could survive the compressive load between the two bones, so there is no true lateral collateral ligament in the elbow. Instead, the lateral ligament inserts on the annular ligament which encircles the radius and is isometric. In about 10% of elbows, there is a slip from the lateral epicondyle to the ulna. This ligament crosses only one axis (the elbow flexion-extension axis) and could be considered to be a lateral collateral ligament, but there is no ligament that goes directly from the humerus to the radius.

Most collateral ligaments are isometric and are not under tension in the normal range of joint motion. If the normal range of motion is exceeded, and particularly if the joint is pushed into an unphysiologic position, that is subluxated or dislocated, these structures are placed under tension, and can tear or rupture. Some ligaments, such as the accessory collateral ligaments in the fingers, are under tension at all times and are dynamic stabilizers of the joints. The accessory collateral ligaments hold the flexor sheath to the bone, and when the finger is flexed are under considerable load (see Fig 5–24).

For centuries, anatomists have defined the three planes by which the location of structures may be described and joint movements defined; the sagittal plane which runs anteroposteriorly, the coronal plane which runs from side to side, and the transverse plane which runs at right angles to the other two planes and is parallel to the ground when the person is standing (see Fig 4–2,B).

There are problems when precision is needed in relating the joints of the hand to conform to the anatomic planes. The classic diagram of a body shows the hands in supination, so that the palms face forward, and it is clear that the palm of the hand lies in the coronal plane. However, the arms are extended obliquely and thus do not lie parallel to the sagittal plane and are cut obliquely by the coronal plane. It was probably intended that the arm and hand should be considered as a longitudinal structure, parallel to the body, and that the fingers should flex in the sagittal plane (see Fig 4–2,C). Furthermore, as we have seen, the axes of rotation of the joints are offset from the anatomic planes, so motion about them is not in the anatomic planes, nor is motion about one joint in a finger in the same plane as the next.

The thumb presents greatest difficulty in matching the hand to classic planes. At rest, the thumb is usually thought of as lying halfway between pronation and supination. This would bring the action which we call flexion into the coronal rather than the sagittal plane. Thus, what we call flexion of the thumb would become, anatomically, adduction.

Hand surgeons have tried to develop systems to standardize the nomenclature of thumb movements, and terms such as "palmar abduction" have been introduced. However, we have not found these terms easy to work with. For the purpose of this book, we will use terminology that defines hand movements in the way most hand surgeons and therapists seem to think about them.

All of the palmar surface of the hand and thumb is to be regarded as the front, as though it were facing forward and in the coronal plane, no matter what its position at the time. The back of the hand is considered to be parallel to the palm, and each digit is to have its planes of motion considered in terms of its own palmar skin and fingernail. Any movement which flexes the palmar digital skin creases may be called flexion, no matter what happens to be the position of the digit in relation to the arm or to the planes of the body. Thus, when the thumb is in the midprone position, adduction and abduction will be movements away from the plane that runs the midline length of the thumb.

THE AXES OF ROTATION OF THE JOINTS OF THE THUMB

The thumb consists of four bones linked by three joints (Fig 4–11). As we have seen, the CMC joint has two axes of rotation, one in the trapezium and the other in the metacarpal. The MCP joint also is a two-axis joint. The flexion-extension axis is in the metacarpal passing under the epicondyles and the abduction-adduction axis passes between the sesamoids just proximal to the beak of the proximal phalanx. The axis of rotation of the IP joint is just palmar and distal to the epicondyles on the proximal phalanx. None of these axes is perpendicular to the bones or to one another.

All of the muscles which cross the distal joints also cross the CMC joint. They tend to have bigger moment arms for the CMC joint than the distal joints. It is interesting to note that the muscles which are abductors of the CMC joint are also abductors of the MCP joint. Those which are adductors of the CMC joint are also adductors of the MCP joint. This means that when the muscles that move the CMC joint into adduction are used, they also move the MCP joint into adduction. In fact, it is very difficult to move one of the joints into abduction or adduction without moving the other in the same direction. This arrangement of the muscles about the abduction-adduction axis of both joints has caused the two joints to be thought of as a single unit, with descriptions for combined movements of both joints such as "opposition" (CMC flexion and abduction with MCP abduction) being common. This is very useful in the normal hand where the motors and the motions of both joints are working in concert, but can cause problems when reconstruction of the joints or transfers for paralysis are contemplated (see Chapter 12).

When a tendon like the extensor pollicis longus (EPL) passes from the wrist to insert on the distal phalanx, it acts on all of the joints, and all of their axes of rotation. A tendon will have an effect on every hinge it crosses, whether or not it is attached directly to the bone nearest the joint. The EPL is dorsal to the flexion-extension axes of all of the joints, and acts as an extensor of all three thumb joints. It is ulnar to the abduction-adduction axis of the CMC and MCP joints, and adducts both joints. The offset of the axes results in the conjoint motion of supination with adduction in both joints, so the EPL supinates the thumb as it adducts. As the thumb is adducted, the EPL can be seen to slide dorsal and ulnar at the CMC joint and ulnar at the MCP joint. This increases the EPL's moment arm for adduction for both joints, and makes possible a trick motion that is quickly learned by patients with intrinsic paralysis who cannot oppose the thumb. They hold objects by an effective sideways squeeze between the first and second metacarpals. If this becomes a habit, patients may continue to contract their EPL for pinch even after a tendon

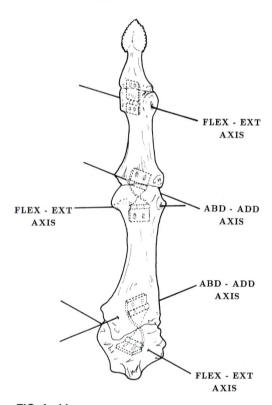

FLEX - EXT AXIS

FLEX - EXT AXIS

ABD - ADD AXIS

ABD - ADD AXIS

FLEX - EXT AXIS

FIG 4–11.
The thumb has three joints, the carpometacarpal (CMC), the metacarpophalangeal (MCP), and the interphalangeal (IP) joints. The CMC and MCP joints both have two axes of rotation, shown here by the *linked, offset pairs of hinges.* The IP joint has one axis or hinge. None of the hinges are perpendicular to the bones or to one another, so there are conjunct rotations with flexion-extension at all of the joints and with abduction-adduction at the CMC and MCP joints.

Joints of the Thumb

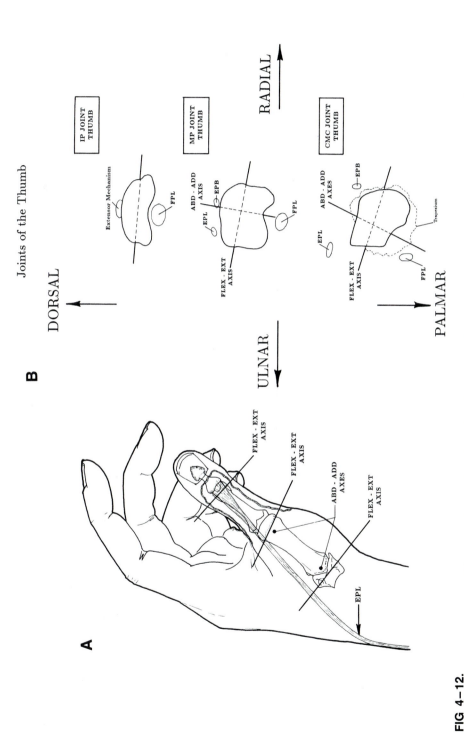

FIG 4–12.
A, the extensor pollicis longus (*EPL*) leaves the thumb on its path to the dorsum of the wrist, and develops moments for adduction of both the carpometacarpal (*CMC*) and metacarpophalangeal (*MP*) joints as it does so. **B,** the cross-sectional diagrams at each joint are for demonstration only, and do not represent comparative numbers for moment arms. The amount of extension-abduction moment at the CMC joint, for example, varies with the position of the joint because there is no sling or retinaculum to hold it at it at that level. *ADD* = adduction; *ABD* = abduction; *EXT* = extension; *FLEX* = flexion; *FPL* = flexor pollicis longus; *IP* = interphalangeal; *EPB* = extensor pollicis brevis.

transfer has been done to restore opposition. In such cases, the old effective trick movement opposes the new transferred tendon and the operation becomes a failure. When I have feared that outcome, I have sometimes rerouted the EPL at the same time as the transfer for opposition, so that the EPL comes to lie over the tendon of the abductor pollicis longus (APL). It then maintains its extensor moment on both the CMC and MCP joints, but loses its adductor–external rotator moment arms. (Fig 4–12,A and B).

AXES OF MOTION OF METACARPOPHALANGEAL JOINTS OF THE FINGERS

When we began to define and measure the way in which muscles and tendons affect the movements of joints, we assumed that MCP joint motion could best be resolved by reference to three axes: (1) flexion-extension, (2) abduction-adduction, and (3) rotation. We assumed that the flexion-extension axis and the abduction-adduction axis would be at right angles to each other, and the rotation axis would be axial or longitudinal to the proximal phalanx. Oblique movement would be resolved by vectors.

It is hard to use the three-axis concept to account for the fact that abduction and adduction are free when the finger is straight, but rotation is extremely limited, while rotation is possible in flexion but not abduction-adduction.

Working with Agee and then with Hollister, I have come to accept their analysis and to realize that all falls into place if one regards the MCP joints of the fingers as two-axis joints. One axis is for flexion-extension and is transverse through the metacarpal head. The other is like a cone axis and runs forward from the metacarpal head with an inclination distally of maybe 30 degrees above a right angle.

An umbrella is rather like a cone that can open and shut. The extended fingers follow an arc like the fabric surface of an open umbrella (Fig 4–13,A). The partially flexed MCP joint in the pinch position allows the proximal phalanx to follow the surface of a partially closed umbrella (Fig 4–13,B). When the MCP joint is fully flexed, the proximal phalanx can rotate but has limited lateral movement. A similar situation is seen in a nearly closed umbrella where the ribs are close to the axis of the umbrella (Fig 4–13,C). Each of the MCP joints has a similar axis, but the index finger, having the most freedom, exhibits it best. The first dorsal interosseus and the first palmar interosseus seem perfectly placed to control the movement around the conic axis.

The proximal phalanx can only rotate, in the true rotational sense, when it lies along the axis of the cone. When the phalanx is extended to form an angle to this axis (Fig 4–14,A), its movement from side to side follows the curve of the cone centered on the same axis. This cone becomes wider as the finger is extended until, when the finger is straight, the sideways movement may appear to be a straight side-to-side tilt. Not so, however. If so-called abduction-adduction is closely observed when the finger is straight, it will be seen to be part of a shallow curve with a wide radius, and it is accompanied by a slight rotation that is constrained to the curve, in much the same way that the thumb rotates as part of its circumduction around its narrower cone.

In order to understand this better, I invite you to carry out a simple experiment on your own index finger as you read on. Bring your index finger and thumb into a normal position of full opposition and pulp-to-pulp pinch with all joints of finger and thumb in enough flexion to produce an open O pinch (Fig 4–14,B). In this position, the cone axis of the index MCP joint passes straight to the point of contact of finger and thumb. This axis is not at right angles to the index metacarpal, but maybe 20 to 30 degrees more open. Now try to move the index finger into abduction and adduction. The tip of the finger will not move, but the PIP joint and the whole curved finger will rock to and fro like a bow turning on a bowstring. Now, without moving the index finger, move the thumb away and repeat the attempt to move the finger laterally and medially. Still the bow will rock, but it will prove impossible to

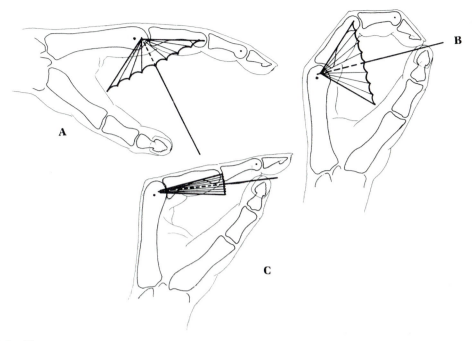

FIG 4–13.
A, the extended finger follows an arc like the fabric surface of an umbrella. The longitudinal axis projects in a palmar direction from the metacarpal head along the line illustrated by the handle of the umbrella. **B,** when the tip of a finger touches the thumb, their pulps embrace the palmar projection of this axis from the metacarpal head. In this position, the intrinsic muscles that appeared to radially and ulnarly deviate the extended finger are now clearly visualized as torsional stabilizers of the entire finger. Note that the proximal segment of the finger continues to follow the surface of a partially closed umbrella. (Also note that a laterally or medially directed force applied to the side of the finger would generate a torsional load as defined by the distance from the point on the side of the finger to the longitudinal axis; it is this torque that the intrinsic muscles have to *shock-absorb* and compete against.) **C,** As the metacarpophalangeal (MCP) joint approaches full flexion, the proximal phalanx approaches and then becomes parallel with the longitudinal axis, similar to the umbrella's ribs embracing the shaft-like handle. (*Note:* When the long axis projects down the proximal phalanx, we take advantage of this position clinically to check the integrity of the collateral ligaments. When the fingers are in this position, a laterally deviating force cannot be easily dissipated and thereby puts the ligaments and bone of the MCP joint at risk for traumatic failure.)

move the tip of the finger sideways. Now flex the MCP and PIP joints a little more so that the fingertip moves an inch or so. Now the same longitudinal axis that projects from the palmar surface of the MCP joint extends *through* the distal interphalangeal (DIP) joint such that an attempt to abduct and adduct the finger will result in the (PIP) joint moving one way and the tip of the finger moving the opposite way while the DIP joint displaces neither medially or laterally, but remains in the same position in space as it rotates about the axis (Fig 4–14,C and D).

Since most of our earlier measurements were made on the assumption of a simple abduction-adduction axis, we will still refer to moment arms for abduction-adduction. However, the figures are only true for the extended position of the MCP joint. The

more it flexes, the greater the significance of the rotatory component of the movement around a narrowing cone (or a closing umbrella).

MECHANICS OF THE INTERPHALANGEAL JOINTS

The IP joints of the fingers are the terminal parts of a chain in which the MCP joint is the key. Position of the finger, mechanisms of application and dissipation of load, and the mechanism of application and dissipation of load, and the mechanics of the muscles are all determined primarily at the MCP joint.

The IP joints are single-axis joints with

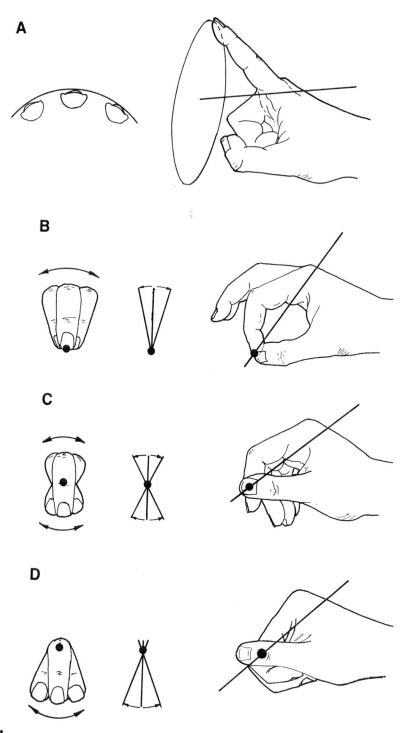

FIG 4–14.
The longitudinal axis is key to understanding how the metacarpophalangeal (MCP) joint works. When the finger is extended **(A)**, the tip is a long distance from the axis, and wide abduction and adduction can occur. As the tip is brought closer to the axis, the tip describes a smaller and smaller arc of motion until, when the tip is along the axis, no lateral motion can occur **(B)**. This is the position of precision pinch, where the action of the interossei can only flex the tip, not laterally deviate it. When you perform precision activities such as picking up a pin or scratching your head, you will notice that the tip of the finger is on the axis of rotation. In lateral pinch **(C)**, the midpoint of the thumb pulp strikes the fingers along this axis. This mechanism greatly reduces the torque needed to stabilize the finger against the lateral thrust of the thumb. When you grab an object such as a hammer in a power grip, you automatically align the center of the shaft with the longitudinal axis. Again, this mechanism decreases the amount of force the interossei would need to stabilize the joint against the lateral force. Finally, in maximal MCP flexion **(D)**, the axis lies just under the proximal phalanx, and tip movement occurs with abduction and adduction, but in the opposite direction as occurred when the finger was extended.

the axis just volar to the origin of the collateral ligaments. The axes are parallel to the flexion-extension creases, and pass just above the dorsal end of the flexion creases of the IP joints. Except for those of the long finger, the IP joints all have carrying angles, which is part of the reason the fingertips close in an arc to point at the volar carpal ligament. The joints have an intercondylar groove on the proximal surface which articulates with an eminence on the distal joint surface and is perpendicular to the axis of rotation. This trochlea gives lateral and rotational stability when the joints are compressed by the flexors and extensors during use. Littler's[8] observation about the ratios of the lengths of the phalanges is important. These ratios allow for the smooth arc which the finger makes in closing and opening, and are also important for normal extensor mechanism function.

WHEN THE AXIS MOVES

All normal transmission of muscle force to the hand is based on the fact that the joints are stable and that the joint axis stays in its relationship to both bones at the joint. However, as a result of injury, arthritis, or paralysis, there is sometimes an alteration in the pattern of movement around an axis, or even a collapse of the whole integrity of a joint that results in great changes in all moment arms and that we must look at here. The axis may change considerably under two circumstances: (1) when the joint no longer glides, and (2) when the joint subluxates or collapses.

FAILURE OF GLIDING

The system of lubrication in a normal joint is so free of friction that there is almost no drag or resistance that is worth even trying to measure (see Chapter 6). However, there are conditions that may roughen the gliding surfaces or that may block gliding because of osteophytes or because of partial subluxation and loss of congruence, or the gliding surface may be roughened when an

intraarticular fracture blocks gliding in one direction.

In any such case, the joint may, at some point, stop being a *gliding* joint and become a *tilting joint*. The functional axis will then immediately move and come to lie at the joint surface. Figures 4–15 and 4–16 show a sequence that demonstrates this.

A wheel on an icy road is rather like a gliding cartilage joint. It spins without movement of its axis (axle). The gliding surface (ice) has a very low coefficient of friction (Fig 4–15,A).

Now if sand is thrown under the wheel, the "joint" no longer glides. It is like a joint in the hand that locks from within because of roughness, irregularity, or lack of congruence. The automobile wheel will now have traction and will begin to roll. The axle will begin to move forward. At that instant the axle ceases to be the axis of the "joint" between the wheel and the road. The definition of axis is "that point that *does not move* in relation to either component of the joint." In a moving wheel, the axis moves from the axle to the road. The axle moves around the new axis (Fig 4–15,B).

As a wheel rolls along, the point in contact with the road constantly changes. Thus the axis is also constantly changing (instant center). Here the analogy is no longer helpful to tilting joints of the hand.

The diagram of a joint shows how that change affects the moments of tendons, capsule, and skin. In attempted flexion, the moment arm of the flexor tendons is very much shortened, and the moment arm of the skin and capsule on the dorsum is increased (or vice versa if extension movement is attempted) (Fig 4–16).

Most seriously, any external force that attempts to move the joint now is resisted by big moments from skin, ligament, and capsule, which now have to lengthen much more for the same angular movement; and the point on the cartilage at which the joint movement is blocked will be subject to very large compressive forces.

All of the above happens at an MCP joint of a rheumatoid patient (Fig 4–17) when it fails to glide and begins to subluxate into the palm. Any attempt to use the long lever of

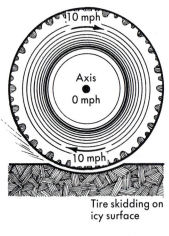

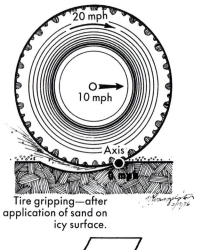

Tire skidding on
icy surface

Tire gripping—after
application of sand on
icy surface.

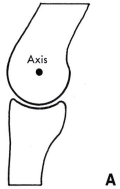

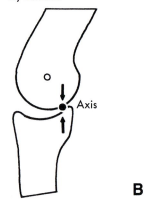

A

B

FIG 4–15.
A, a wheel spinning on its axle (axis) on an icy road is rather like a convex joint surface turning with a free gliding motion on a concave joint surface. **B,** if, with sand on the road, the gliding-sliding is prevented and the tire grips the road, the wheel moves foward and now the point of contact with the road becomes the axis around which the wheel turns. In the joint, if gliding is blocked and the joint tilts without sliding, the effective axis moves at once to the joint surface where the tilting occurs.

the finger to extend the MCP joint results in the axis becoming localized to the dorsal edge of the proximal phalanx, where it impinges on the head of the metacarpal. The result is pain and also absorption of the dorsal lip of the base of the phalanx. Note also how the moment arms of tendons and skin become totally changed.

Therefore, if there is any question of failure of gliding of a joint, the full length of the phalanx should never be used as a lever to restore angular movement. The phalanx should be held near its base and congruence and gliding restored first; then, with the gliding restored and the axis in its right place, angular movement may be carefully restored.

JOINT COLLAPSE

Synovial joints are not the only types of joints between bones. The axes are lost altogether when the joint surfaces are destroyed or when a joint becomes fractured, dislocated, or collapsed. When a joint is replaced by a fibrous arthroplasty, it no longer moves about axes of rotation. If the fibrous tissue is fairly thick, it deforms when the loads of muscles and external load are applied. The mechanics for this type of joint are very different from those for a normal or even subluxated joint. There is no fixed pattern for the movements, since small changes in the direction and magnitude of the forces will markedly change the nature of the deformation. Since increasing movement in a given

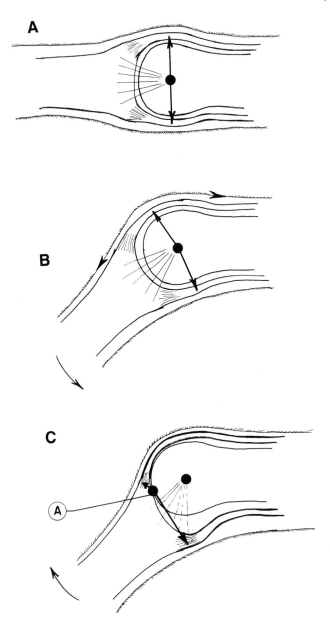

FIG 4–16.
A and **B,** joint axis in normal place. Moment arms for flexion and extension are about equal. **C,** in partial subluxation the axis *A* moves to the surface of the joint, which no longer glides but tilts. The moment arm of the extensor is now very small, and that of the flexor is very large.

direction produces increasing deformation, there is increasing resistance from the joint to the movements away from the resting position. This type of joint works well in settings where it is designed to gently compensate for dissimilar movements between large portions of the body, such as at the pubic symphysis.

This condition can occur at the CMC joint at the base of the thumb. The APL can pull strongly, but since the axes are gone, all that happens is that the whole metacarpal and thumb move proximally bodily, with no abduction. Opposing muscles may contract simultaneously around a shattered or dislocated joint, and the only result may be shortening of the limb or digit. In all such gross cases, the objective of the surgeon is

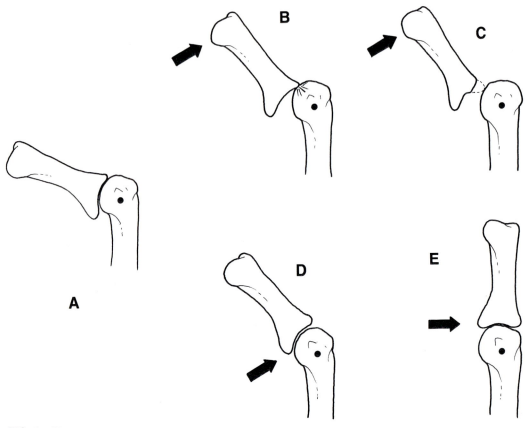

FIG 4–17.
In rheumatoid arthritis, if a phalanx begins to subluxate in flexion **(A)**, it is tempting to use passive external force to reduce the joint. If the force is applied distally **(B)**, it results in tilting, and that results in high stress on the dorsal lip, which becomes absorbed **(C)**. If the external force is applied at the base of the phalanx **(D)**, it tends to restore joint congruence and correct the subluxation **(E)**.

to restore and stabilize the axes or create a new one or perform an arthrodesis on the joint so that at least the muscles crossing it may maintain their length and be effective at other joints proximal or distal to the arthrodesis.

The dramatic change in the mechanics of the CMC joint with introduction of a fibrous joint affects not only the joint in question, but all of the joints in the chain. We shall see (Chapter 5) how changes in the joint moments by loss of muscles can lead to a variety of deformities. If one performs a fibrous or Silastic arthroplasty on a basal finger or thumb joint, one changes the way the joint moves, hence the axes of rotation and therefore the moment arms for all of the muscles at that joint. This can lead to an imbalance in the moment arms similar to that seen in paralysis, and to deformity in the distal joints. Extension deformity of the thumb MCP joint, as in Jeanne's sign, is common distal to severe arthritis with subluxation of the CMC joint or following Silastic or fibrous CMC arthroplasty. There is nothing wrong with the MCP joint itself. The deformity results from the destruction of the mechanics of the basal joint which all of the thumb muscles cross.

REFERENCES

1. Agee, J., Hollister, A.M., and King, F.: The longitudinal axis of rotation of the metacarpophalangeal joint of the finger, J. Hand Surg. 11(A):767, 1986.
2. Fick, A.: Die Gelenke mit sattelförmigen Flächen. Z. Rationelle Med. 314–321, 1854.
3. Fick, R.: Handbuch der Anatomie und Mechanik der Gelenke unter Berücksichti-

gung der bewegenden Muskeln, 1908, Vol. 2, Spezielle Gelenk- und Muskelmechanik, Jena, 1911, Gustav Fischer.

4. Hollister, A., et al.: The axes of rotation of the thumb carpometacarpal joint, J. Orthop. Res. 10:454–460, 1992.

5. Hollister, A., et al.: The axes of rotation of the thumb interphalangeal and metacarpophalangeal joints, J. Orthop. Res. (submitted).

6. Hollister, A., et al.: The axes of rotation of the knee, Clin. Orthop. (in press).

7. Inman, V.T.: The joints of the ankle, Baltimore, 1976, Williams & Wilkins.

8. Littler, J.W.: On the adaptability of man's hand, Hand 5:187, 1973.

9. London, J.T.: Kinematics of the elbow. J. Bone Joint Surg. 63A:529–535, 1981.

10. Singh, A.K., et al.: The kinematics of the ankle: A hinge axis model. Foot Ankle (in press).

11. Wright, V., and Johns, R.J.: Quantitative and qualitative analysis of joint stiffness in normal subjects and in patients with connective tissue diseases, Ann. Rheum. Dis. 20:36–46, 1961.

CHAPTER 5

Transmission

THE HAND AS A WHOLE

We have discussed the way muscles work, and have noted that the output of a muscle may be defined by two variables—tension and excursion. We have also discussed the way joints move, and have seen the way in which muscle tension is translated into joint movement around one or more axes. Now we need to address the way in which it all comes together, as a limited number of muscles control many segments of bone moving around multiple axes to control the posture and movement of the hand as it meets the demands of external loads.

LEVERAGE OR MOMENT ARM

In this chapter, we address the question of how the force output of a muscle is used to move bones and joints. We have already noted that a muscle has two variables—tension and excursion. Now we introduce a new variable—leverage or *moment arm*. This variable determines not the output of the muscle but rather the effectiveness of the muscle tension at a joint. The total effect of a muscle at a joint is called the *moment* or *torque*. It is the product of two parts: (1) the

force or *tension* from the muscle and (2) the leverage, or *moment arm*.

$$\text{Moment} = \text{tension} \times \text{moment arm}$$

is the same as

$$\text{Torque} = \text{force} \times \text{lever arm}$$

The moment arm of a muscle-tendon unit at a joint is measured (Fig 5–1) by the perpendicular distance between the axis of the joint and the tendon as it crosses the joint. The key word is *perpendicular*. It is the distance at right angles to the tendon. It is the shortest distance between the axis and the tendon. If the tendon does not actually cross the joint, the line of action of the tendon may be extended on a diagram and the moment arm is then measured from the axis to the extended line of the tendon (see Fig 1–5,C).

In the measurement of torque, it is not necessary to worry about the angle of the tendon to the bone (angle of attack) or any other factor that engineers use. If we use moment arm alone, we shall not be far wrong. A tendon may cross more than one joint; it may change direction after crossing one joint on its way to another. The tension is the same along the length of the tendon (unless

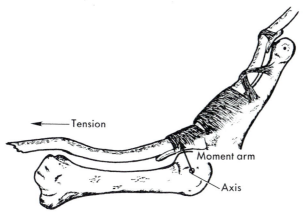

FIG 5–1.
The effect of a tendon at a joint is determined by two variables: the tension of the tendon and the moment arm or perpendicular distance from axis to tendon. (*Note:* This tendon will affect other joints distally. That does not change its effect on this joint unless the action of the other joints changes the tension of the tendon.)

it is held back by adhesions). It does not have to be shared or divided between joints. At each joint, the moment arm will be different and that alone will determine the relative effect of the one muscle at each joint. For example, a certain flexor digitorum profundus (FDP) tendon has a moment arm (Fig 5–2) of 1.25 cm at the wrist joint, 1.0 cm at the metacarpophalangeal (MCP) joint, 0.75 cm at the proximal interphalangeal (PIP) joint, and 0.5 cm at the distal interphalangeal (DIP) joint of a finger. If the muscle now contracts with a tension of 2 kg, it will exert a moment at each joint (Fig 5–3). These moments will be:

Wrist joint $2 \times 1.25 = 2.5$ kg·cm
MCP joint $2 \times 1.00 = 2.0$ kg·cm
PIP joint $2 \times 0.75 = 1.5$ kg·cm
DIP joint $2 \times 0.50 = 1.0$ kg·cm

Many other forces and moments from other muscles will affect these same joints, so that finally not all joints will actually move. However, that one profundus muscle cannot in any way alter the ratio of its effectiveness at those four joints. It can only increase or decrease its tension. The moment arms determine the way the tension is used.

MECHANICAL ADVANTAGE

On any lever the ratio of the moment arms of the force and the load determines the mechanical advantage. If the load is very heavy, it is good to have it close to the axis of the lever and the force is applied farther away. The load then moves very little while the force moves much more. The mechanical advantage is high. This is called *low gear* in a car or bicycle. If there is plenty of available force, and it is desirable to move the load fast or far, then the force is applied near the axis and the load further away. This is called *high gear*. High gear has a low mechanical advantage; low gear has a high mechanical advantage (Fig 5–4).

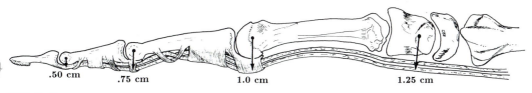

.50 cm   .75 cm   1.0 cm   1.25 cm

FIG 5–2.
A typical but hypothetical profundus tendon showing how moment arms tend to be larger as the joints are more proximal. The figures are approximate and rounded off, but realistic.

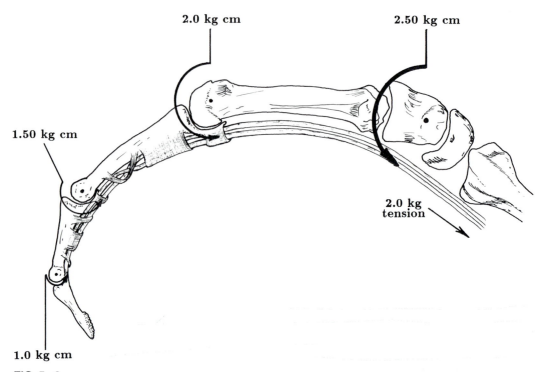

FIG 5–3.
Based on the moment arms from Figure 5–2 and on a tendon tension of 2 kg, the actual torque or moment at each joint is shown.

Mechanical advantage

$$= \frac{\text{Distance moved by force}}{\text{Distance moved by load}}$$
$$= \frac{\text{Distance of force from axis}}{\text{Distance of load from axis}}$$

In the body there are no low-gear transmissions, no high mechanical advantage ar-

rangements. The highest mechanical advantage is probably at the masseter muscle in reference to the molar teeth. Here the ratio is 1:1 (Fig 5–5). The muscle is right beside the load and the moment arms are the same. In the arm most of the work is done out at the end of the hand where the load is carried. The force is applied much closer to the

FIG 5–4.
If the little fellow is in control, he has a high mechanical advantage. If the big fellow takes charge, his mechanical advantage is low.

axes of the joints. The average mechanical advantage is 1:5 or less for each bone. Therefore, a muscle often has to deliver five times more force than the opposing force or the load it needs to move at the end of the bone.

EFFICIENCY

Now let us consider the way in which work may be done efficiently with various placements of tendons in relation to an axis and a bone. When we analyze the forces around a joint, it is useful to simplify the system by assuming that at a joint the proximal of the two bones remains stationary and the distal bone moves around it. In fact, the distal bone may be fixed, the proximal may move, or both bones may be moving. Within the limited triangle of bone-muscle-bone, we will consider A the axis, AB the unmoving bone, AC the moving bone, and

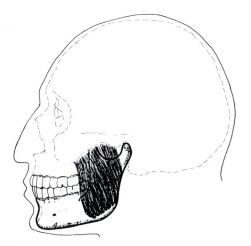

FIG 5–5.
The masseter may be the only muscle in the body that exerts an effective action with a 1:1 mechanical advantage. The back molar teeth are right beside the anterior fibers of the masseter so both have an equal moment arm for the temperomandibular joint. In the hand, the adductor pollicis may exercise almost a 1:1 mechanical advantage if something is squeezed between the head of the first metacarpal and the neck of the second metacarpal while the skin of the thumb web is pushed aside.

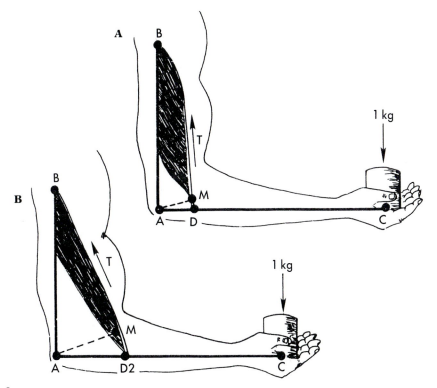

FIG 5–6.
Two hypothetical arrangements for an elbow flexor, using about the same bulk of muscle. **A,** with insertion close to the axis, a small moment **(M)** arm allows a small mechanical advantage. **B,** a large moment arm **(M)** allows a larger mechanical advantage.

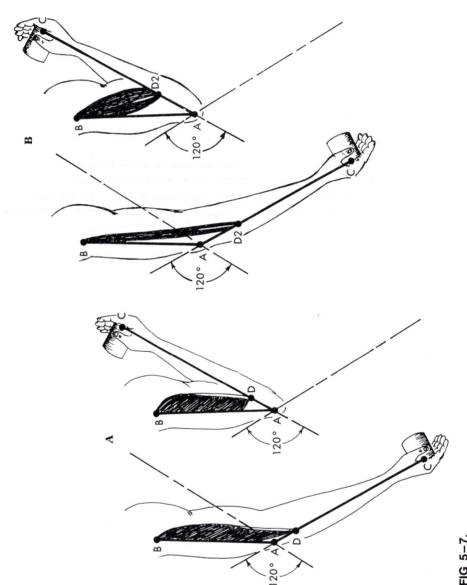

FIG 5–7.
A, a shorter moment arm requires a stronger muscle but can use short fibers to complete the range of motion. **B,** a longer moment arm can use a weaker muscle and needs longer fibers to complete the range of motion.

D the insertion of the muscle that forms the third side of the triangle. If *A* is the elbow, *B* is the upper end of the arm, *AC* is the forearm with the hand beyond it, and *D* is the attachment of the biceps tendon. *AM* is the perpendicular distance from the axis to the tendon. Let us further suppose that the function of the muscle is to flex the joint, lifting 1 kg at the end of *AC*. With the joint at right angles, the muscle contracts with a tension *T*. Now consider two alternative placements of the tendon on the bone *AC*; the attachment *D* (Fig 5–6,*A*) is close to the axis and *D2* (Fig 5–6,*B*) is further from the axis. Which is the more efficient?

Clearly *D2* has a bigger moment arm, so the muscle can support 1 kg using less force. But when the joint moves through 120 degrees, the muscle has to lengthen and shorten through a greater distance (Fig 5–7,*B*) than when the insertion is closer to the axis (Fig 5–7,*A*). Thus the only way a muscle could accomplish this would be to have long fibers running from the origin to the insertion of the muscle. Even then the fibers would have to bunch up maximally in flexion and stretch to their limit in extension.

If the tendon is at *D*, the muscle needs to exert twice as much tension to overcome the low mechanical advantage (see Fig 5–7,*A*). However, the muscle insertion only has to move a short distance between flexion and extension for 120 degrees of movement. Thus the muscle could be composed of many short fibers running from bone *AB* to a long tendon of insertion. By having many short fibers, the muscle has a larger cross-sectional area of its fibers and thus can exert more tension without using a greater bulk of muscle tissue.

It should now be clear that for a given task, like lifting 1 kg through a 120-degree elbow movement, a short-fibered muscle using a short moment arm may do exactly the same amount of work with the same efficiency as a longer-fibered muscle with a longer moment arm, provided the same total bulk (mass) of muscle is used. Given such a choice, the body has usually selected the shorter moment arm for at least two good reasons. First, big moment arms mean thick limbs. If the biceps were attached to the midforearm, a web of skin would cross from shoulder to forearm (Fig 5–8,*A*). Slim arms have more mobility and can surround an ob-

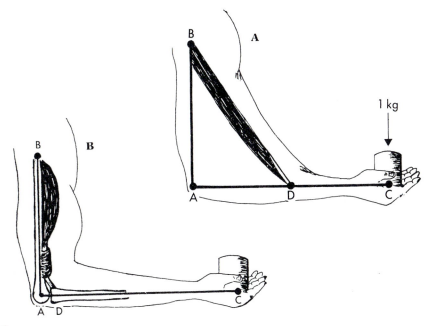

FIG 5–8.
A, a really big moment arm would give enormous power but a very limited range of motion and a thick webbed arm. **B**, a sling to hold the tendon to a short moment arm and a thick head of bone to hold the tendon away from the axis results in a slender limb and constant moment (or torque).

ject much more satisfactorily without bow-stringing tendons. Second, big moment arms mean variable moment arms. The moment arm is big when the elbow is flexed. It would become small with the elbow in extension. If a sling or pulley is used to keep the tendon close to the axis, then the moment arm will be shorter and also more nearly constant (Fig 5–8,B). In many joints in the hand, the thickness of the bone at the joint keeps the tendon away from the axis when the joint is extended, and a sling or pulley prevents much increase in moment arm while in flexion. The result is an almost constant moment arm.

It is worth noting at this point that the moment arms of external loads are also variable (Figs 5–9 and 5–10). As the elbow moves from a right angle to the extended position, the biceps tendon moves closer to the axis, but so does the line of action of the weight carried in the hand. As the moment of the biceps is diminished, so, in this case, is the moment of the load diminished.

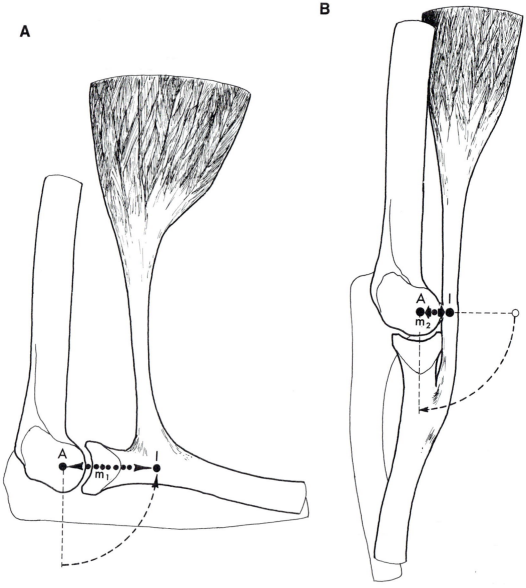

FIG 5–9.
The biceps tendon has a larger moment arm in flexion **(A)** of the elbow than in extension **(B)**.

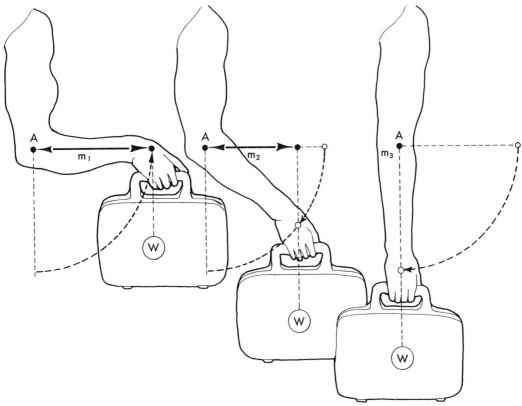

FIG 5–10.
A load (suitcase) also has a larger moment arm when the elbow is flexed. A = axis of elbow; W = weight (load) of suitcase; m^1 = moment arm of load, elbow flexed; m^2 = moment arm of load, elbow extending; m^3 = zero, elbow fully extended.

OPTIONS FOR THE SURGEON

The original designer of the body may have had a wide open choice of fiber lengths and moment arms, but the surgeon can only use existing fiber lengths and modify some moment arms. However, he or she does have some limited choice about which muscle to transfer, and in that choice it is wise to take fiber length into account, especially if the choice of moment arm is limited.

MEASUREMENT OF MOMENT ARMS

In view of the fact that moment arms are a vital factor in the transmission of force in the hand, we have made a study of ways in which a surgeon may measure or estimate the moment arms of tendons before and after tendon transfers. Direct measurement is never possible in the intact hand, because the axis of a joint is never exposed. There is one relationship that we can use and we recommend it from our experience.[1] It is simple and practical and covers both fresh cadaver dissections and actual surgical operations. It is the measurement of tendon movement (excursion) related to a measured angular movement of a joint. There is a rule of geometry that connects these two. When a lever rotates around an axis, the distance moved by every point on the lever is proportional to its own distance from the axis. Thus if a point on a lever that is 5 cm from the axis moves just 1 cm, then a point that is 10 cm from the axis will move 2 cm (Fig 5–11). There is one specific instance of this rule that must be emphasized: *When a lever rotates around an axis through an angle of a radian, every point on the lever moves through a distance equal to its own distance from the axis* (Fig 5–12).

The angle of 57.29 degrees is called a ra-

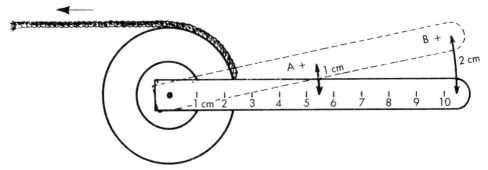

FIG 5–11.
Point *A*, 5 cm from the axis, moves 1 cm. Point *B*, 10 cm from the axis, moves 2 cm.

dian. It is the angle at the center of a circle that has an arc equal in length to the radius (Fig 5–13). The model in Figure 5–14 shows the connecting link between a lever and a pulley wheel. If the wheel is fixed to the lever, a cord that runs around the pulley wheel will serve as an extension of the lever system, but with a constant leverage (or moment arm) whether the cord is pulled vertically, horizontally, or obliquely.

The value of this model is that it demonstrates the way in which the lengthwise movement of a tendon may be used to measure its moment arm at any joint. When the lever moves through an angle of 57.29 degrees, the length of cord that runs off the pulley wheel is equal to the moment arm of the cord (see Fig 5–14). In the same way, if a finger segment is flexed or extended 57.29 degrees while the movement of a tendon is measured, the distance the tendon moves will be exactly equal to the moment arm of

Radian = 57.29 degrees

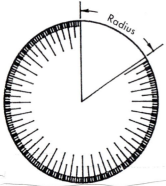

FIG 5–12.
A radian is the length of a radius, measured on the circumference, joined to the center by two radii.

the tendon at the joint. The beauty of this system is its simplicity and universality. It is applicable to the ideal joint-pulley arrangement, as at finger joints, where the head of the proximal bone is very like a wheel (sphere) (Fig 5–15) and where the tendon hugs the skeletal plane. It is also applicable to joints where the tendon is free to bowstring and is subject to changing moment arms. In such a case the excursion of the tendon will represent the *mean* functional moment arm throughout the measured angular joint motion.

Warning

This system of estimating moment arms by measuring tendon excursion related to joint motion is valid *only if the tendon is free to move.* It is *not* applicable to a tendon that is limited by adhesions so that it is not free to move fully with the joint. It is thus *not* applicable to the extensor digitorum at the PIP when the MCP joint is hyperextended, since the tendon is unable to exert its tension distally when it has used up its excursion proximally.

However, even in these cases the tendon excursion for a given joint movement will measure the *effectiveness* of the tendon in relation to that joint movement. It will show that the tendon behaves as if it had a moment arm of only 5 mm (though its actual moment arm may be 8 mm) because its restricted movement makes it less effective. In some cases, a tendon will be free to move through 30 degrees of joint movement and then be blocked by adhesions for the last 30 degrees of joint motion, suggesting a mo-

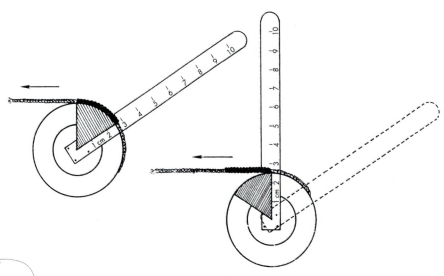

FIG 5–13.
The lengthwise movement of a tendon may be used to measure the moment arm of a joint. If the joint moves 57.29 degrees, the length of rope that runs off the pulley must be equal to its moment arm at the joint (i.e., the radius of the pulley).

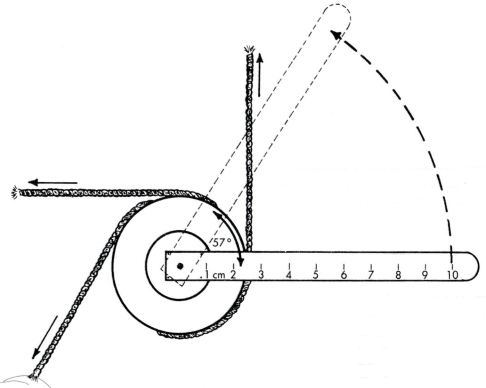

FIG 5–14.
The rope (or tendon) will have the same moment arm and same excursion for each angular change so long as it runs off the convex wheel (like a tendon that runs on the surface of the head of a bone at a joint).

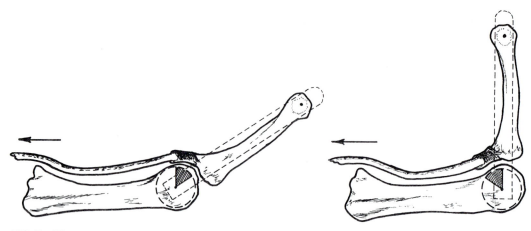

FIG 5–15.
Joint-tendon arrangement in which the tendon is hugging the skeletal plane.

ment arm of only half the actual moment arm. However, it should be obvious to the surgeon that the tendon only moves during part of the range, and he or she should then recheck the excursion for the 30 degrees that it actually moves, and a fairly good analysis of the probable effectiveness of the tendon and of the cause and extent of its limitation will be obtained.

In actual practice, during operations we do not attempt real accuracy. It is quite satisfactory to use a standard triangle for joint-angle measurements (Fig 5–16). We use a 30-60-90-degree triangle of aluminum and a metal millimeter scale for measuring tendon

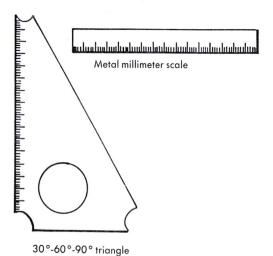

Metal millimeter scale

30°-60°-90° triangle

FIG 5–16.
A 30-60-90-degree triangle. The corners are cut off to allow the triangle to be tucked into web spaces where the actual joint axis is deep in tissue.

movement. We record the tendon movement for a 60-degree joint motion as if it were a radian (Fig 5–17,*A*), or we sometimes move a joint 90 degrees and call it 1.50 radians (Fig 5–17,*B*). If we have a free tendon that may be subject to bowstringing, we measure the tendon excursion that occurs with two 30-degree ranges of motion, one near extension and one near flexion, and call each 0.5 radian (half a moment arm). Thus if the 30-degree range near extension gives tendon motion of 5 mm and the 30-degree range near flexion gives 8 mm of tendon excursion, then we assume that the mean moment arm near extension is $2 \times 5 = 10$ mm (Fig 5–18) and the mean moment arm near flexion is 16 mm, and that this difference is caused by bowstringing (Fig 5–19).

Example at Surgery

If a flexor tendon graft has been placed in a finger at surgery and the pulleys have been reconstructed, there remains a doubt about the tightness and effectiveness of the pulleys.

The proximal end of the graft may be pulled while the finger movement is measured with a triangle. We hold the finger straight at all joints while the tendon is pulled proximally by a stitch and measured against a millimeter scale (Fig 5–20,*A*). Now the finger is flexed through 60 degrees at the MCP joint only. The tendon may move 15 mm (Fig 5–20,*B*). Then we hold

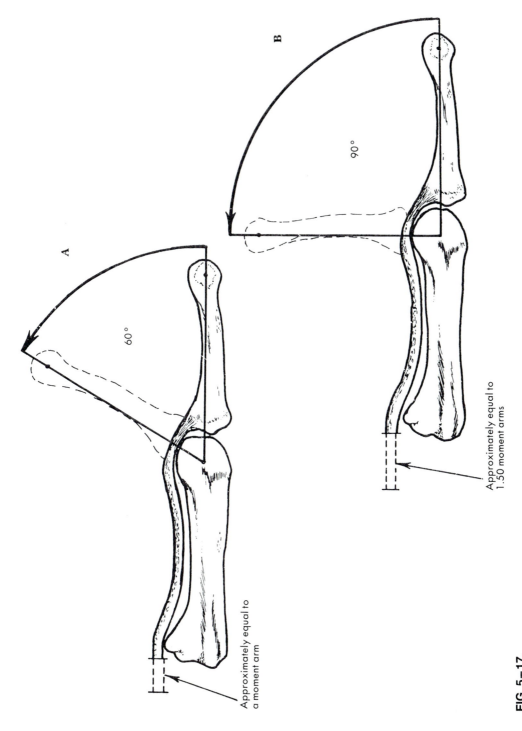

FIG 5–17.
A, tendon movement is approximately equal to a 1.0-moment arm. **B,** tendon movement is approximately equal to 1.50 moment arms.

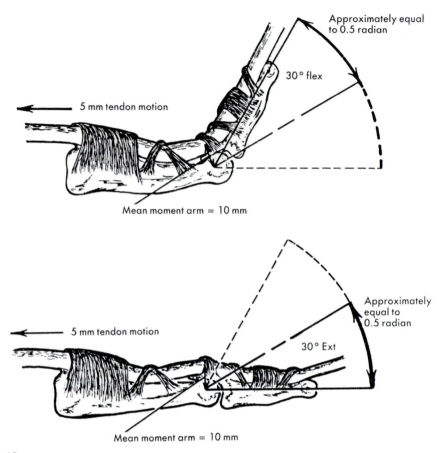

FIG 5-18.
Normal finger joint shows little difference between tendon excursion near flexion and near extension.

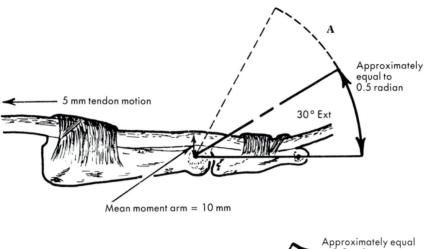

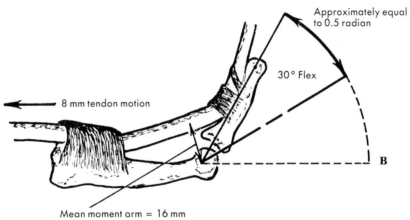

FIG 5–19.
A, tendon with excised sheath shows unchanged tendon excursion as joint is near extension. **B,** mean moment arm near flexion is 16 mm as a result of bowstringing.

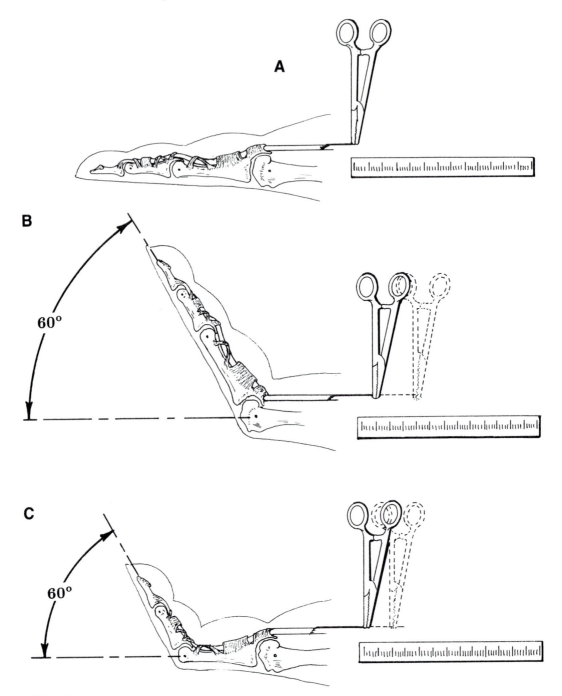

FIG 5–20.
Measurement of tendon and pulley placement at surgery. **A,** the finger is straight at all joints. **B,** the finger is flexed only at the MCP joint through 60 degrees. **C,** the MCP joint is straight and only the PIP joint is flexed through 60 degrees.

the MCP joint straight and flex only the PIP joint through 60 degrees. The tendon may move 8 mm (Fig 5–20,C). Now we know that our pulley at the MCP joint is too loose or too far from the axis. We know it because

the average moment arm figures for adults are available to us and we know that the moment arm for a flexor profundus tendon at the MCP joint should not be more than 3 mm greater than at the PIP joint. Another way to check the pulley is to measure the tendon excursion that matches the 30-degree range near flexion and the 30-degree range near extension as noted above (see Fig 5–19). If the first is much greater than the second, this indicates bowstringing of the tendon, so we go back and place a new pulley or tighten the old one and check again.

This measurement may take 5 minutes. Failure to identify this common error of a poorly placed proximal pulley will result either in failure or in weeks of extra effort after operation to get the finger moving. This is because a bigger moment arm at the MCP joint makes the tendon more "effective" at that joint. It also uses up more excursion. Thus, when the patient attempts to flex his finger after surgery, the MCP joint will flex easily. The muscle uses up its peak effectiveness on the MCP joint and leaves only the weak tension of an already contracted muscle for the PIP joint. If tendon motion is limited by proximal adhesions, it is even more likely that all available tendon movement will be used up at the MCP joint.

ONE MUSCLE: TWO INSERTIONS

Sometimes, following extensive paralysis, a surgeon may want to transfer the tendon of one good muscle and attach it to two or more tendons of insertion, because there are not enough muscles to go around. This is often a useful device to balance a hand, but there are hidden problems that must be understood. The first is rather obvious. The surgeon must recognize that if one muscle is to have two insertions, in parallel, both will always move together. If one is held back, the other will be held back. The surgeon should consult with the hand therapist and with the patient and make it clear that certain movements will be linked in the future and make sure that there is no incompatibility between the linkage and the patient's functional needs.

One example of this is in high median and ulnar palsy when a wrist extensor such as the extensor carpi radialis longus (ECRL) may be used to power all the finger flexor tendons. After this, if one finger flexes, all will flex, and if one is prevented from flexion

by stiffness, then none will be able to flex. A stiff finger should not be included in a multiple-finger transfer of a single motor. The second problem is more subtle and is often unrealized by the surgeon: unless both tendons have about the same moment arms at the joints they cross, the final effect may be unfortunate.

A good example is wrist extension. There are three wrist extensors that are all paralyzed in a radial nerve palsy. Two are also radial deviators, the extensor carpi radialis brevis (ECRB) and the ECRL, and one is an ulnar deviator, the extensor carpi ulnaris (ECU). Commonly in a radial palsy, a single motor, the pronator teres, is used to restore wrist extension. This is reasonable, because there is a severe shortage of muscles that can be used.

What is not reasonable is the advice given in some standard texts on hand surgery, that the pronator teres tendon should be attached to both the ECRL and ECRB tendons of insertion. These two tendons have different moment arms. Since the ECRL has a smaller moment arm for wrist extension, it does not move as far as the ECRB tendon unless radial deviation also occurs. Thus the patient will be forced into full radial deviation every time he or she tries to extend the wrist. It is better to put the pronator only into the ECRB, which has the least effect on radial deviation. Other surgeons, ourselves included, have tried in the past to compensate for the radial deviation by using two insertions, one on each side of the axis for deviation of the wrist. This was a good idea except that we used the insertions of the ECRB and the ECU. This was before we had studied the mechanics of the hand and before we knew that the ECU has a very small moment arm for wrist extension (and none at all when the wrist is in pronation) (Fig 5–21).

Even if we had known, we might still have naively assumed that, offered two moment arms, the muscle would make use of the bigger and more effective one. Not so. When ropes from two pulley wheels, fixed to the same axle, are joined and then pulled together, the one with the *smaller* moment arm is effective while the other rope be-

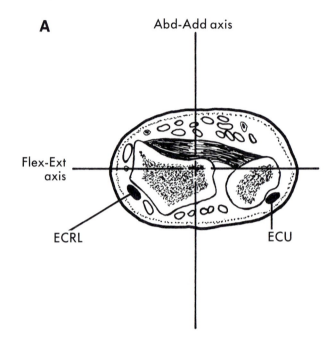

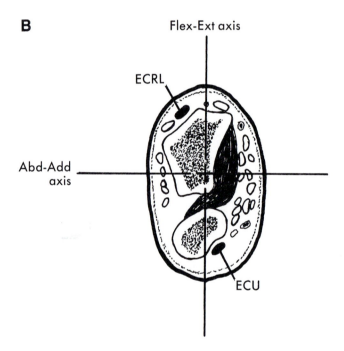

FIG 5–21.
A and **B,** during pronation and supination, the wrist and hand rotate with the lower end of the radius. The ulna does not rotate but rather serves as an anatomic axis around which the radius and wrist rotate. However, since the axes for flexion-extension and abduction-adduction of the wrist stay with the lower end of the radius, the ulna seems, in a wrist function sense, to rotate within the wrist complex. The only forearm muscle tendon that stays with the ulna bone is the extensor carpi ulnaris *(ECU).* This tendon changes its relation to wrist axes during pronation and supination. On supination it is a wrist extensor and ulnar deviator. In full pronation it is still an ulnar deviator but loses its extensor moment arm and may even become a wrist flexor. *ECRL* = extensor carpi radialis longus.

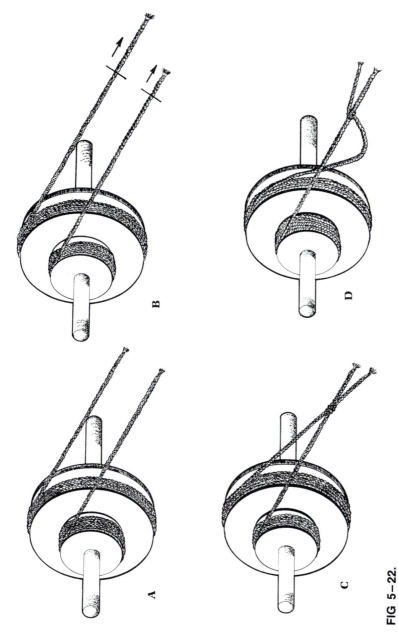

FIG 5–22.
A, two pulley wheels fixed to a common axle (axis). **B,** for a given range of wheel motion, more rope (tendon) comes off the wheel with the larger moment arm. **C,** both ropes are now fixed together (common muscle). **D,** pulling on the common origin makes the rope on the larger wheel loose and useless. Only the small moment arm remains effective.

comes loose (Fig 5–22). If one moment arm is so small that it has no effect, the total movement is nil. Thus the net effect of putting the pronator into the ECRB and ECU is to produce very weak extension, using only the smaller moment arm. Or it may produce stronger extension, but only in full ulnar deviation.

To achieve a good balanced wrist extension, the following may be tried. The tendon of the ECRL is detached from its insertion and withdrawn proximally at the level of the musculotendinous junction. It is then tunnelled distally and attached to the base of the fourth metacarpal. We have checked that the moment arm for wrist extension of a tendon at that level is almost the same as that of the ECRB, while the moment arm for ulnar deviation is about the same as that of the ECRB for radial deviation. Now the tendon of the pronator teres may be attached to both tendons proximally. When the pronator contracts there will be balanced wrist extension, with no radial or ulnar deviation.

Another example is the linking together of the movement of opposition of the thumb and the opposition of the fifth ray in order to achieve a cupping of the palm and to avoid the loss or reversal of the metacarpal arch, when all intrinsic muscles are paralyzed. It has seemed a good idea to use one tendon, perhaps a flexor superficialis, and split it so that one slip opposes the thumb and one the little finger. This has been done in the past, but the problem is that the tendon excursion needed for thumb opposition is at least double that needed for the fifth ray. Thus the thumb will move only half as far as it should. It would be possible in theory to attach the slip to the thumb close to the proximal end of the metacarpal so that its required excursion would be reduced to match that of the fifth metacarpal slip. However, it would seem a pity to reduce the moment arm of an important action of the thumb to match a less important action that might not always be wanted at the time the thumb was being used for pinch.

Whenever a surgeon is planning an unconventional redeployment of tendons, it is a good idea to place a monofilament nylon thread or a wire in the exact position and with the same attachments as the proposed tendon. Then the thread can be pulled and its excursion measured while the hand is put through the movement that is desired. Then if two or more insertions are planned for one single motor, the surgeon will know exactly how much limitation any one insertion will place on the others.

ONE TENDON: TWO OR MORE JOINTS IN SERIES

When one tendon is split to affect two joints in parallel, then both joints have to move together or neither will move. However, if one tendon crosses two joints one after the other, in series, one may flex while the other fails to move or even extends. Thus one may have the false idea that the tendon is able to selectively affect one joint more than another.

In fact, the influence of the tendon is always the same at each joint it crosses. Its tension is the same. Its *effect* depends on its moment arm at each joint and on other muscles that may add to or oppose its action at various joints. It also depends on the moments caused by external forces. We consider some of these factors in Chapter 8, when we study the balance between internal and external forces.

PULLEYS AND SLINGS

Most of the tendons in the hand are restrained to some extent by sheaths or retinacula, so that they stay close to the skeletal plane and maintain a relatively constant moment arm rather than bowstringing across joints. However, no sheath is so rigid and snug that some mobility is not possible. Even flexor tendons in the digital sheath have a slightly larger moment arm when the finger is flexed than when it is extended. In the muscle-by-muscle tables in Chapter 12, the figures quoted for moment arms at joints are averages. They are accompanied by a graph for each tendon at each joint showing how the moment arm varies according to the angle of the joint. There are three other mechanical factors associated with pulleys and slings that should be noted.

1. The pressure between tendon and sheath on the bend is so great that high friction and drag might be expected. In fact, however, the contiguous surfaces of tendon and sheath on the concave side of a curve are never attached to any paratendinous tissues but rather are free-gliding surfaces lubricated with synovial fluid. There are paratenon tissues or vincula on the convex curve of the tendon and in areas where it is free of friction under pressure.

2. There are no sharp or angular bends in the course of any normal tendon. The sheaths or pulleys always allow for a smooth curve. This minimizes local points of high pressure between tendon and sheath (Fig 5–23,A). On the contrary, when surgeons place a pulley to restrain a tendon, they sometimes use a narrow strip of tendon or fascial graft. This is bad for two reasons: (a) a lot of force is concentrated on that narrow band; it may break or become stretched; (b) the narrow band distorts the tendon and may cause an obstruction to gliding (Fig 5–23,B). We have sometimes used a stout monofilament nylon sling at operation, just to test the placement of the tendon and sling, before putting in the definitive tissue

sling. We have noted the way a narrow sling can deeply indent a tendon, making a swelling on each side of the sling so that it is difficult to start the tendon moving. Fortunately, the body will usually compensate for a narrow sling by filling in around it with scar tissue as it heals. However, it is better to use wider pulleys or multiple pulleys that will allow smaller changes of tendon direction at each (Fig 5–23,C). Then the scar that fills in the gaps will leave a smooth curve.

3. When a pulley occurs at a joint and the joint is flexed, the tension in the tendon will tend to pull the pulley away from its attachment or to pull the bone that forms that attachment away from the joint. If all the tissues are strong and the joint is stable, there is no problem, but if the joint is unstable, as in rheumatoid arthritis, there may be a danger of severe subluxation. This danger is widely recognized. What may not be realized is how great these subluxating forces may be.

When a joint (and its flexor tendon) is flexed 60 degrees, the two limbs of the tendon will have an angle of 120 degrees between them (Fig 5–24,A). At that point, the tension in the restraining pulley system must be equal to the tension in the tendon. When the joint is flexed at 90 degrees, the tension in the pulley will be nearly 50% more than the tendon tension. If the attachment of the pulley is to the base of a phalanx (Fig 5–24,B), which is then suspended to the head of the proximal bone by a ligament, this ligament may have to resist up to the whole of this dislocating force. We say "up to" the whole, because, of course, the bony configuration of the joint should give some resistance to any subluxating force, especially if there is a longitudinal force to impact or hold the joint together.

THE TRANSMISSION OF EXTERNAL FORCE

The hand is for work. Too often muscle balance is thought of as if flexor strength had to balance extensors. In fact, the major use of muscles is to oppose external loads. The

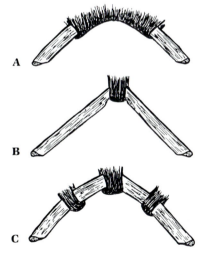

FIG 5–23.
A, there are no sharp or angular bends in the course of any normal tendon. **B,** a narrow strip of tendon or fascial graft. **C,** it is best to use wider or multiple pulleys.

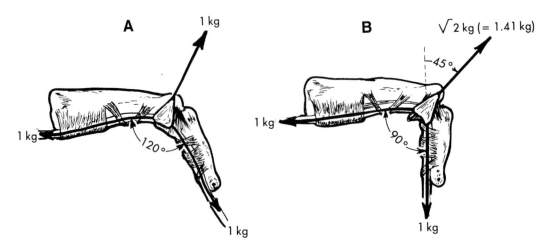

FIG 5–24.

A, at 120 degrees, the tension in the restraining pulley system is equal to the tension in the tendon . **B,** at 90 degrees, the tension in the pulley system needs to be greater than the tendon tension. It works out at the square root of twice the tendon tension.

interplay of flexors and extensors is coordinated in a rhythmic dance where each alternates with the other to produce the harmonious ballet whose theme is to defy gravity and to make loads seem light and inertia turn to fluid motion. Thus no diagram of force and balance within the hand is really complete until some arrows are included to indicate the size and direction of external forces or loads.

The big difference between the transmission of external and internal forces is in the moment arms. It has already been pointed out that most moment arms of muscle-tendon units are small and are related to the thickness of the bone or joint, whereas external moment arms are large, and are related to the length of the bone. But there is a greater difference still when we consider a series of joints. Whereas internal muscle forces affect every joint that is crossed by a tendon in proportion to a separate moment arm at each joint, external force applied to the tip of a finger may have a cumulative moment arm that summates as it affects more proximal joints. Consider the diagram of a flexor profundus tendon from fingertip to forearm and the effect of an opposing force at right angles to the tip of the same finger (Fig 5–25).

Profundus moment arms at the four joints from fingertip to forearm are about 0.5,

0.75, 1.0, and 1.25 cm at the DIP, PIP, MCP, and wrist joints, respectively. The moment arms of the external force at the same joints might be 2.0, 5.5, 10.5, and 20.0 cm. It might be possible for the profundus tendon alone to flex the finger and hand in a smooth coordinated movement so long as the hand was unloaded. The muscle would only have to overcome the weight of the hand and the soft tissue drag at each joint. However, as soon as an external load of 2 kg is applied to the fingertip, the profundus would need a tension of 8 kg for equilibrium at the DIP joint (ratio of moment arms of external to internal force = 2.0:0.5), but would need a tension of 32 kg to hold the wrist in equilibrium (moment arm ratio 20.0:1.25). Since the tension in a tendon has to be the same along its length, there is no way that this can be accomplished by one tendon alone. It can be done only by adding other muscles. In the above example, the 2-kg external load will produce an extensor *moment* at each joint as follows:

DIP joint $2.0 \times 2.0 = 4$ kg·cm
PIP joint $2.0 \times 5.5 = 11$ kg·cm
MCP joint $2.0 \times 10.5 = 21$ kg·cm
Wrist joint $2.0 \times 20.0 = 40$ kg·cm

To achieve equilibrium, muscles must therefore produce flexor moments to equal and oppose these extensor moments. It may be done thus. A flexor profundus tension of

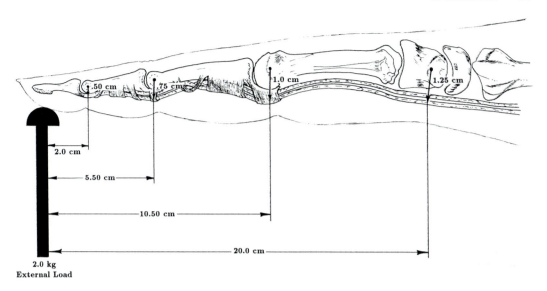

External Force Moment Arm

FIG 5-25.
In Figures 5-2 and 5-3, we now introduce an external load and show the moment arms of the external load at each joint. Whereas the tendon moment arms increase moderately at each proximal joint, the moment arms of the external load increase enormously.

8.0 kg (Fig 5-26) will give the following moments:

DIP joint $8.0 \times 0.50 = 4.0$ kg·cm
PIP joint $8.0 \times 0.75 = 6.0$ kg·cm
MCP joint $8.0 \times 1.00 = 8.0$ kg·cm
Wrist $8.0 \times 1.25 = 10.0$ kg·cm

Now the flexor digitorium superficialis (FDS) is needed to make up the deficit of 5

kg·cm at the PIP joint. It can do this by a tension of 6.66 kg. The same superficialis will also add flexion moment to the MCP and wrist joints:

PIP joint $6.66 \times 0.75 = 5.00$ kg·cm
MCP joint $6.66 \times 1.00 = 6.66$ kg·cm
Wrist $6.66 \times 1.25 = 8.33$ kg·cm

Now the MCP joint has 14.66 kg·cm (8.0

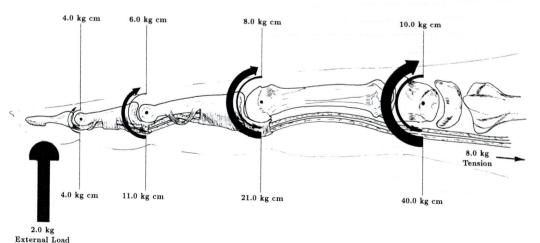

FIG 5-26.
External force opposed by the tendon of flexor profundus. A composite diagram of moment arms and force gives actual moments or torque at each joint. The flexor torque is from the tendon; the extensor torque is from the external load. The upper figures are for flexor torque (or moment). The lower figures are for extensor torque. (*Note:* The tendon tension has been set at 8 kg to keep the distal joint at equilibrium. The proximal joints cannot be balanced unless additional tendons and muscles are added.)

+ 6.66) and needs 6.33 more. So the intrinsic muscles must supply a flexor moment of 6.33. The wrist now has a flexor moment of 18.33 kg·cm (10.0 + 8.33) but needs a total of 40.0. So a wrist flexor, such as the flexor carpi radialis (FCR), with a moment arm of 2.0, needs a tension of 10.66 kg. Joints, moments, and muscle tensions needed to support 2 kg at a fingertip are shown in Table 5–1.

These figures are only approximations and they take no account of other muscles and of joint position. However, they do indicate how the need for muscle tension is much greater at proximal joints than at distal joints when the load is distal. They also demonstrate the impossibility of a single tendon controlling a number of joints in series.

The tendons of the flexor superficialis are often used for transfer, because they are readily available and easy to reeducate. In removing the superficialis for other duties, surgeons often forget that its loss from its own finger is serious. Once the superficialis is gone, there is only one flexor, the profundus, for the last two joints of the finger. As the hand gets back into active use, it is bound to be required to oppose distal loading. Distal load always results in a much higher torque at the PIP than at distal joint. One tendon cannot itself control two joints. If there is no flexor superficialis, there will not be enough flexor torque at the PIP joint. If we look at the previous example, the profundus generates a flexor torque of 4.0 kg·cm at the distal joint to offset the 4 kg·cm torque of the external load of 2 kg. This external load produces an extension torque of 11 kg·cm at the PIP joint, while the profundus only produces 6 kg of flexor torque at this joint. If more tension is added to the FDP to balance the PIP joint, it will produce too much flexor torque at the distal joint. Without a superficialis, the PIP joint will go into extension and the DIP joint will flex (Fig 5–27). As time goes on, these unbalanced forces result in stretching of the dorsal capsule of the distal joint, often accompanied by stretching of the volar plate of the PIP joint. The deformity is more serious in very mobile fingers, but may develop more slowly even in thicker, stiff fingers.

We call this tendency for the PIP joint to hyperextend and the distal joint to flex the superficialis-minus finger. It is seen in a number of situations where the superficialis is absent, such as following harvesting of a superficialis for a transfer, a free tendon graft to restore finger flexion, or a superficialis-to-profundus transfer. In the case of a free tendon graft, the flexion deformity of the distal phalanx can be minimized by concentrating attention on gaining motion at the PIP joint and protecting the distal joint from flexing. Good early motion at the distal joint is likely to end in a flexion deformity which is secondary to the lack of superficialis flexion torque at the PIP joint. In any finger left with only a profundus tendon in which the terminal joint begins to go into progressive flexion, with or without PIP hyperextension, the deformity should be recognized as progressive and irreversible. It can be corrected by tenodesis of the profundus tendon (or its split half) to the middle phalanx (Fig 5–28).

When the external force opposes finger

TABLE 5–1.

Muscle Tensions and Moments Needed at Each Joint (Distal Interphalangeal [DIP], Proximal Interphalangeal [PIP], Metacarpophalangeal [MCP], and Wrist) to Support a Load of 2 kg at a Fingertip

Tensions	Moment or Torque (kg·cm)			
	DIP	PIP	MCP	Wrist
Profundus—8 kg	4	6	8	10
Superficialis—6.66 kg	—	5	6.66	8.33
Intrinsics	—	—	6.33	—
Wrist flexors—10.66 kg	—	—	—	21.66
Totals	4	11	21	40

A

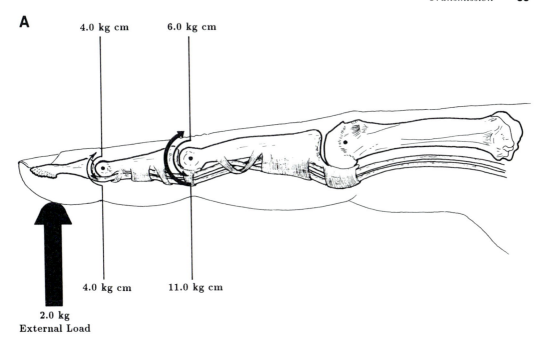

4.0 kg cm   6.0 kg cm

4.0 kg cm   11.0 kg cm

2.0 kg
External Load

B

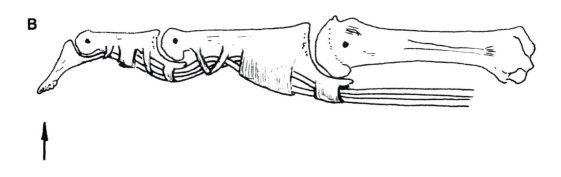

C

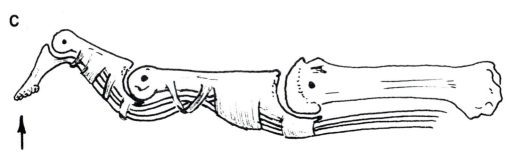

FIG 5–27.
A, the flexor digitorum profundus (FDP) generates a flexor torque of 4 kg·cm at the distal joint to resist the external load of 2 kg. However, it can only produce a flexor torque of 6 kg·cm at the PIP joint where the external load has an extensor moment of 11 kg·cm. **B,** without the added flexor torque of the flexor digitorum superficialis at the middle joint, the distal joint will flex and the proximal joint will extend. **C,** with time, hyperextension deformity of the PIP and fixed flexion deformity of the DIP joint wil develop.

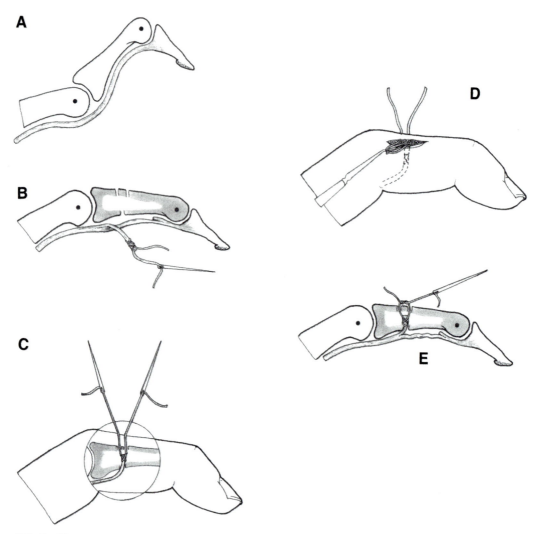

FIG 5–28.
A–E, the split half of the FDP is pulled out and sutured to the middle phalanx. The tendon is sutured with the PIP joint in slight flexion.

flexion directly, the wrist needs independent wrist flexors to support it. However, if the hand grasps an external object and circles it, the fingers oppose the palm of the hand or the fingers oppose the thumb; then the tension of finger flexors may produce unwanted flexion of the wrist and must be opposed by extensors at the wrist level. One more example will demonstrate a familiar problem—the thumb with ulnar nerve paralysis. We shall consider only the IP joint and the MCP joint and explore the reason for Froment's sign (i.e., the flexion of the tip of the thumb when pinch is attempted).

Flexion of the thumb is accomplished in

the normal hand by the flexor pollicis longus (FPL) (median nerve) and the adductor–short flexor intrinsic muscle complex, usually ulnar nerve–supplied. In Figure 5–29, the pulp of the thumb is opposed by 2 kg force, which may come from the opposing index finger in pinch. The moment arms of the FPL at the IP and MCP joints are about 0.75 and 1.0 cm, respectively. The point of application of the opposing force on the thumb pulp is 2.0 cm from the axis of the IP joint. It is 6.0 cm from the axis of the MCP joint. Now to hold the equilibrium of the IP joint against 2 kg of external load acting through a 2-cm moment arm, the FPL must

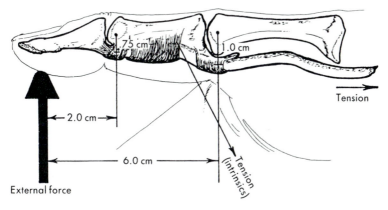

FIG 5–29.
The thumb in ulnar palsy. Moment arms of the flexor pollicis longus (FPL) at the IP and the MCP joints are 0.75 and 1.0, respectively.

provide a moment of 4 kg·cm (2.0 × 2.0). Since the tendon has a moment arm of 0.75 cm, the tension it needs is 5.33 kg (Fig 5–30). At the MCP joint the moment of the external force is 2 kg × 6 cm = 12 kg·cm. The FPL at 5.33 kg of tension and 1-cm moment arm can produce less than half the needed moment. In the normal hand the intrinsic muscles will provide the extra flexion moment needed to stabilize the MCP joint. If the intrinsic flexors are paralyzed, the

MCP joint will begin to bend backward in response to the 12-kg·cm moment caused by the external force and the inadequate 5.33-kg·cm flexor moment provided by the flexor longus tendon. Now one of two things can happen. The MCP joint, if it has limited passive range and a tough volar plate, may be stabilized by the static force of volar plate tension and will not hyperextend. Or, more probably, the FPL will respond to the sense of weakness and commencing hyperexten-

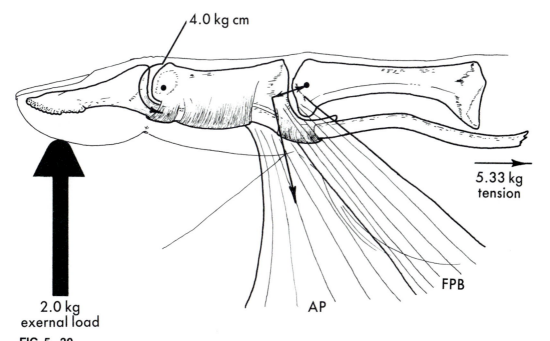

FIG 5–30.
For equilibrium at the IP joint, a 2-kg load on the thumb pulp may require 5.33 kg of tension in the flexor pollicis longus tendon. At the MCP joint the added moments of the adductor pollicis *(AP)* and the flexor pollicis brevis *(FPB)* maintain equilibrium.

sion and will contract more strongly until it stabilizes the MCP joint and prevents further extension. For this it will need a tension of 12 kg to give a 12-kg·cm moment at a moment arm of 1.0 cm (Fig 5–31).

In doing this, the tension of the FPL will become 12 kg at the IP joint also, and this will give a flexion moment of 9 kg·cm to oppose the extension moment of 4 kg·cm caused by the external force. The IP joint will now be unbalanced and will flex, inevitably and helplessly, producing the typical Froment's sign of ulnar palsy* (Fig 5–32). When the IP joint starts to flex, the line of action of the external force will move closer to the axis of the IP joint, thereby shortening the moment arm and further increasing the disparity between the moments caused by external force at the IP and MCP joints. Thus the patient has no way to correct the deformity.

Some patients with ulnar palsy soon realize that they cannot control their MCP and

---

*Froment's sign deformity is caused by a deficiency on the flexor side and cannot be corrected by adding to extensor tendon strength It needs extra flexion at the MCP joint.

IP joints simultaneously in pinch, and they learn a trick movement. They hyperextend their IP joint in advance and pinch down onto the end of the thumb with the index finger. This brings the line of action of the external force *behind* the axis of the IP joint and in front of the MCP joint (i.e., the force of pinch runs down the length of the thumb rather than across it) (Fig 5–33). This pinch is stable and very useful so long as the volar plate of the IP joint can hold up against the stretch of hyperextension or so long as the patient can adjust his or her thumb position so that the moment arms of the external force stay small and about equal at the two joints.

EXTERNAL FORCE AROUND AN AXIS OF ROTATION

A skillful wrestler gains control of his opponent by applying rotational forces to an arm by the use of a crank action. Once the position is achieved, there is no way in which the victim can oppose that external force, which has a large moment arm, because all rotational forces generated by mus-

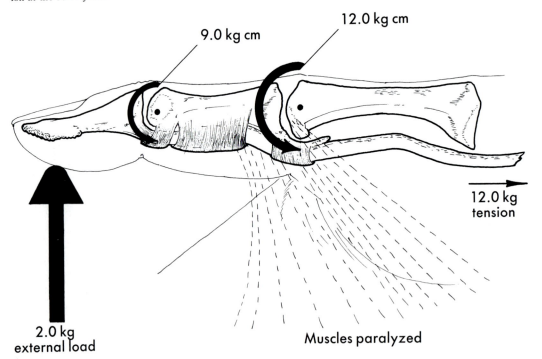

9.0 kg cm

12.0 kg cm

12.0 kg tension

2.0 kg external load

Muscles paralyzed

FIG 5–31.
For equilibrium at the MCP joint, a 2 kg load on the thumb pulp needs a 12.0 kg tension in the FPL tendon if other muscles are paralyzed.

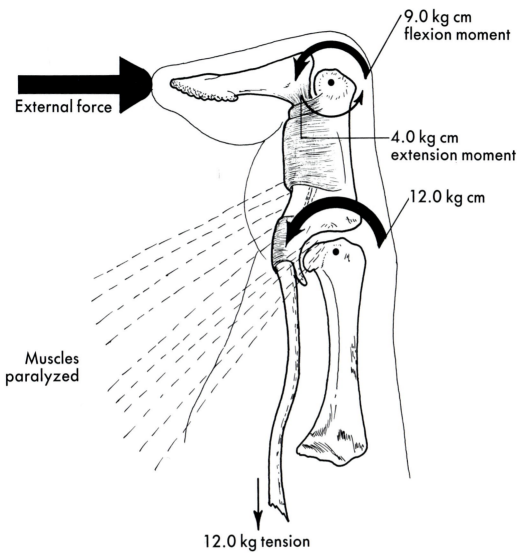

External force

9.0 kg cm
flexion moment

4.0 kg cm
extension moment

12.0 kg cm

Muscles
paralyzed

12.0 kg tension

FIG 5–32.
The 12.0-kg tension that is needed to stabilize the MCP joint is much too great at the IP joint. The latter goes into sharp flexion. (*Note:* Once the IP joint goes into flexion, the moment arm of the load is shortened and may even run through the axis. This is the mechanics of Froment's sign for ulnar palsy.)

cles inside the body have small moment arms. Typically the wrestler will flex his victim's arm at the elbow and use the forearm as a crank to twist the humerus. Here the external force will have a moment arm of about 30 cm while the rotators of the humerus have moment arms of about 3 cm or less. There is no contest.

Surgeons and therapists often fail to recognize the same overwhelming disparity in the hand when rotational forces are concerned. An obvious example is the thumb. Opposition of the thumb is a complex inter-

action of circumduction powered by abductors and adductors. So long as the thumb is straight at its MCP and IP joints, the muscles that move it toward opposition will do so through small moment arms. External forces that oppose them also will have small moment arms.

If the thumb is flexed at the MCP joint and extended at the IP joint (Fig 5–34,A), the distal part of the thumb can be used as a crank to reverse the intended opposition just as the forearm can be used to rotate the humerus when the elbow is flexed. This

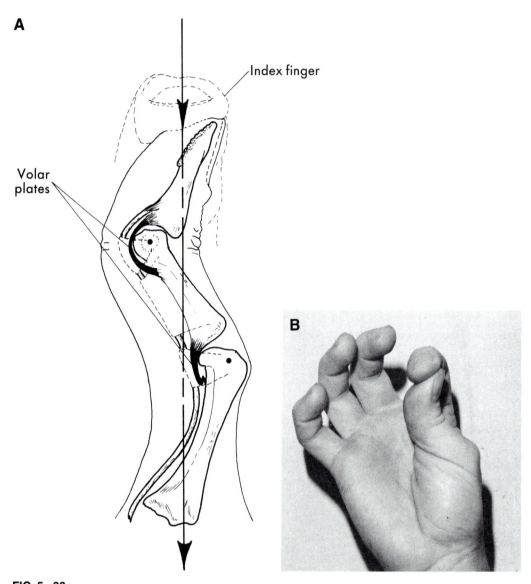

A

Index finger

Volar
plates

B

FIG 5–33.
A, in ulnar palsy, in order to avoid the weak pinch with acute flexion of the IP joint (Froment's sign), some patients learn to pinch down onto the tip of the thumb, so that the index finger locks the IP joint in hyperextension, leaving the FPL to control the MCP joint. This places a strain on the volar plate of the IP joint but is quite an effective pinch. **B,** pinching on the end of thumb—a stable solution to the weakness of ulnar palsy.

happens very frequently following tendon transfer for low ulnar-median paralysis of the intrinsic muscles of the thumb. This is how it works. A tendon (perhaps a flexor superficialis) is transferred to oppose and abduct the thumb. It is attached to the IP extensors. Or the surgeon is unable to correct the acute flexion of the IP joint (Froment's sign) by his tendon transfer, so he does an arthrodesis of the IP joint to hold it in extension.

The FPL then tends to hold the MCP joint in flexion as the IP is fixed in extension. In either case, the patient now goes to pinch with an incompletely opposed thumb with MCP flexion and IP extension. The index finger meets the thumb on the ulnar side of the pulp (Fig 5–34,B).

This is a *crank action*. The index finger pushes the thumb from the side. The metacarpal is held back by the thumb web. The

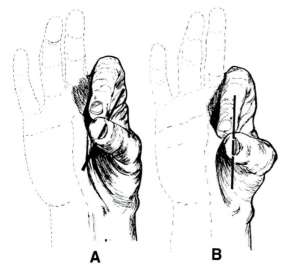

FIG 5-34.
Pinch with an incompletely opposed thumb with MCP flexion and IP extension. Crank action rotates the thumb metacarpal.

A **B**

distal half of the thumb becomes a crank that strongly rotates the metacarpal into supination (see Fig 5–34,B). The newly transferred tendon, which is supposed to produce medial rotation (pronation, opposition), may try to oppose the external force, which is producing lateral rotation (supination), but it fails. The moment arm for rotation into pronation is internal and small, whereas the moment arm of the external crank is large. The result of this unequal struggle is that the thumb progressively goes into chronic external rotation and supination. A frequent final position is with the supinated metacarpal lying beside the palm, and the flexed distal half of the thumb projecting forward. The surgeon who does not understand the mechanism mutters, "I know this thumb was working just after the operation. It must be that the patient did not exercise properly." So either the patient or the therapist gets the blame. This is not an uncommon end point of surgery for intrinsic paralysis of the thumb, but it is recognized only at late follow-up. To avoid it, the following hints may be useful:

1. Do not accept incomplete opposition of the thumb at surgery. Even though normal thumbs are rarely fully opposed (square pinch), they can get away with it because of good intrinsic muscle support. If thumbs with intrinsic paralysis are not fully op-

posed, so that the pinch is pulp to pulp, a tip pinch should not be attempted and a key pinch should be the aim of surgery. Full opposition may require a dorsal webplasty to allow full pronation.

2. Patients must be taught to pinch pulp to pulp, not pulp to side of the thumb.

3. Avoid arthrodesis of the IP joint of the thumb, if possible, in cases of intrinsic palsy.

4. Consider rerouting the extensor pollicis longus (EPL) (an external rotator) to lie over the abductor pollicis longus (APL), where it is no longer a rotator.

5. If the MCP of the thumb is chronically flexed, consider an arthrodesis of the MCP joint more nearly straight with some medial rotation of the phalanx on the metacarpal.

EXTERNAL FORCE TOWARD ULNAR DRIFT R.A effect/adaptat".

In rheumatoid arthritis, many patients have problems with a pulp-to-pulp pinch. They do not usually make the mistake of fingertip-to-side-of-thumb pinch (Fig 5–35,A), producing crank action of the thumb, but they do very commonly use the pulp-of-thumb-to-side-of-finger pinch. If they do this on the side of the tip of the index finger, this is on the longitudinal axis of the MCP joint and should be stable. However, in rheumatoid arthritis the axis may have lost

FIG 5–35.
Pinching by the thumb against the side of the tip of the index finger makes a crank action that rotates the proximal phalanx of the index finger. True key pinch uses full flexion of the middle phalanx of the index finger at the PIP and DIP joints. This results in less rotation effect but more ulnar-deviation effect of the index finger.

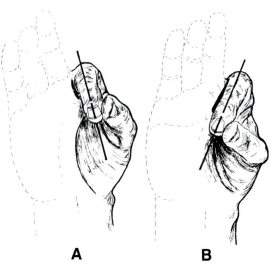

A   **B**

stability and may allow the force of pinch to enhance ulnar drift. If the patient uses a true key pinch against the middle of the index finger when it is fully flexed at the PIP and DIP joints, the MCP joint is more nearly extended and the proximal phalanx needs the support of the first dorsal interosseus to oppose the ulnar deviation.

The key pinch is so useful that it may be better to accept it even though it tends to deviate ulnarward and try to reinforce the index finger by teaching the patient to pinch against the index supported by the other fingers together (Fig 5–36). Also, the first dorsal interosseous muscle may be strengthened by exercise or by reinforcement by another tendon. We mention the ulnar drift problem here because it is an example of a deforming force made more effective by a long external moment arm. However, the internal muscle that opposes the force of key pinch is also a very strong muscle. The first dorsal interosseus has a tension fraction of 3.2, which is stronger than most of the flexor

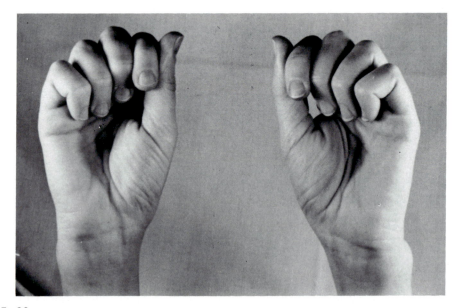

FIG 5–36.
Key pinch does not demand much intrinsic muscle support of the index finger provided it is supported by all the fingers flexed together.

profundus and flexor superficialis muscles and indicates that key pinch is provided for in the original design of the hand.

REFERENCES

1. Brand, P.W., et al.: Tendon and pulleys at the metacarpophalangeal joint of a finger, J. Bone Joint Surg. 57A:779–784, 1975.
2. Kapandji, I.A.: Biomechanics of the thumb. In Tubiana, R., editor: The hand, Philadelphia, 1981, W. B. Saunders Co., pp. 404–422.

CHAPTER 6

Mechanical Resistance

Paul W. Brand, M.B., M.S., F.R.C.S.

David E. Thompson, Ph.D., F.R.C.S.

We have considered the way in which a muscle, when activated, will shorten and create tension in its tendon. The tendon is pulled toward the origin of the muscle, thereby transmitting its tension to the hand and digits distally. Not all the force generated by the muscle will reach the distal joints. Some will be dissipated on the way by passive resistance of structures around the muscle and tendon. In the case of a newly transferred tendon or tendon graft surrounded by scar, most or even all of the muscle power may be lost in this way. More muscle balance operations fail because of scar, adhesions, and joint stiffness than because of poor selection of muscles to transfer, yet little emphasis has been placed on the study of these passive structures that resist motion.

This neglect is due in part to the lack of basic knowledge about resistance forces, or stiffness, in the body. We cannot surgically explore scarred structures without modifying or adding to the scar. The resistance to movement is complex and dependent on many factors. Nevertheless the importance of the subject demands attention, and we must find an objective method of monitoring the problem in individual patients so that we may know which treatment methods are effective. To do so, we must gain an appreciation for the underlying effects that produce these losses.

The losses produced by the sliding of ten-

dons in their sheaths, the compression and distention of soft tissues, and other associated actions are discussed in this chapter. The second law of thermodynamics states that real processes have associated losses. There are no exceptions to this principle in the human body. In the hand, the dissipation of forces and torques is a consequence of the nature of the soft tissues that transmit the forces, and the types of interface between the tissues—fixed and free. The losses are modified by the vitality of the tissues, the possible presence of scar tissue, and many other factors. Some of these factors are described here along with some basic principles on the mechanical behavior of soft tissues. The schematic finger shown in Figure 6–1 is used to portray the areas where the forces of muscles are lost to friction, viscous drag, elastic effects, and scar. Each of these is discussed subsequently along with measurement techniques to evaluate their effects.

The complexity of soft tissues makes it difficult to reduce their mechanical qualities to simple terms that can be quantified. The most vexing of these tissues' qualities is that they are not homogeneous. The way in which the various required qualities are blended into a single heterogenous tissue, while it frustrates the analyst, excites the admiration of the user and the appreciative observer. An example is the pulp of a finger tip. A finger tip is often the interface for

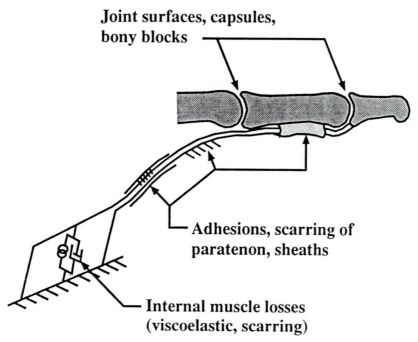

Joint surfaces, capsules, bony blocks

Adhesions, scarring of paratenon, sheaths

Internal muscle losses (viscoelastic, scarring)

FIG 6–1.
A schematic of the losses in the muscle-tendon-skeletal chain.

strong application of force in lifting a rock, pulling a rope, or for the gentler action of stroking a baby to sleep. These need the mechanical qualities of (1) compliance to the shape of an external object, such as a rock, (2) low elasticity for compression, and (3) elastic resistance against the pull of an object or a tool when tension might pull the skin away from the skeleton of the digit.

The elasticity of the pulp in compression is provided by little globules of fat which are almost liquid at body temperature and thus would be pushed aside by strong localized thrust. However, these fat globules are each contained and restrained by fibrous envelopes, serving as the skin of a balloon and creating a tissue that feels soft and exhibits elastic behavior in compression. The whole tissue is further subdivided by strands of collagen which run from skin to skeleton as septa. When a rope is grasped and pulled, these septa are pulled into tension and hold the whole tissue firmly to the bony skeleton. The compartmentalization of load-bearing soft tissues by their fibrous septa provides the benefits of prevention of the spread of infection and creating viscoelastic support structures. As the pressures within the tis-

sues alternate under applied external loads, the fluid within each compartment is forced to move from regions of high pressure to regions of low pressure.

Tissue fluid flows sluggishly around and between the fat cells and fibroblasts and fibers. It is this fluid which, by being subject to a viscous flow under pressure, permits even more compliance to the shape of external objects under sustained pressure. When a finger tip is examined after a ball point pen has been pressing on it for several seconds, a marked hollow remains in the pulp for some minutes while the fluids that were displaced now gradually return.

Any attempt to analyze and quantify each component of the finger tip separately would be difficult and inappropriate. What we can attempt to do is to understand separately the mechanical behavior of fat in general, and that of collagen, and of interstitial fluid. Then, as we study the mechanics of the hand we may recognize factors that combine in various proportions to modify the forces of muscle contraction, or the impingement of external force. This may help us to know how to avoid loss of useful force or compensate for the lack of it.

VISCOUS ELEMENTS

When a fluid layer flows over a surface or over another fluid layer, the faster-moving molecules of the fluid exchange their energy with those of the slower-moving fluid layer or the surface. The greater the relative velocity, the greater is the resulting energy transfer. This effect is seen within soft tissues when a quantity of fluid such as that produced by edema is forced from one compartment of the finger to another as a joint is flexed. The resulting drag forces which retard the fluid's motion are transmitted back through the fluid to the surface and this is felt as a resistance to their motion. For example, when an external pressure is applied to the finger pad, the longer the pressure is present, the greater the amount of fluid which moves from segment to segment. This provides one means by which the pulp tissues adapt or conform to an external surface. Once the majority of the unbound fluids have migrated, however, the solid components of the tissues begin to accept and transmit a greater portion of the load from surface to bone. When the external load is removed, very small restoring forces are in place and the surface tissue shape may be restored to its original shape very slowly as the fluids return to their resting distribution. This return to the resting state is slower that the rate of fluid flow under pressure, and results in the characteristic hysteresis in the stress-strain curves of soft tissues (see Fig 1–7,B). The more fluid available to be moved, the larger the difference between the loading and unloading portions of the curve.

ELASTIC PROPERTIES

Elastic tissues allow the soft tissue to be deformed and provide the forces to restore their resting shapes. To better understand the nature of elastic tissues, let us consider several types of materials and their elastic nature. One way to do this is with a stress-strain curve, where stress (the force applied per unit of cross-sectional area of the material), is plotted against strain (the change in length that results from the stress).

METALS

Metals have the simplest stress-strain curves of any of the materials. Steel has a characteristic behavior as shown in Figure 6–2. In *curve A*, the initial relation between stress and strain is linear. The slope of this portion of the curve is termed *Young's modulus*, or the elasticity of the material. When engineers design buildings and cars using steel, they always use a sufficient amount of the material to keep the stresses (force per unit area) well within region 1. Young's modulus for steel is on the order of 30×10^6 lb/in.^{2} As the load is increased, the curve is no longer linear, reaching the proportional limit at the extreme of region 2. In region 3, the metal will suffer permanent deformation. "Plastic change" has occurred when internal microscopic damage results in the material failing to return to its original length when it is unloaded. Only when the metal is stretched to the end of region 3 does it fail catastrophically.

Aluminum, as shown in *curve B*, does not exhibit a simple linear region and it therefore does not have a simple modulus of elasticity. There are published values of the elasticity for this metal, but these values are normally just an approximate stress at a 10% elongation. Aluminum does have a lower ultimate strength than steel, but a much greater maximum strain or deformation. If two identical bars of aluminum and steel are clamped to a bench and then loaded in bending, the steel is found to offer much greater resistance than the aluminum. The aluminum not only requires less force to bend, it also deflects more before breaking than does steel.

ELASTIC BIOLOGIC MATERIALS

The stress-strain curves of human tissues are much more complex. Most normal soft tissues contain fibers of elastin and collagen. Some of these tissues are specialized to transmit tension (tendon) or stabilize joints (ligament and fascia). These have a high per-

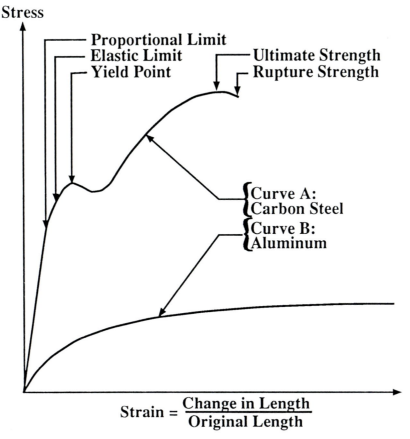

FIG 6–2.
The stress-strain curve of high carbon steel *(A)* and aluminum *(B)*.

centage of collagen which is arranged as parallel polymeric fibers; most of these force-transmitting structures allow 10% or less of total lengthening under physiologic tension. The length-tension curves of these tissues have a classic elastic profile (Fig 6–3). Soft tissues such as paratenon are designed to permit free movement and behave more like nylon stretch fabrics (Fig 6–4).

Elastin, a common biologic building block of soft tissues, is an extremely compliant, linearly elastic material. In terms of elasticity, it has a Young's modulus of 200 lb/in². The Young's modulus of steel is 10,000 times higher, yet elastin has 2,000 times as much compliance, in that elastin will accept 200% lengthening without deformity, whereas steel accepts only 0.1%. Although elastin is reported to make up only about 5% of the fat-free dry weight of the soft tissues, its contribution to the compliance of human soft tissues is important.

Collagen is a more abundant material within soft tissues, contributing an average of 77% of the fat-free dry weight. These fibers are 20 times stiffer than elastin, having a Young's modulus of around 15,000 lb/in.² This rigid material only permits strains (lengthening) of around 10% before the first internal tearing occurs. The distribution of collagen and elastin within the body is not uniform. In tendons and ligaments, the collagen content is very high and the elastin content low. In paratenon, this distribution is reversed.

THE BEHAVIOR OF POLYMERS

The properties of human tissues are similar to those of polymers because both polymers and human tissues are mainly composed of long hydrocarbon chain molecules.

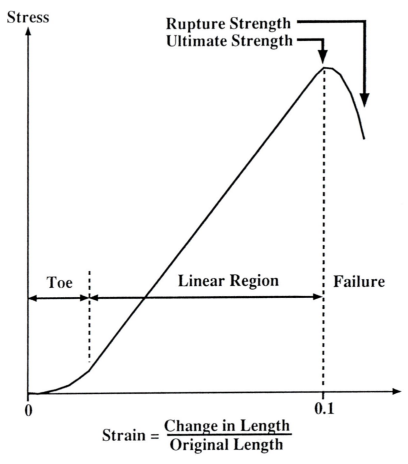

FIG 6–3.
Stress-strain curve for tendon showing the nonlinear *(Toe)*, linear, and failure regions.

Random Coil   Fully Extended Chain

FIG 6–4.
Limiting conformations of polymer chains. Each "zig" denotes a structural unit. *(Redrawn from Schultz, J: Polymer materials science, Englewood Cliffs, N.J., 1974, Prentice-Hall, 1974.*

To better understand human tissues, consider the following description of the properties of polymeric compounds.

Polymers are composed of long-chain molecules with a repetition of structural units along the chain. Usually these chains consist of hydrocarbon monomers bonded from head to tail. The "configuration" of the chains refers to the arrangement of units along their axes. Different polymers can be made by arranging the same chemical units in different configurations. The bonds between these repeating polymer chains are usually flexible enough to permit some rotational freedom. This allows the molecules to be arranged in a variety of physical arrangements, or "conformations." These conformations can be classified as random coil, fully extended, or crystalline states.

The random coil and fully extended states are isolated strands (Fig 6–4). These structures have relatively few cross-connections between chains. Loosely organized soft tissues such as paratenon are examples of living material in this state. The crystalline state exists if a number of similar chains si-

multaneously become extended and aligned. This conformation is created by cross-linking or hydrogen bonding side to side between the structural units of adjacent chains. Tendons are the best examples of this material state. If stress is applied to a polymer in a direction not aligned with the principal polymer axis, deformation occurs in four stages (Fig 6–5).

When we begin to study the elastic response of polymers to stress, we recognize a new factor—*time*. An important difference between the deformation of metals and polymers is that the deformation of polymers varies with variations in the length of time over which the force is applied. Consider the response of an entangled linear polymer when a load is instantaneously applied and held for a time. If this time of loading is very short, the chains in the amorphous regions will not have time to straighten and no deformation will occur. As the time over which the load is applied is increased, the chains will be able to uncoil, allowing the polymer to lengthen. At the same time, this stretching of the chains produces an elastic restor-

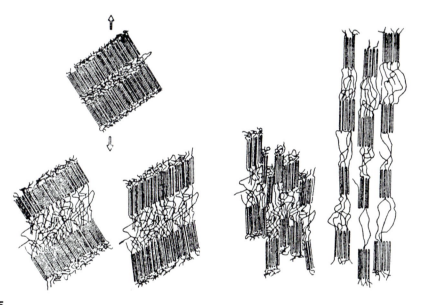

FIG 6–5.
The four stages of polymer deformation *(counterclockwise from upper left)*. *First stage:* Lammellar ribbons slip rigidly past one another. The ribbons lying parallel to the tensile axis cannot do this and the strain is accommodated almost entirely by the interlamellar amorphous layer. *Second stage:* Here the tie chains are highly extended and deformation occurs by the slip-tilting of the crystalline lamellae. *Third stage:* Blocks of crystal are pulled out of the ribbons. The blocks are still attached to each other by tie chains. *Fourth stage:* The blocks become aligned along the tensile axis and slip past each other. *(From Schultz, J: Polymer materials science, Englewood Cliffs, N.J., 1974, Prentice-Hall. Used by permission.)*

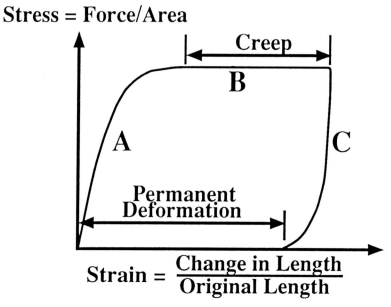

FIG 6–6.
A schematic representation of the yield and plastic deformation of a polymer. **A,** the load is increasing. **B,** the creep deformation when the load is held constant. **C,** the load is removed.

ing force. If the load is applied slowly enough, the chains will uncoil to a state in which the restoring force exactly equals the applied load.

However, polymers have both elastic and viscous characteristics. When a polymer is loaded, the elastic and viscous properties oppose one another. The viscous aspects of the polymer let the microfibril chains slip past one another. This allows the chains to recoil and relaxes the stress within the material. Thus if the polymer is held at a constant length for enough time, the entire lengthening will occur by slippage of the chains on one another. No load will be required to maintain the deformation.

These two aspects of time-dependent mechanical behavior are termed, respectively, *creep* (the uncoiling of the chains) and *stress relaxation* (the slipping of the chains past one another). Creep is the continued deformation of a material under constant stress. Stress relaxation is the continued decrease in stress needed to maintain a given deformation. Figure 6–6 presents the stress-strain behavior of a polymer, showing both creep and stress relaxation phenomena. The kinship of human tissues to polymers would normally allow us to draw some conclusions

from similar situations in these simple materials. In living tissues, however, the time response of tissues to sustained pressures is quite different.

BIOLOGIC STRUCTURES

What we have developed so far is a set of elementary concepts which let us begin to assemble and analyze more complex structures. The length-tension curves (Fig 6–7) of composite biologic tissues have four definable parts:

1. Unfolding: a compliant phase where the internal elements of the tissues adapt or unfold without significant stresses.
2. Alignment: a phase where the internal elements gradually align themselves with the predominant stresses.
3. Stiffening: the internal elements are essentially aligned with the stresses and each increment in load results in a smaller and smaller lengthening.
4. Failure: internal tears which spread and finally result in total separation.

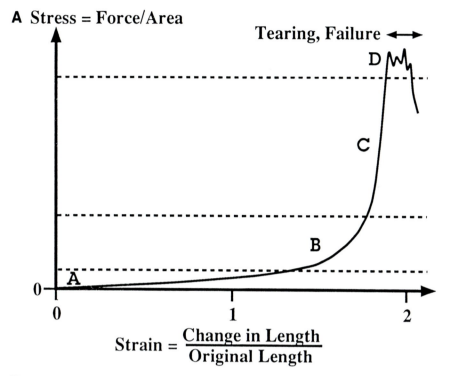

A Stress = Force/Area

Tearing, Failure ⟷

D

C

B

A

0

$$\text{Strain} = \frac{\text{Change in Length}}{\text{Original Length}}$$

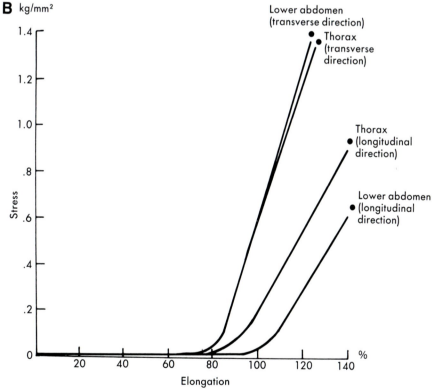

B kg/mm²

Lower abdomen
(transverse direction)

Thorax
(transverse
direction)

Thorax
(longitudinal
direction)

Lower abdomen
(longitudinal
direction)

Stress

%

Elongation

FIG 6–7.
A, the elastic behavior of soft tissues. *A,* unfolding; *B,* alignment; *C,* stiffening; *D,* failure. **B,** human skin from various parts of the body. Each curve shows up to 80% lengthening with very little tension. Then it curves up into a steep straight section typical of elastic response. These curves stop short of failure. *(From Yamada, H. In Evans, F.G., editor:* Strength of biological materials, *Baltimore, 1970, Williams & Wilkins Co.)*

Skin and many soft connective tissues are really a fabric of loosely woven strands of elastin and collagen. Figure 6–7,B is the length-tension curve for skin in situ. It has a long flat segment before it begins to turn upward into a typical elastic curve. In this flat segment, very little tension results in a large amount of elongation. This feature allows great freedom of motion in the neutral position of joints which can be produced by very small forces. In the early part of this flat segment the skin is not really being stretched; its elastic elements are merely being *unfolded* and realigned along the principal direction of the forces. As the collagen fibers become more and more aligned with the direction of the stresses, the stiffness of the tissues becomes greater and greater. Only when these two stages are complete are the actual individual tissue elements being stretched. That is when the curve turns upward as more and more solid elements are elastically stretched to a greater and greater extent.

A simple analogy is a fabric, like a piece of cloth. If it lies loosely on a table, it will be wrinkled, and the first pull will undo the creases. This apparent lengthening requires almost no tension. If the fabric is pulled on the bias, rather than parallel to the fibers (Fig 6–8,A and B), the weave will be distorted. The fabric will lengthen in one direction while narrowing or shortening in the other. All the fibers gradually become

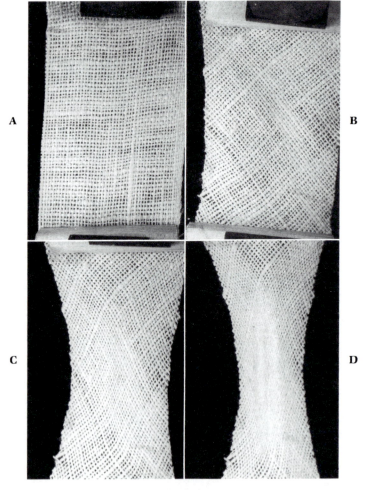

FIG 6–8.
A strip of burlap is shown in an Instron testing machine. **A,** tension parallel to burlap fibers. **B,** burlap cut on the bias without tension, **C,** with light tension, **D,** with strong tension. In **C** and **D** the increase in length is through alteration in the fiber directions rather than by actual fiber lengthening.

aligned more nearly parallel to the line of the tension (Fig 6–8,C). This rearranging of the weave may allow lengthening with slightly greater tension. Finally, when most fibers are parallel to the line of pull, it takes a much larger force to produce very little lengthening (Fig 6–8,D), as every individual fiber has to be stretched. Finally, under the highest tension, the cloth will tear and the length-tension characteristics will alter completely.

As in the fabric example, pulling skin in one direction will cause lengthening of the tissue with a shortening or narrowing at right angles to the line of pull. This narrowing of the tissues is actually a mechanism by which the stresses are spread out to adjacent tissues which aid in carrying the total applied load.

Other tissues, like paratenon, have to stretch in many directions with little resistance. A simple analogy for paratenon is pantyhose. Regular nylon fabric will lengthen very little when stretched. "Stretch" nylon, in stockings and pantyhose, has much more capability of lengthening due to the fact that strands of nylon are coiled or folded on themselves producing an intertwining set of elastic fibers. The elasticity that the wearer feels is that of the coiling and uncoiling, not of the actual lengthening of the strands of nylon fiber. If it lies loosely on a table, it will be wrinkled (Fig 6–9,A), and tension will straighten the creases. This apparent lengthening requires almost no tension. The length-tension curve of pantyhose thus has a large adaptation capacity in which it unfolds to approximately 200% of its resting length. At this point all of the nylon fibers are fully aligned, and the curve turns upward approaching the length-tension curve of the fully extended nylon fibers (see Fig 6–19,D). The elasticity of ligaments is sufficient to allow approximately 10% lengthening under stress, providing a protective "give" at the limits of joint motion. There might be damage if the ligament were nonextensible, as demonstrated in the following analogy.

When steel cables first became popular on ships, ropes were thrown away, and cables were used on boats and ferries to tie up to the wharf. It was not long before ropes were back on the job, because steel cables were found to tear out the cleats and bollards to which the boats were moored. The twisted coils of hemp and manila fiber that make up ropes behaved like misaligned collagen fibers as they took the strain to bring a ship to rest. First they altered their internal arrangement and tightened the twist to align themselves to the load. During this realignment, the interfiber forces increased and frictional resistance aided in resisting the elongation of the rope. Then, as the fibers approached an alignment parallel to the rope, the ultimate creaking pull produced large forces and fiber lengthening. By this time, however, the ship would have almost stopped (Fig 6–10). Steel does not have the low elasticity and large deformation capacity necessary to give the ship time to slow down and stop. The forces of deceleration then become too high, and something has to give.

LIVING TISSUE

Length-tension characteristics of fresh cadaver and living human soft tissues are similar. Since it is difficult to make accurate measurements in an intact living hand, most scientists and bioengineers use fresh cadavers. Tissues can be excised and tested in machines. However, differences between living and dead tissues are so fundamental that we cannot accept the results of cadaver studies until we compare them with studies using measurements in the living hand. However crude and inaccurate these clinical measurements may be, they are *real* and must be the basis on which we determine plans for treatment and evaluate progress.

CREEP AND GROWTH

A typical difference between living and dead tissue is demonstrated by a comparison between the concepts of *creep* and *growth*. When a composite tissue (e.g., skin) is excised and subjected to tension in a machine, it exhibits typical length-tension changes. At higher levels of tension, when the stress is

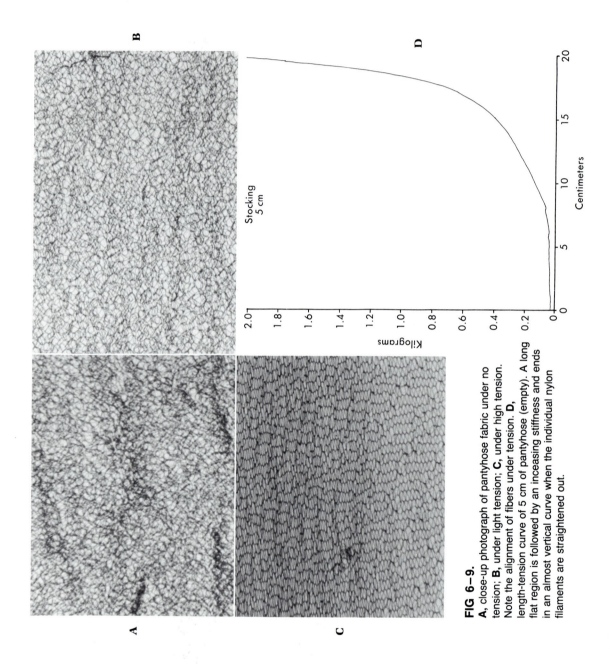

FIG 6–9.
A, close-up photograph of pantyhose fabric under no tension; **B,** under light tension; **C,** under high tension. Note the alignment of fibers under tension. **D,** length-tension curve of 5 cm of pantyhose (empty). A long flat region is followed by an inceasing stiffness and ends in an almost vertical curve when the individual nylon filaments are straightened out.

maintained at constant tension, the skin sample will continue to elongate over time. When the tension is relaxed, the tissue will not return to its original length, even though it has not obviously been torn or otherwise changed. This gradual irreversible lengthening is called *creep*. It is caused by a slippage of short collagen fibers on one another within the tissue. Some fibers may be ruptured while others just slide on each other. The skin as a whole has become *stretched* permanently. This is termed *plastic* behavior.

This knowledge about cadaver skin (or joint capsule) may suggest that similar high tension will help to loosen tight skin and tight joint capsules that restrain joint movement in the intact living finger. This is not necessarily so. If extra force is used to stretch skin or joint capsule beyond its normal elastic limit, creep would occur in an intact living hand. However, at the same time, the microscopic tearing of fibers and cells might result in inflammation and possibly small hemorrhages. The inflammation may cause fibrinogen to be laid down, which

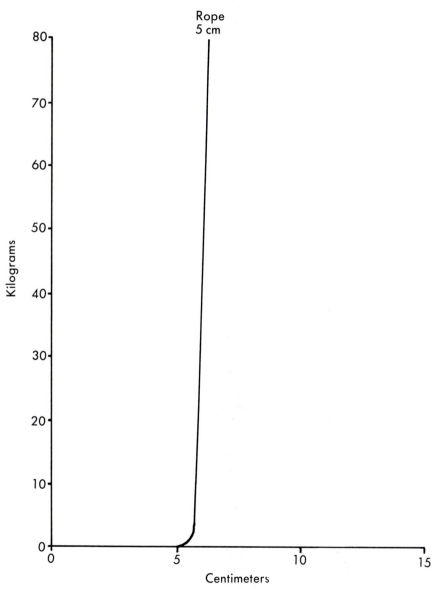

FIG 6–10.
A rope of hemp, tested under tension in an Instron machine shows about 10% lengthening under stress.

ultimately may become an interstitial fibrosis that could permanently limit the mobility of the skin. Thus the final contracture may be worse than before. Then again, living tissue might experience pain, resulting in active behavior that would attempt to escape from the tensile load. By contrast, if the same intact living skin is held in only a slightly lengthened position, within its elastic limit, for a period of hours or days, the living cells will sense the strain and the collagen fibers will be actively and progressively absorbed and laid down again with modified bonding patterns with no creep and no inflammation. This is *growth* rather than *stretch* and it occurs only in *intact living tissue*.

Having introduced the terminology associated with the material behavior of living soft tissues, it may be helpful to categorize them with respect to their response to stress over time. Figure 6–11 attempts to do this

visually. The x-axis represents the time after the stress is initiated and the bar length represents the duration of the tissue response. It must be remembered that the type of stresses and responses are very different. Figure 6–11 is an attempt to symbolically depict the relationship between the stresses applied to living tissues and their response. The three dimensions depicted are those of stress, time of application of stress, and the body's response measured qualitatively as hypertrophy and atrophy. This figure assumes that the stress being applied is not constant, but rather an alternating or repetitive stress. The maximal value of the stress is shown in the figure.

If a single large damaging blow is delivered to tissues, the damage occurs almost immediately. This may result in minor internal structural cellular damage (process 1–2) or the stresses may exceed those at 2 and produce an open wound or broken bone.

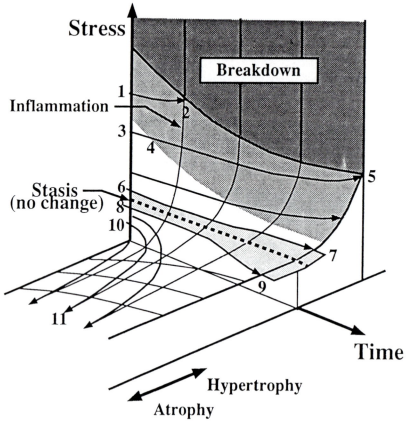

FIG 6–11.
The time response of tissue to mechanical stress.

Thus, any stress applied at or above the level indicated as A in Figure 6–7 will produce damage sufficient to result in a frank wound. The aftermath of this damage is not addressed in this figure. What we will now discuss are the numbered paths which indicate the sequence of events following the application of different stress levels.

If the applied stress was lowered to point *(1)* and repeated at that level, the inflammatory response of the living tissues, combined with the continuance of the stress, would result in damage, inflammation, and eventual breakdown of the tissue. If the levels were lowered to the point *(2)* where there was a minor inflammatory response, the same stress level would eventually produce inflammation *(3)*, more damage, and finally breakdown *(4)*. At some point *(5)*, the stresses are low enough to result in an active response by living cells which recognize new demands and respond by hyperplasia or hypertrophy, or both. This process may be accompanied by mild inflammation and produce a tissue that, over time, has an improved ability to resist mechanical stress. In the region in or around the normal stress levels of the tissue, marked *Stasis*, the body exhibits no real change to a continued, repetitive application of stress. If the stresses are constantly less than this, however, owing to disuse of a hand, or to the adoption of protective devices or orthoses, the tissues will slowly atrophy and eventually reach a state of equilibrium (processes 8–9). Absence of any significant stress results in greater and greater atrophy (processes 10–11).

WORK AND RESISTANCE

Work is normally defined as the product of a force and the distance through which it acts. If one considers a whole finger or a wrist, the work required of the muscles to produce joint angle changes depends upon the external load and the distance that it has to be moved or lifted. However, in the absence of external loads, work will be required in an amount proportional to the product of the pressure of the tissues and their volumetric change during angulation. The power this same angulation consumes is time-dependent and is proportional to the rate of work required. Thus, whenever the volume of the tissues surrounding a joint swell during edema, joint angle changes produce increased tissue pressures and large volume changes. The muscles must work very hard to move the unloaded hand. This effect was experienced by the NASA astronauts upon donning their space suits. They found that the gloves acted much like a second skin and the internal pressure within this skin was much higher than the surroundings. They quickly became fatigued doing simple tasks during early space activities. This problem was further exacerbated in later missions since a higher internal pressure was used for the space suits. Attention was subsequently focused on the necessity to design joints in the wrist and hand that minimized the volume change during joint angulation. The example shows how edema and other viscous effects increase the power required of the musculature. Edema has two negative influences: (1) it increases the volume change with joint angulation and (2) its viscoelastic character dramatically increases the pressure within the tissues.

The only noninvasive way to study the mechanical qualities of the tissues that resist movement is to apply a series of increasing forces to a joint, *torque* or moment, and measure the way joint angles change in response to the changing torque. The angular change in the joint is related to the change in length of the restraining tissues. Although this cannot define the mechanical qualities of each tissue that may be involved in drag, it does produce a curve which is a summation of their overall effect. If a joint becomes constrained by a tissue problem, the torque-angle curve of that joint will be typical of the restricting tissue, because the other tissues are too loose to affect the curve. It is a method that grows in usefulness as it is used, because although the curve is a composite, it has relevance to the actual situation of each joint or tendon. It is also repeatable. If a curve produced today is different from one made last week, it represents a real change in the involved tissues and quantifies the value of the treatment modalities used. For a surgeon or therapist starting to use the system, he or she must test the passive motion of a number of normal joints and note how various joints have similar torque-angle profiles. Normal metacarpophalangeal (MCP) joint torque-angle curves are quite different from normal proximal interphalangeal joint (PIP) curves. The MCP joint moving into extension has a long, gradual curve, whereas the PIP joint curve ends steeply as collateral ligaments and palmar plates become tense (see Fig 11–6). Then one must compare the torque-angle profiles of the same finger joints with the wrist flexed and with the wrist extended to demonstrate the influence on the finger joints of the passive tonus or viscoelastic behavior of the long flexor and extensor muscles. As experience is developed, this method can be used to differentiate joint stiffness caused by adhesions around tendons from stiffness caused by tissue effects at joints and to differentiate elastic from viscous restraints. Finally, it is possible to use a torque-angle curve to determine therapy recommendations or whether conservative treatment is appropriate. The basic torque-angle curve is made from five readings of joint angle vs. five increasing levels of torque.

METHOD OF TORQUE-ANGLE MEASUREMENT

In order to simplify and speed up the measurement of torque-angle curves, we make the following suggestions (see also Chapter 11).

1. Because the measurement of torque is important in a relative sense rather than in an absolute sense, and because it is difficult and time-consuming to measure the exact moment arms or lever arms around a joint, we use a *standard moment arm* that is easy to define and do not attempt to measure it. For the PIP joint, the standard moment arm is the flexion crease of the distal interpha-

langeal (DIP) joint and the extensor torque is applied there. The convexity of the head of the middle phalanx is used for applying flexion torque. For the MCP joint, the proximal of the two creases at the PIP joint (Fig 6–12) is recommended. Any recognizable anatomical marker will do, as long as it is consistent. It is the *shape* of the curve that matters, not the level.

2. The force is applied through a narrow applicator for accuracy and repeatability. A regular sling is too wide to apply tension because it may exert its effect through an edge rather than the middle, thus changing the lever arm. A loop of nylon filament will allow great precision in placement, but it may be too narrow and cause pain or cut into tender skin at high tension. A narrow tape or cord about 2 mm in diameter is accurate and safe. An example of a *pusher* to apply flexion force to the dorsum of the hand may be an object similar to the eraser on the end of a pencil, both in texture and size (narrower is better).

3. To measure the angle, the goniometer must be light and have low friction. It must be possible for a therapist to hold it firmly on one segment of the hand (or digit) and follow the changing angle by moving the other arm of the goniometer by the touch of a finger. It may also be attached to the finger by rubber bands. We have used a number of devices of varying degrees of simplicity and cost (see Figs 11–4 and 11–5).

4. The skill of the therapist in making the evaluation lies in his or her ability to hold the goniometer or angle-measuring device with one hand while applying the torque with the other hand and writing down the figures with the third hand. In order to avoid the need for a third hand, use a tape recorder so that the therapist may rapidly apply a series of five tensions to the digit and read aloud the angles produced by each tension without stopping or relaxing the tension. Then the figures are recorded directly on the graph paper as five dots to outline a curve.

5. Depending on the size of the joint, the forces used may be 100, 250, 400, 550, and 700 g, or 200, 400, 600, 800, and 1000 g. Since the moment arm for a PIP joint is

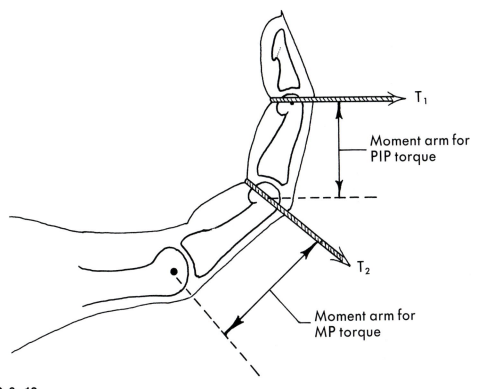

FIG 6–12.
Force T_1 is applied to extend the proximal interphalangeal *(PIP)* joint by tension at right angles to the middle segment of the finger through a narrow sling at the distal crease. Force T_2 is applied to the metacarpophalangeal *(MP)* joint at the PIP crease at right angles to the proximal segment.

about 2.5 cm, the numbers for tension, when multiplied by 2.5, would give torque. Although we do not do this, bear in mind that when we record 200 g, we really mean 200 x, where x is the moment arm at which the force is applied. For an average PIP joint, 500 g cm of *torque* would result from a *tension* of 200 g applied 2.5 cm from the joint axis.

GRAPH OF TORQUE-ANGLE CURVES

For therapists who are short of time and daunted by the need to plot curves, the range of motion of each joint at two different torques may be recorded. The joint may be pulled open with a tension of 200 g while the angle is measured. Then the tension may be increased to 500 g and the angle measured again. The *difference* between the two angles relates to the quality of the restraint. We suggest that for the continuing *longitudinal* study of a single, stiff joint a torque-angle curve should be plotted at the start of treatment, and the second and third curves should be plotted *on the same graph* at various stages of progress. Not only will the curve move bodily along as the range of motion improves, but the shape will change[4] and thus provide as much information as the movement (Fig 6–13).

TORQUE-ANGLE LOOP

Some of the graphs in this chapter show a loop rather than a simple curve. These graphs were made with a more sophisticated device that is not necessary for routine evaluation. It measures the torque that is needed first to move a joint toward its maximum range and then to hold it or restrain it on its way back to the relaxed state. It is thus a measure of moment-to-moment variables, such as viscous change in the tissue and friction that may resist movement in both directions. Since viscous changes are time-related there should be a standard rate of testing so that the return loop may be

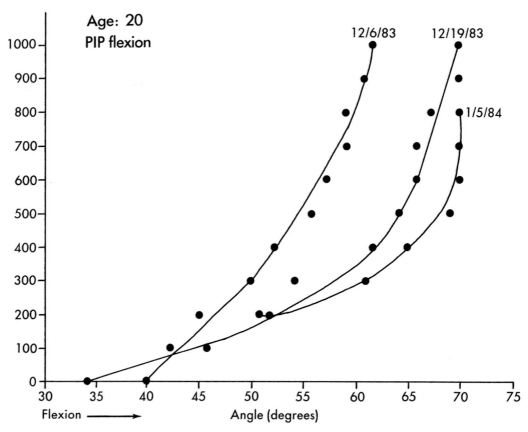

FIG 6–13.
This record of a patient's stiff finger shows three torque-angle curves with 2 weeks' treatment between each. The last curve shows a vertical segment at high torque. This led to a decision to discontinue treatment at that point. PIP, = proximal interphalangeal. *(Courtesy of K.R. Flowers, Phoenixville, Pa.)*

noted as starting 30 seconds after the testing is initiated and only the first test loops compared. The significance of these changes is discussed later in this chapter. If one wants to plot the entire loop, it can be performed with nonelectronic devices simply by recording angles as torque is increased and then by reducing torque, stage by stage, and repeating the measurement of angles on the return trip. Engineers call this graph a *hysteresis loop* (Fig 6–14; also see Fig 1–7,B).

TISSUES DESIGNED TO PERMIT MAXIMUM LENGTHENING AT MINIMUM TENSION

The function of tendons demands that they move. Most tendons do not move through synovial sheaths; they move through connec-

tive tissue and part of the tissue moves with them. This specialized tissue is *attached* to the tendon surface and also to surrounding structures that may be immobile. This paratenon has to carry the blood supply and permit movement. Paratenon is composed of a few elastic fibers and a few folded and coiled collagen fibers and blood vessels, all of which are attached at one end to the tendon and at the other end to the static tissues that line the pathway of the tendon. The bulk of paratendinous tissue is made up of ground substance that is semifluid yet elastic. It is too complex to study here, but it is fascinating to watch under the high power of an operating microscope. The length-tension curve of the paratenon of a tendon such as the palmaris longus shows a flat long section in which the tissue seems to exercise no restraint on tendon movement. At the end of the available range of motion, the paratenon

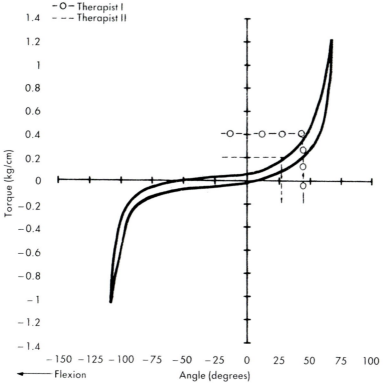

FIG 6–14.

Two therapists measured the range of extension of this metacarpophalangeal joint of an index finger. One *(open circles)* came up with 18 degrees more extension than the other *(dotted line)*. Subsequently this torque-angle curve showed that one had been using twice as much torque as the other.

almost suddenly comes to the end of its permitted lengthening and curves into a rather steep elastic restraint. There is no inorganic material that has such a convenient stretch pattern, giving great freedom within a certain range and then firm resistance beyond those limits. However, the pattern is more familiar if we do not call it a *stretch* pattern but rather a *looseness* pattern or an unfolding. The skin on the back of fully extended knuckles and elbows demonstrates a folded tissue (Fig 6–15). The commencing flexion of a knuckle simply unfolds the dorsal skin but then when the joint is flexed beyond about 60 degrees, the skin is fully unfolded and has to actually lengthen. Finally, when the skin blanches, it may be exerting quite a significant elastic resistance to further stretch and further joint flexion.

There are some important practical considerations of the concept of the composite length-tension curve of normal compliant connective tissue and skin. These tissues

may, under certain circumstances, cause significant drag or resistance to movement. One such circumstance relates to use and disuse.

USE AND DISUSE

In tissues designed to allow stretch and movement and yet which also limit motion, we see the same length-tension curve. In Figure 6–7, the A section is rather flat, the B section curves into a steeper slope, the C section is steep and straight, and the D section shows gradual failure rather than a sudden break. In life, these tissues are not static. All of the elements of the composite tissue are constantly being renewed and adjusted. Collagen in particular is being absorbed and then laid down again with updated length, strength, and new bonding patterns. The body does not maintain a range of motion or a redundancy of loose tissue if it is never used. When a finger or a

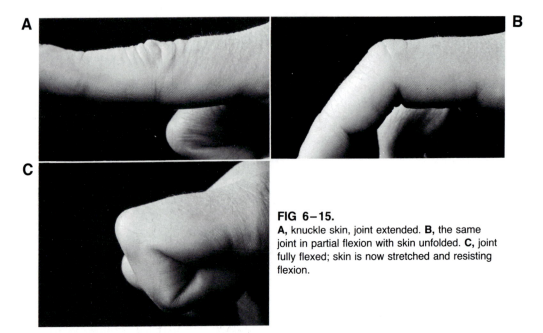

FIG 6–15.
A, knuckle skin, joint extended. **B,** the same joint in partial flexion with skin unfolded. **C,** joint fully flexed; skin is now stretched and resisting flexion.

limb is placed in a plaster cast for a few weeks, it becomes stiff and takes time to regain its full range of motion.

One factor in disuse stiffness is the simple shortening of structures that need length to allow tendon and joint movement. Refer again to the skin on the knuckle. Before the joint is immobilized in extension, the skin is deeply folded and creased. After several weeks of immobilization, the creases have disappeared and slight flexion puts the skin under tension. One may think this is a change in the elasticity of the skin, but it is an actual shortening. If the postimmobilization skin were tested as an isolated tissue excised from the body, it would not show much change because the quality of the tissue is unchanged; there is just less of it.

We labor this point because surgeons often talk about *stretching* skin that is tight. The word *stretch* is appropriate to the short-term lengthening of an elastic tissue. If we assume that contracted skin needs to be stretched, there is a danger that we may try to force the joint and stretch the skin to the point of creep. It is better not to use the word *stretch* for what should be long-term *growth*. If we want to restore normal length to a tissue that has shortened after disuse, we need to reverse the process and apply the stimulus of activity, or better, the stim-

ulus of holding the tissue in moderately lengthened position for a significant time. Then it will *grow*. We do not know how much tension is needed to stimulate growth, but from experience, when skin is held gently and continuously on the stretch to the point of early blanching, it is stimulated to lengthen. Time is the important factor. It is not a minute-by-minute change; growth is a matter of days, and the stimulus needs to be continuous for hours at a time to be most effective. Thus these compliant normal tissues adapt to the requirement of their pattern of use. They become shorter when immobilized in a slack position. They grow longer fastest when held constantly in a slightly stretched position.

NONCOMPLIANT SOFT TISSUES

A group of tissues composed of parallel collagen fibers are tendon, fascia, and ligament. They are designed to resist stress rather than to permit movement. Their fibers are not woven on the bias and there is no flat *A* segment to their length-tension curves (see Fig 6–7). These tissues, tested by Yamada[13] and others, show less than 10% lengthening under maximum stress. The greater part of the length-tension curve is a steep straight line like the length-tension curve of steel wire, not living tissue (Fig 6–16). The greatest contrast between these parallel fibered structures and the loosely woven structures is not just in their response to mechanical stress as dead excised tissues. It is in their long-term response to stress in the intact living hand, especially in their behavior after injury. These tendon-

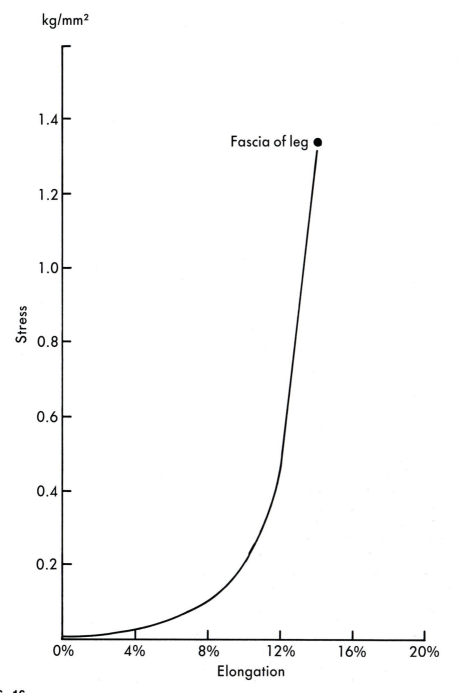

FIG 6–16.
To compare the shape of this curve of fascia with that of skin or other compliant tissue, one should note the scale along the bottom. The total lengthening of this tissue is only about 12%. Compare steel, 0.2%; Manilla, 10% (see Fig 6–11); and skin, 100% (see Fig 6–9).

like tissues do not respond to repetitive tension by growing longer. They are programmed to resist stretch and have a tendency to become thicker and stronger as they are subjected to repetitive stress over many days and weeks.

SCAR IN CONTINUITY WITH TENDON

In the growth and development of an area like the palm of the hand, every normal child has millions of fibroblasts laying down tough inextensible bands of collagen fibers that are oriented longitudinally. Some fibers are bound firmly to one another to form tendons and are in continuity from forearm to fingers; yet at the interface between tendons lying side by side there is a cleavage that is absolute. Even before babies have the wit to move their fingers separately, they are assured of perfect gliding between their crowded tendons. Before they are strong enough to apply longitudinal force, there is a mechanism that will tend to restore continuity in a tendon if it should be ruptured or cut. Whatever the nature of the blueprint that the fibroblasts recognize and obey, the result is that the tendon keeps the tough parallel-fibered collagen inside and the compliant, extensible tissues outside. This may all be disrupted by injury or by the knife of the surgeon.

Once the surface of the tendon is disrupted, some qualities that maintain the strength and mechanical integrity of the tendon seem to be communicated to the new scar outside the tendon. The work of Potenza[8] has demonstrated that tendon healing in a sheath may be associated with the migration of fibroblasts from any locally available connective tissue. The more recent work of Lundborg and Rank[5] has shown that tendon repair may also occur from the cells within the cut ends of the tendon. Whatever the source of the fibroblasts, most authors agree that once the new scar forms between the cut tendon ends, the new collagen fibers become oriented in parallel fashion. They tend to shorten and draw the tendon ends toward each other. This new tendon-like scar responds to repeated tensile stress by hypertrophy and further shortening. The scar that forms off to the side of the tendon, outside of its intact surface, remains disoriented in its fiber pattern and may respond to tendon movement by gradual relaxation or by undergoing some absorption, leaving lighter more areolar tissue that may allow more tendon movement.[4]

The objective of the hand surgeon is to repair the wound in such a way that the dense parallel-fibered scar is confined to the tendon and to effect the repair or the attachment at the new insertion. The remainder of the wound must heal by scar, but that scar should not prevent the movement of tendons and joints. Gelberman et al.[4] have demonstrated how tendon movement during healing makes a dramatic difference in the development of a plane of gliding around a tendon scar. Immobilized canine flexor tendons healed by ingrowth of connective tissue from the sheath and became surrounded by adhesions (Fig 6–17,A). Similar tendons that were mobilized after tendon suture healed by proliferation of cells from the epitenon. The mobilized tendons healed more rapidly, there was freedom from adhesions, and the strength of the repair was enhanced (Fig 6–17,B).

The contrast between those results was caused by mechanical factors, not chemical factors. It encourages us to believe that fibroblasts are responsive to mechanical stimuli and that we should be able to control many of the variables in healing by carefully graded mechanical input monitored by mechanical evaluations of the stress-strain qualities of scar and the changing drag on tendon movement.

Tissues with a long or shallow length-tension curve have a better prognosis for long-term shift of the curve in response to imposed tension. Tissues with a steep length-tension curve not only are noncompliant on a short-term basis but are resistant to changes imposed by splinting and exercises (Fig 6–18). Because the shape of the length-tension curve (and therefore of the torque-angle curve that derives from it) has prognostic significance, one needs to recognize the types of curves and refer to them by descriptive names. However, there are no clear definitions, and we tend to refer to

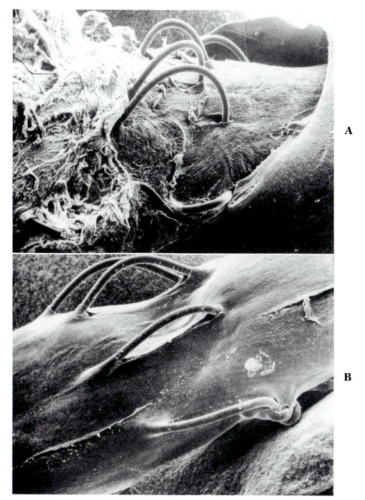

FIG 6–17.
The experimental surgery of Gelberman et al. on canine flexor tendons. **A,** immobilized tendon after 10 days. Surface features show adhesions to the left of the sutures. The surface of the tendon beneath the injured sheath (to the right of the sutures) is smooth and free of adhesions (×60). **B,** mobilized tendon after 42 days. Surface features show a slight depression at the repair site. However, the repair surface is smooth and free of adhesions (×60). *(From Gelberman, R.H., et al: J. Bone Joint Surg. 65A: 70–80, 1983, Used by permission.)*

them as *shallow* or *gentle curves* if they have a long, shallow rise at first and then go gradually into a steeper curve. The tendon-like tissues that may block joint movement and that have parallel fibers have *steep curves*, or simply *tendon-like curves*.

MASKING OF A STEEP CURVE

When joint movement is blocked by a tenodesis or other noncompliant tissue, there is sometimes pain or discomfort associated with the sudden block to flexion or extension. The patient therefore rarely uses the last few degrees of motion short of the block. When a range of motion is not used, it is gradually lost. The skin and other soft tissues take up the slack and no longer permit the use of those last few degrees. Now if a torque-angle curve is plotted, it may show a gentle curve typical of disuse contraction of soft tissue. If a therapist now works with the joint and brings all the available range into use, the compliant tissues will lengthen while the tenodesis remains solid. Therefore the *shape* of the torque-angle curve will change and become more square (shallow at first and then suddenly steep). The range of

Age: 28
Fracture of radial head (8 years ago)

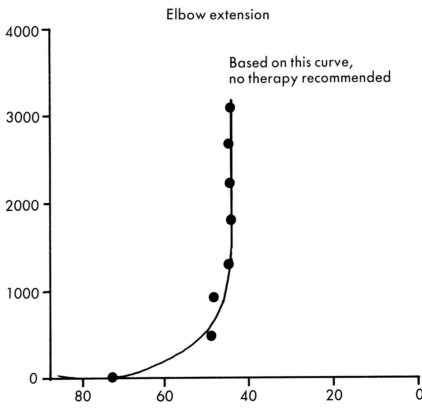

FIG 6–18.
Torque-angle curve of a stiff elbow, typical of bony block or of tenodesis. *(Courtesy of K.R. Flowers, Phoenixville, Pa.)*

motion at the high torque may not have changed while the range at low torque will have increased (Figs 6–13 and 6–19).

The therapist has *unmasked* the real problem. Now the surgeon may decide to excise the unmasked tenodesis.[2] The post-operative torque-angle curve may jump after surgery (we sometimes measure it at surgery) to an improvement of 15 to 25 degrees and will now show compliant resistance at high torque and therefore a good prognosis for more improvement with active therapy. This unmasking may occur in other ways. For example, a tenodesis or tight muscle-tendon unit may cross more than one joint and result in a loss of range of motion at a distal joint. This is composed of the restraint of the tendon plus, after a time, the restraint

of the soft tissues around the joint that have shortened from disuse. These two factors may be analyzed by flexing the proximal joint and testing the distal joint with a torque-angle curve, then extending the proximal joint and retesting (see Fig 11–8). In this way, one may determine the significance of each problem separately.

TISSUES IN SERIES

Sometimes scar occurs at some position along the length of a tissue. The scar may have a noncompliant profile, and the normal tissue may be compliant. For example, an ulcer on the palmar surface of a finger may

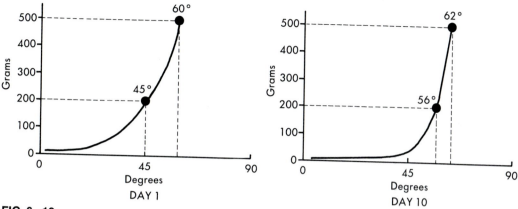

FIG 6–19.
The torque-angle curve of this contracted PIP joint shows a rather gentle, nonspecific curve. After 10 days of therapy, the angle of extension at 500 g has only increased 2 degrees while the angle at 200 g has increased 11 degrees. The shape of the curve has changed to one typical of a tough tenodesis with a steep straight terminal section. The 10 days of therapy improved the problem of disuse contracture of the skin and fat, but made no difference to the underlying tenodesis. The presence of the tenodesis has been unmasked.

heal by scar, pulling the skin together and resulting in a flexion contracture. A torque-angle curve for the joint will be shallow because the residual skin will stretch enough to give a compliant curve, even though the scar itself may not stretch more than 5%. Such a digit may improve with serial castings (as predicted by the shallow curve), but it will be the residual skin that has lengthened while the scar may have even shortened. In clinical practice, it is often less important to define the progress of an individual tissue than to monitor the total change in resistance by a torque-angle curve.

Torque-angle curves are always composites. The problem tissue may be either in parallel or in series with normal tissue. It is rarely possible to define the contribution of each. However, the torque-angle curve does give a real picture of a total joint situation. It is also numeric, repeatable, and educational, and it has prognostic significance.

VISCOUS ELEMENTS IN DRAG

EFFECT ON TORQUE-ANGLE CURVES

A truly viscous response from a tissue would be one in which the tissue, after being deformed by stress, has no recoil or tendency to return to its previous shape. Just as

pure elasticity is rare in biologic tissues, so pure viscosity is rare. Many tissues are viscoelastic. Within a fabric or matrix that is mainly elastic, there are fluids and free cells that are able to move around within the tissue and adapt to imposed stress. In the palm of the hand, a sudden localized pressure, such as from a pencil eraser, is accommodated and resisted by elastic pulp tissues. If the pressure is immediately released, the tissues spring back. If the stress is maintained for a minute, the viscous elements in the palmar pulp tissues flow around the deforming object and adapt to the imposed shape. When the stress is removed, there will be a residual hollow in the pulp skin that will take time to level out. This change over a short period of time is characteristic of viscous responses. Shock absorbers in a car are designed to modify the *elastic* response of springs and damp out the movement to a slower pace. It is the springs that restore the shock absorbers to their original shape. It is the elastic portion of the soft tissues that restores the original shape after pressure is removed.

If a stiff joint is moved through its range of motion, and if its angles are measured under *sustained* moderate torque twice in succession, the second measurement will often be different from the first. The difference reflects the viscous element of the drag (Fig 6–20). Since most normal soft tissues have

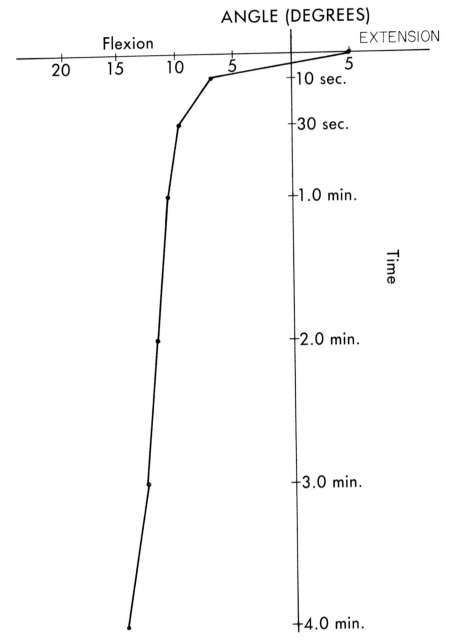

FIG 6–20.
Viscous drag: a time-angle curve at constant torque. A constant tension of 350 g was applied at the level of the IP joint, pulling the MCP joint into flexion. At this constant torque, the angle of the MCP joint was measured at intervals for up to 4 minutes. Note that the gradual change of angle in this swollen hand is a direct effect of edema.

some viscous element, a small variation of joint angle under constant torque is normal. This is one reason we usually take our measurements quickly. A rapid run through the angles saves time and measures the elastic tissue restraint. Some stiff joints respond to repeated torque or to sustained torque by a very significant progressive increase in their range of motion. This demonstrates a large viscous modulus. This may alter our treatment program significantly. It is important to be able to evaluate the presence and the role of the viscous element of total resistance and to use our evaluation to identify it.

Simple edema is a collection of water and

electrolytes in the tissues. Tissue fluid may accumulate in the hand because of dependency or diminished muscle action. Simple edema restricts movement partly by serving as a cushion of passive material that takes energy to move. Also it limits the longitudinal movement of fibers by a reorientation in a transverse direction. Weeks and Wray[10] have pointed this out in relation to skin mobility. Loose connective tissue is rather like a fabric woven on the bias, and it is a three-dimensional fabric. If it is pulled lengthwise, it shortens from side to side and also perhaps from the top downward. If it is held stretched from side to side, lengthening may be restricted. Empty pantyhose have a good longitudinal range of stretch. When pantyhose are filled with leg, the lateral distention limits the possibility of lengthening. Collagen fibers in paratenon are attached to the tendon at one end and to the parietal tissues at the other. The tissue is slack from side to side and thus has freedom longitudinally. If fluid distends the area, it forces the

parietal tissues away from the tendon and the fibers of the paratenon become transversely oriented. The freedom for longitudinal movement then becomes limited. This applies to edema in and under the skin. It lifts the skin away from joint axes forcing connective tissue fibers to become oriented in a perpendicular direction (Fig 6–21), limiting longitudinal skin movement.

Edema also has direct effects on joint movement by changing the moment arms of skin (on the extensor side) and by direct obstruction (on the flexor side). For example (Fig 6–22), the skin on the dorsum of the PIP joint of a finger may ordinarily lie 8 mm behind the axis of the joint. During normal joint flexion through 90 degrees (about 1.5 radians), the skin must be lengthened by 8 × 1.5 = 12 mm to permit the flexion. The 12 mm of lengthening may occur by perhaps 8 mm of unfolding of the loose dorsal skin plus 4 mm of actual stretch of the skin. Now suppose a 5-mm thickness of edema fluid distends the back of the finger and pushes

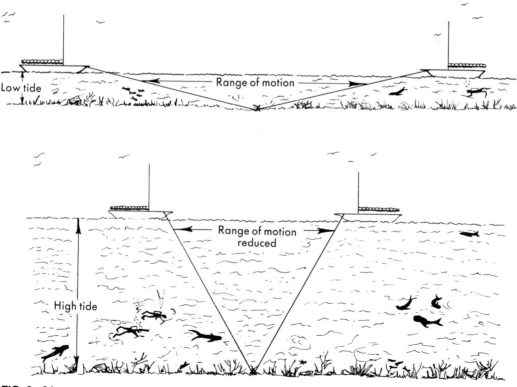

FIG 6–21.
Every sailor knows that a ship at anchor has a wide range of mobility at low tide and much less mobility at high tide. This is because the water (edema) has lifted the boat. The anchor chain (connective tissue fibers) is now oriented vertically, limiting its range horizontally.

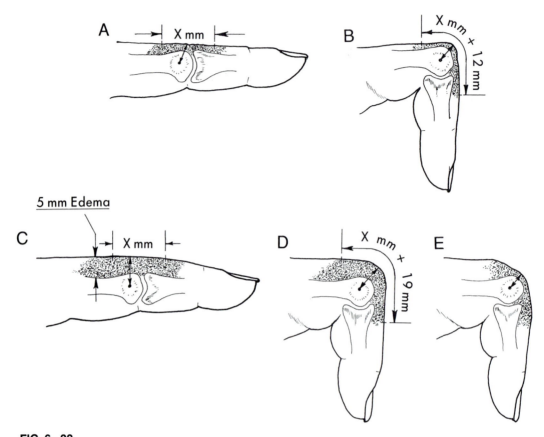

FIG 6–22.
A and **B**, dorsal skin requires 12 mm of lengthening for 90 degrees of flexion. **C** and **D**, with 5-mm thickness of edema, skin requires 19 mm of lengthening for 90 degrees of flexion. **E**, with continuing torque, slowly applied, the edema fluid moves around permitting the skin to cross closer to the joint axis and require less stretch.

the skin back 5 mm more from the joint axis. Now the skin will lie 8 + 5 = 13 mm from the axis and 90 degrees of flexion will now need skin lengthening of 13 × 1.5 = 19.5 mm. In addition, the edema will have produced tension in a transverse direction to the skin by an increase of the circumference of the finger, thus reducing the freedom to stretch longitudinally.

The need for the same skin that previously only needed to lengthen 12 mm to now lengthen 19 or 20 mm may result in limitation of flexion or it may overstretch the skin. More commonly, flexion is accomplished by forcing some edema fluid away from the knuckle so that the skin can cross the joint closer to the axis. This will take time to do, so the finger will flex only slowly and after repeated attempts. This will be clearly demonstrated by the contrast between a fast torque-angle curve and one that

is tested slowly with sustained torque. On the flexor side of a finger, edema acts as a direct barrier or cushion of fluid to block flexion. Here also, repeated active movement will gradually disperse and redistribute fluid to areas where pressure is less.

When tissues are injured or infected, there is usually an inflammatory response. This inflammation brings extra blood supply to the part and also an infiltration of cells whose function may be to handle the infection or to repair damage. A feature of inflammation is the exudate. This viscous, cell-containing fluid escapes from distended capillaries and contains large protein molecules such as fibrinogen.

This inflammatory edema has the same effect as simple edema. It forces a transverse reorientation on the fibers of connective tissue, but its high viscosity makes it more difficult to move and redistribute. It also has another more sinister potential. Under appropriate conditions, the fibrinogen in the fluid becomes organized and becomes fibrin, finally forming new fibrous tissues. It forms a sort of interstitial scar. This new scar tissue forms a web or fabric among the original tissue. It is nonspecific and is not designed for movement but for repair and fixation. The following analogy is a gross oversimplification but may dramatize this serious situation. Haberdasher's elastic, used in dressmaking, comes in strips made of loosely woven cotton or polyester with strands of rubber running through it. The strip may be pulled out and stretched until the loose cotton weave becomes tense; the cotton then limits further stretch of the rubber. This is rather like the mix of elastic fibers and collagen fibers in stretchable connective tissue. If in the slack position, glue is poured into the fabric, it will set and bind the strands of cotton side to side. Now that the cotton can no longer uncoil or unfold, the rubber strands are locked in a nonextensible fabric.

Although the analogy is fanciful and inexact, it does indeed give a vivid picture of what really goes on when inflammatory edema invades a digit. Previously mobile tissues become thickened and stiff. The additional scar tissue is not a separate definable entity that could be excised but has become a part of the existing tissue. It has changed its nature and its mechanical qualities. The length-tension curve of such scarred tissue will have lost the long low *A* section of its profile (see Fig 6–3) and requires real stretch for every millimeter of gain in length.

Some patients are more liable than others to produce this kind of scar. People who develop generalized stiffness of a hand following injury or surgery have demonstrated that they are likely to develop this kind of scarring. It may be a quirk of their metabolism or a feature of their sympathetic nervous system or just a very vigorous and hyperactive defense mechanism. If it happens once to an individual, it may happen again to the same person on another occasion, especially in the same hand and digit, and perhaps with increasing sensitivity to its proximate cause. This kind of patient needs extra care in handling joints and tissues, to avoid inflammatory exudate and stiffness.

EVALUATION OF VISCOUS RESISTANCE AND SWELLING

Quantitative.—Since every type of swelling of the hand is potentially harmful and may cause stiffening, we need to be constantly alert to it. It creeps up on the patient in many ways. The lazy patient develops gravity edema by leaving his hand hanging down. Eager and compulsive patients develop inflammatory edema by overuse or strong stretching of stiff and healing tissues. As much as 50 mL of edema fluid may accumulate in a hand and still be unnoticed by a busy therapist dealing with a succession of patients. For this reason we recommend the regular use of a volumometer for every postoperative hand. It is a quick test, simple and repeatable. It is sensitive to levels of change of volume far smaller than can be observed even by an astute surgeon or therapist. Patients can observe and understand the volume record and relate quickly to its objective evidence. If they have not been keeping their hand elevated, they may now see the evidence that just 30 minutes of active exercise with the hand held elevated produces a dramatic shift in the graph of volume and also improves their range of motion (Figs 6–23 and 11–24).

Qualitative.—The main qualitative analysis the hand surgeon needs about a swollen hand or finger is an answer to the question, Is it simple edema or is it inflammatory? This means, Is it thin or thick?, or in practice, Does it flow fast, slow, or not at all? The standard test of pitting on pressure is valid but only if there is a large quantity of fluid.

Torque-Angle Curve.—The torque-angle curve is profoundly affected by the viscous element of drag. It may be demonstrated by applying just one level of torque (e.g., 500 g at the distal joint crease), holding it there for a full minute while taking successive angle measurements every 15 seconds. A joint

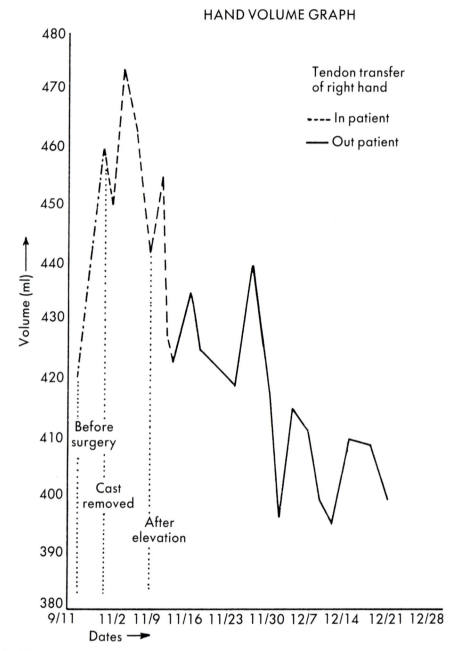

FIG 6–23.
Record of hand volume for a patient after tendon transfer taken 1 month after cast removal and before volume returned to normal levels.

that has only a structural stiffness will have little change after the first 15 seconds. An edematous finger may continue to change angles for the full minute. The change will be faster for simple edema and slower for inflammatory high-protein edema. Another way to evaluate this is to use the hysteresis loop technique in which a series of five sequential torques are applied in increasing order of magnitude while the angles are recorded and then applying the same five torques in decreasing order while the new angles are recorded. If the loop is narrow (i.e., if the angles recorded while torque is increasing are close to the angles recorded for the same torque while decreasing), then the stiffness is structural and elastic. If the loop is wide, there is probably a large viscous element.

"Rheumatic" or "Old Age" Stiffness of Joints.—We use these words in quotation marks because we do not understand the etiology. However, the evaluation is important so that we may learn about it. We have all noticed how old people tend to be stiff in the morning. After injury or surgery it is more difficult to get old joints moving and one frequently finds that the increased motion that is achieved after a session of therapy is lost by the time the patient returns for the session the next day. It is also typical of rheumatoid arthritis that patients wake up in the morning with their joints "all locked up." They have to go through their own set of mobilizing exercises before becoming free enough to go about their daily tasks. Wright

and his associates[11, 12] in Leeds, England, have done studies of torque-angle curves on rheumatoid joints and normal joints. They have shown that even normal persons go through a diurnal change of joint mobility and strength, being relatively stiff in the early hours of the morning and mobile in the afternoon (Fig 6–24).

What does all this mean? It seems that the day-to-day variation cannot be caused by a change in the cellular fabric nor can it be a change in the elastic structural tissues such as collagen bands. These structures are too solid. It is unlikely that they would be absorbed and laid down again on a diurnal basis. This is especially true since the problem is worse in old age when there is the least flexibility in cell multiplication and metabolism. Thus it has to be caused by a change in the more labile fluid elements of the tissues. We visualize it as an infiltration of ligaments

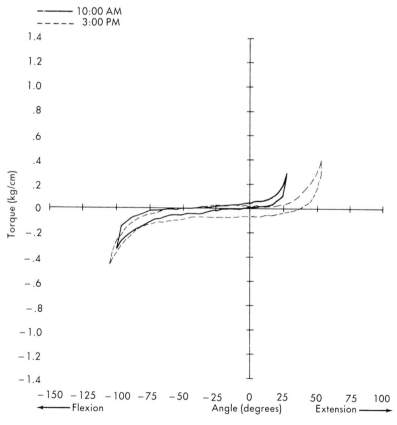

FIG 6–24.
Torque angle curve showing 25 degrees better extension at 3 P.M. than at 10 A.M. The data are for the PIP joint of an index finger.

and joint capsules by viscous fluid, resulting in a change of shape and a relative distention of these structures during rest. Then, when motion starts, the act of movement has first to squeeze out this fluid and redistribute it by the internal pressures of joint motion until the capsular structures have restored their shape, length, and mobility. In the case of postinjury and postoperative stiffness of the joints in older people, there may be a similar process occurring with an added acute inflammatory element. The immediate instinct of a scientist investigating these phenomena would be to try to extract tissue for histologic, or fluid for biochemical, study. Our plea is that while we await better understanding based on histology and biochemistry, we should relate our treatment of stiffness directly to the mechanical analysis of improvement or worsening. To do this we have to do more than range-of-motion studies and more even than torque–range-of-motion curves. We have to do time-related, sequential torque-angle measurements and exercise-related graphs. These are not difficult. A single torque may be applied to a joint for 1 minute in the early morning and the angle measured at the beginning and end of the minute. If the angles are different, the same test may be done later in the day or after a period of therapeutic exercise. If the change of angle under sustained torque is fast, it suggests a thin fluid. If the change of angle is slow, it suggests a fluid of high viscosity. If the angles are about the same, there is no significant viscous element to the stiffness. Finally, in analysis of the mechanical profile of viscosity, let us not forget that the fluid may or may not be part of an active inflammatory or healing process. Attempts to mobilize the part must never be allowed to exacerbate an infection or inflammation. Therefore, along with torque-angle-time curves, we suggest taking a skin temperature of the joint (see Chapter 17) before and after treatment to make sure we are doing no harm. Another way in which "old age" and rheumatoid arthritic stiffness may be caused by fluid is in the variations of fluid lubrication, which we now consider.

FRICTION

FRICTION BETWEEN SURFACES: GLIDING

We like to restrict the term *gliding* to the movement of two surfaces over each other when no tissue intervenes. Tendons may seem to glide through the forearm, but in fact they carry with them an investment of paratendinous tissue that allows movement only by its ability to stretch. The same tendons in another part of their path may truly glide in synovial sheaths. The movement in true gliding depends on smooth surfaces, is encouraged by good lubrication, and is hindered by friction.

When tendons glide within their sheaths and through other tissues, they do so on a synovial membrane lubricated with synovial fluids. Tendons are not, however, free to move large distances within their sheaths because their nutrient blood supply arises from the synovial vincula and is delivered through small vessels contained within the paratenon. Because the coefficient of friction between tendon and synovium is so small and the elastic modulus of elastin so low, the viscous and elastic resistance to motion is almost negligible in normal anatomy. In inflammation, the change in volume of these tissues deprives the tendon of its protective layer of fluid and forces the tendon and surrounding tissues together, thereby raising the frictional losses.

True gliding takes place only in synovial sheaths of tendons and between hyaline cartilage surfaces at joints. We shall not attempt to study the mechanics of these surfaces in detail because their efficiency is so great and their coefficient of friction so low that most bioengineers have found that friction within the joint contributes minimally to joint stiffness. However, there are several points that require attention.

When Sir John Charnley,[1] working toward the design of total hip replacements, began to try to imitate the qualities of normal joint cartilage by using metals and plastics, he invited the collaboration of the Cavendish Laboratories in Cambridge,

England. The engineers there, when first exposed to mammalian joint lubrication systems, were at a loss to explain how animal joints had coefficients of friction as low as one fifth of the best levels that could be achieved by substituting mechanical bearings using metals and plastics that were engineered and polished to the finest tolerances. Animals use a system of lubrication that is relatively new in engineering technology—hydrodynamic lubrication.

The term *hydrodynamic* is used in contrast to *conventional boundary lubrication*. The latter is the common type that is used when a shaft runs through a bearing that is slightly larger in diameter and is highly polished. When oil or grease lubricates this interface it forms an intimate boundary layer, almost a bond, between itself and each metal surface. Each surface presents the opposing surface with a film of oil that helps it to glide smoothly. In hydrostatic lubrication, more oil or other lubricant is introduced into the interface with sufficient pressure to actually lift the surfaces apart, so that they float on the lubricant and do not make any metal-to-metal contact. In a similar manner, in hydrodynamic lubrication, the eccentric-ity of the static inner shaft and the moving outer bearing coupled with the viscous action of the fluid causes fluid to be forced into the wedge shape between the shaft and journal of the bearing. This cannot happen unless the joint surfaces have room to move apart. They must be engineered to fit just close enough to one another to provide the proper eccentricity without clearance to permit the fluid pressures to equalize.

The body uses hydrodynamic lubrication in human joints by two mechanisms, both of which depend on the fact that the two joint surfaces are not congruent and have different radii of curvature. A convex articular head (Fig 6–25,A) glides on a concave socket that has a larger radius. This results in a wedge-shaped cavity that is filled with synovial fluid (Fig 6–25,B). As the joint rolls or glides, it moves faster than the viscous or gelatinous (thixotropic) fluid can escape, so the fluid is trapped and remains to lift the surfaces apart. This same phenomenon occurs when a smooth automobile tire rolls along a wet road on which an actual layer of water is resting. At high speeds the rolling tire rides over the water faster than the water can escape, so the tire is lifted off the

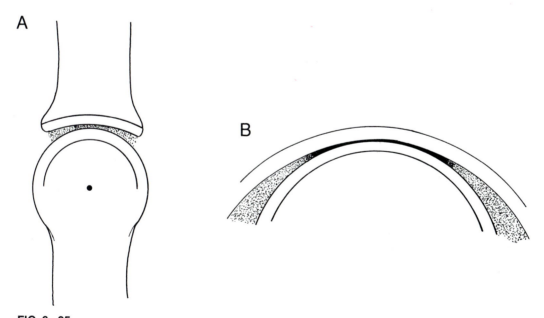

FIG 6–25.
The lubrication of joint surfaces is by both hydrodynamic and hydrostatic means. When the joint is moving **(A)**, the articular surfaces separate. When the joint is loaded and relatively stationary **(B)**, the rolling compression releases synovial fluid from the intracartilage.

road (hydroplaning) and floats. It has negligible traction or braking power, and the car goes into an uncontrollable skid. (Therefore never allow tires to lose their tread and become smooth, and never travel fast when the rain is falling faster than it can run off the road.) In a joint the lubrication becomes hydrodynamic only when the joint is actually moving. As soon as it comes to rest, the free fluid escapes into the joint cavity leaving only a film of fluid for boundary lubrication.

Another feature of cartilage related to hydrostatic lubrication depends on the incongruity and compressibility of the joint surfaces. As the joint moves, the area of contact moves and different areas of cartilage are compressed. This seeming inefficiency of design allows a constant change of shape of the cartilage and forces flow of lubricant in and out of the substance of the cartilage. Joint cartilage is like a firm collagen sponge with a contoured surface. When it is relaxed, it is full of synovial fluid that is a gel in the deeper parts of the cartilage. When the joint moves, the opposing cartilage surface rolls and glides across and compresses the cartilage in contact, thus squeezing fluid out of tiny canals. The thousands of miniature fountains of synovial fluid squirt out of the surface under pressure literally lifting the cartilage surfaces off each other, so that the joints are floating. As the joint continues to move, the cartilage under stress is relieved of pressure, expands, and reabsorbs the loose fluid in the joint. It is then ready to lubricate the joint again. As soon as joint motion stops, the whole system changes. The area of cartilage under pressure remains under pressure. All its fluid is extruded, leaving only gel in the deeper part, and the fluid leaks out from between the opposed surfaces to the open parts of the joint. Now the hyaline joint surfaces are in actual contact, solid surface to solid surface, and there is only boundary lubrication until movement occurs.[3, 6, 7]

Human joints are efficient while moving but may lose their efficiency at rest. If a joint is kept immobile for a long time, it takes both motion and time to get the system working correctly. If a joint is kept still in a plaster cast for many weeks, the little patch of cartilage that has been under long-term pressure may actually lose some of its texture and its gel and take longer to move freely.

If, during immobilization, an area of cartilage is under abnormally high pressure, the damage to the cartilage may be severe, permanent, and result in pain and stiffness later. Joints remain healthy when they move. Long-term immobilization is harmful for systems of lubrication. Long-term immobilization may be necessary and *is* necessary for various reasons, but it is better avoided if possible. When immobilization is required, there is often value in finding ways to interrupt it at intervals to permit carefully controlled and guarded joint motion.

Consider the common situation when a digital joint has a limited range of motion, because of tightness of skin or capsule while it has good friction-free joint surfaces. If that joint is immobilized continuously to keep shortened tissues on the stretch *(good)*, it will harm the joint cartilage and lubrication system *(bad)*. However, there is a compromise. Healthy joint lubrication systems do not need *constant* movement to keep healthy. We suggest holding a digit (such as the above) in its immobilized position for the greater part of the day, even 23 hours, but to insist on active or passive movement of the joint for half an hour daily or at least on alternate days. Thus the tissues get periods of continuous stretch needed to stimulate growth and intermittent movement needed by the cartilage.

HIGH PRESSURE ON CARTILAGE

As long as joint surfaces are free and moving concentrically around their axes, joint friction is so low that it may be neglected in attempts at evaluation (Fig 6–26,A). Most stiffness results from ligaments and capsules and skin and scar. Sometimes, however, when ligaments are diseased and a joint may be partially subluxated or when its surface has been made rough or irregular by severe arthritis or by fracture, the joint surfaces themselves may block movement. If that happens, the joint

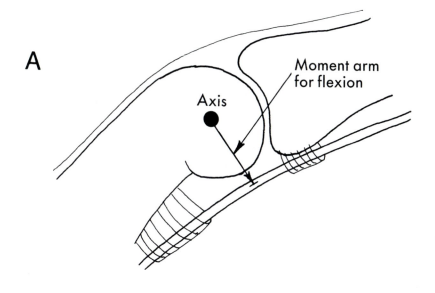

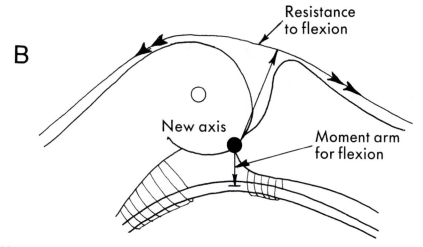

FIG 6–26.
A, when the functional axis is the same as the anatomic axis, the flexor tendon has good moment arm for flexion.
B, when gliding is blocked at the joint surface, the functional axis for flexion moves to the point at the joint surface at which the joint now tilts (not glides). Immediately the moment arm for flexion is reduced to the point that the flexor tendon loses its ability to effectively flex.

immediately changes its mechanical profile. It no longer glides but tilts on an edge. The axis of rotation (instant center) moves at once to the point on the joint surface that blocks movement, and all moment arms change. Figure 6–26 shows the effect of this change on moment arms. This may cause pain and in a normal sensitive person it may inhibit further attempts at movement. If sensation is absent or if movement is im-

posed from outside by a dynamic splint or therapist, this block to gliding results in a joint *tilting* on the edge of the articular surface. Serious damage may result from localized high pressures at this new axis and to the ligamentous restraints opposing distraction of the joint surfaces. The mechanics of this change from gliding to tilting is portrayed in Figure 6–25,B.

The work by Salter et al.[9] on constant

passive motion in joints demonstrates that not only lubrication but also structural integrity of joint cartilage is improved by motion. This is true of injured joints and of the healing of intraarticular cartilage defects.

FRICTION IN TENDON SHEATHS

Tendons are designed to transmit tension. The tension stress remains within the structure that transmits it so long as it runs in a straight line. Thus a straight tendon has no mechanical effect on what is around it except for the need to lengthen the fibers of the paratendinous tissues that may be attached to it.

As soon as a tendon turns around a bend, some of the tension generates a new type of stress, i.e., pressure or compression that acts at right angles to the tendon against any structure that resists the tendency of the tendon to straighten out again. When a tendon turns around a bend, friction could become a very important factor. The smaller the radius of curvature, the greater the pressure. The greater the pressure, the greater the friction. This is why normal tendons always lie in synovial sheaths whenever they turn around a concave bend. The pressure on the concave side of a curve would be too great to allow the survival of soft paratenon tissues. The shear stresses generated would tear them apart.

When a transferred tendon is passed around a bend where there is no sheath, it first becomes attached to tissues around it. When it begins to move, those tissues may begin to stretch and move with it. When strong movement takes place, shearing stress on the concave side may quickly break down the tissues, forming a bursa. This happens only if movement is taking place. Finally, after weeks or months, the tendon seems to develop a true synovial sheath. Synovial tendon sheaths secrete a lubricating fluid much like that in a synovial joint. However, the method of lubrication is different. Joints in motion (see above) use hydrodynamic lubrication. This is possible because of two mechanisms: (1) the convex surface overrides a wedge of viscous lubricant, and (2) the cartilage is compressible, like a sponge, and exudes fluid from its surface. *Neither* of these mechanisms is possible for a tendon in a sheath.

It is noteworthy that the mechanics forcing the tendon against its sheath produce much smaller interface pressures than are seen in joints. The lubrication in the sheath is simple boundary lubrication. Although less efficient than in the joint, it produces little drag in the normal hand. Unlike the joint, this resistance is not lessened by continuous motion. A secondary mechanism, viscous shear stresses, occurs during fast, strong, continuous reciprocal motion. The shear stresses act directly on the paratenon, a tissue with little ability to resist such stresses. This sometimes results in tenosynovitis, but rarely in joint synovitis. This is aggravated by factors which increase the interface pressure between tendon and sheath. This is one reason to avoid taking tendon transfers around a bend and to try very hard to avoid acute bends. A person who tends to get a tenosynovitis in flexor tendon sheaths from strong repetitive grasping stress should be advised to use a larger grasp involving less acute joint flexion.

DIAGNOSIS OF FRICTION DRAG

Since the drag from friction is low in most normal and disease conditions, it scarcely affects the torque-angle curves in most cases. In a full torque-angle cone the horizontal segment between flexion drag and extension drag usually shows a gap of a few gram-centimeters between the two lines. It takes a small torque to move the joint in either direction, even when no elastic tissues are on the stretch. The gap contains an element of friction, but most of it is caused by the energy needed to move tissues on one another in the composite hand. True friction in the joint and sheath is miniscule. However, when the loop is wider, it may include a significant element of frictional drag and should receive attention. One way to get an impression of joint friction in a joint that is stiff primarily from capsular causes is to hold both segments of the digit firmly and slide them

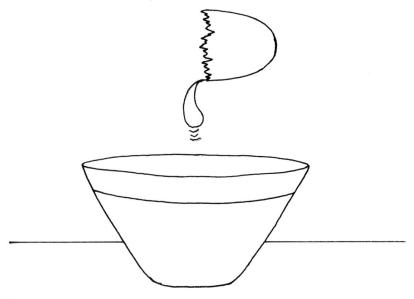

FIG 6–27.
Thixotropic fluid is tenacious and resists pouring (like egg white). The last drop bounces up and down as if on an elastic band. This quality is important in synovial fluid.

transversely on each other with no angulation. Thus the examiner will feel the cartilage surfaces sliding even though the joint is not turning. If grating or crepitus is felt, there is irregularity of the cartilage surface. This sliding of joints without angulation is called *joint mobilization* by those who use it as a therapeutic modality. Frictional problems in a tendon sheath may similarly be felt as a crepitus when the examiner lays a finger over a tendon sheath during movement.

THIXOTROPY OF SYNOVIAL FLUID

Viscous fluids flow slowly. The word *thixotropy* refers to a special quality of the molecules of some organic fluids, like synovial fluid, that causes them to cling together and resist change of shape. Thus mucus and raw egg white form strings of sticky fluid when they begin to pour (Fig 6–27). This thixotropic quality is vital to hydrodynamic lubrication in joints, as it holds a blob of fluid together so that the joint rides over it rather than pushing it away. It has been shown in rheumatoid arthritis and also in old age that the thixotropic quality of synovial fluid may be lost. This may be an important factor in the increase of friction and pain.

REFERENCES

1. Charnley, J.: Arthroplasty of the hip: a new operation, Lancet 1:1129–1132, 1961.
2. Flowers, K.R.: Personal communication, 1984.
3. Fung, Y.C.B.: Biomechanics: mechanical properties of living tissues, New York, Springer-Verlag, 1981.
4. Gelberman, R.H., et al.: Flexor tendon healing and restoration of the gliding surface: an ultrastructural study in dogs, J. Bone and Joint Surg., 65A:70–80, 1983.
5. Lundborg, G. and Rank, F.: Experimental intrinsic healing of flexor tendons based upon synovial fluid nutrition, J. Hand Surg. 3:21–32, 1978.
6. Malcolm, L.L.: Frictional and deformational responses of articular cartilage interfaces to static and dynamic loading, PhD dissertation, University of San Diego, La Jolla, Calif., 1976.
7. McCutchen, C.W.: Sponge-hydrostatic and weeping bearings, Nature 184:1284–1285, 1959.
8. Potenza, A.D.: Flexor tendon injuries, Orthop. Clin. North. Am. 1:355–373, 1970.
9. Salter, R.B. et al.: The biological effect of continuous passive motion on the healing of full-thickness defects in articular cartilage: an experimental investigation in the rabbit,

J. Bone and Joint Surg., 62A:1232–1251, 1980.

10. Weeks, P.M., and Wray, R.C.: Management of acute hand injuries: a biological approach, St. Louis, Mosby–Year Book, Inc., 1973.

11. Wright, V., Longfied, M.D., and Dowson, D.: Joint stiffness: its characterization and significance, Biomed. Eng. 4:8–14, 1969.

12. Wright, V.: Stiffness: a review of its measurement and physiological importance, Physiotherapy 59:107–111, 1973.

13. Yamada, H.: Strength of biological materials, In Evans, F.G. editor: Strength of biological materials, Baltimore, 1970, Williams & Wilkins Co., p. 226.

CHAPTER 7

External Stress: Effect at the Surface

PRESSURE AND SHEAR STRESS ON THE SURFACE OF THE HAND

The purpose of all the activities of the muscles, bones, and joints of the hand is not just to position the hand in space; it is to impose the will of the body on the environment. This is done through the skin. It is the skin that actually touches the tools and materials that we work with. It is the toughness of the skin that determines how much force may be applied to an external object, and it is the sensibility of the skin that keeps the central nervous system informed of all the pressures and temperatures to which it is being subjected, so that these may be modified by further action or relaxation of the muscles.

The typical immediate effect of hand activity is pressure on the palmar or pulp surface of the digits. A good way to get a visual image of this is to grasp a clear glass tumbler or cylinder in the hand and view it through the open end of the tumbler (Fig 7–1). A very light grasp will bring every segment of every finger, as well as the thumb and the palm, into contact with the glass. As the grip slowly tightens, the rounded pulp surfaces are gradually flattened so that larger areas of skin come into contact with the glass. With increasingly strong grasp, the most prominent parts of the finger pulps become blanched and then a widening area becomes compressed until the whole palmar surface of the hand may be white and bloodless. All

of this feels quite comfortable. It is very difficult to grasp this type of large smooth cylinder so tightly that it causes pain or damage at the surface of the hand. However, if even a moderately firm grasp is continued for a very long time, discomfort is felt and then the pain of ischemia. It is worth noting that in a normal hand this type of grasp applies approximately equal pressure over the whole palmar surface. Each segment of each finger begins to blanch at about the same time, which means that the force applied through that segment is proportional to the area of the segment. Some of this precision may be controlled by feedback from skin, but the normal balance of grasp clearly is based on the coordination of muscle contraction that results in an orderly sequence of internal and external moments.

In the case of partially paralyzed hands, the pattern of grasp viewed through a tumbler is quite different (Fig 7–2). In the case of ulnar palsy with intrinsic muscle paralysis, the tips of the fingers and the metacarpal head area of the palm get all the pressure while the major part of the palmar surface of the fingers does not even touch the glass. The parts of the hand that do touch the glass get blanched very quickly and are dead-white before much force is used. The reduction of the contact area increases the pressure that results from the same force. Increase of the force of grasp results quite soon in discomfort or pain in the fingertips if they have normal sensibility.

If a normal hand grasps an irregular ob-

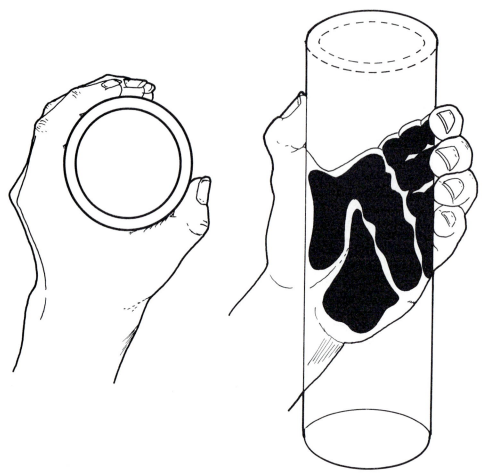

FIG 7–1.
Normal hand grasping a transparent cylinder. Areas of skin contact are marked in *black*.

ject, such as an oval or fluted or square handle, there may be some adaptation of the positions of the finger joints to maximize the area of contact, and there is also a marked adaptation that results from a redistribution of the substance of the tissues of the palm. This occurs in part because of compression of the elastic elements of the pulp tissues by the unequal projections of the object that is grasped. It also results from a rearrangement of the viscous elements of the pulp, especially the tissue fluids that flow around the projecting prominences of the object and allow hollows and pits to be created in the palmar pulp to fit any shape that is imposed on it. We shall not attempt to quantify or analyze the contributions of elastic and viscous elements to the adaptability and compliance of the palm. However, we must point out that any loss of pulp substance by injury and scarring results also in loss of compliance, especially in loss of viscous adaptability. This results in an apparent loss of strength, because the patient is unwilling to exert his or her full strength in situations in which noncompliant skin presses on rough, irregular objects, producing points of unacceptably high pressure. In such circumstances, insensitive skin that is scarred will quickly become injured and ulcerated. Hand surgeons and therapists need to be aware of surface pressure and shear stress as modalities of hand mechanics for several reasons, one of which is the danger of damage from stress.

1. Excessive stress may cause direct damage.

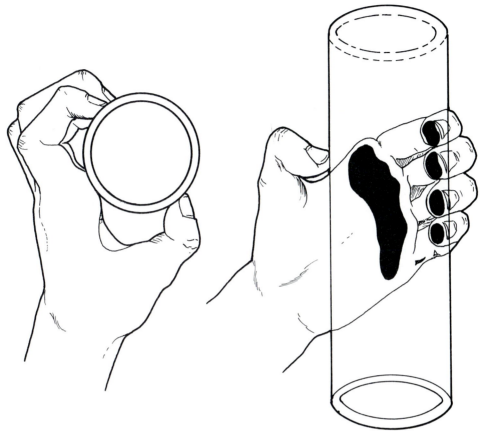

FIG 7–2.
Clawhand (paralyzed intrinsic muscles) grasping a cylinder. Reduced areas of skin contact result in very high pressures on the fingertips.

2. Continuous stress may cause ischemic necrosis.
3. Repeated stress may cause progressive inflammatory change, finally resulting in damage.

DIRECT DAMAGE

In his book *Strength of Biological Materials*, Yamada[10] reproduces stress-strain curves for skin and quotes a figure of 1.6 to 2.5 kg/mm^2 for the ultimate tensile strength of skin of the upper limb. In a normal hand it is very unusual for a person knowingly to apply enough pressure or shear stress to a hand to break the skin or to tear it. This is not because of thoughtful control but because intolerable pain results in inhibition of the intended action before damage is caused. If skin is damaged by direct force, it is either accidental, as a result of sudden un-

expected pressure, or the hand must be insensitive. Only rarely will a person knowingly allow direct damages to skin. A classic exception to this occurred in India when a woman tried to escape from a tiger by climbing a tree. The tiger caught her by the legs and dragged her away. Afterward, the skin of her palm was found still attached to the branch, showing that even pain will not inhibit maximal muscle contraction when life is at stake.

Dedicated rock climbers often finish a climb up a rock face with skin abrasions and even full-thickness wounds of the skin of their knuckles from stress that they accept as part of the constant interplay between danger and pain and the joy of achievement that is, to them, the essence of their adventure (Fig 7–3).

Within the limits of ordinary activity a very wide range of stresses are accepted by

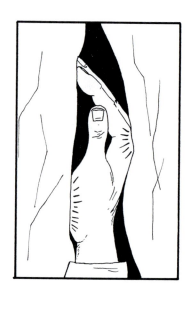

FIG 7–3.
A, rock climber using metacarpophalangeal flexion to trap his hand in the crack of a rock. **B,** detail showing how localized stress on the knuckle is used to fix the hand that may then serve to suspend the body (gloves should be worn).

the skin of the hand and it is really quite extraordinary that (1) it is not damaged more often and (2) most people never give a thought to the actual limits to the strength of their skin. Even engineers are at a loss if suddenly asked how many pounds per square inch of hand surface will be required on a standard wrench to loosen or tighten the nuts while changing a tire or to turn the little knob that turns on the bedside light in a hotel room. They are surprised when we tell them that the latter is often much higher than the former and is closer to the limits of tolerance of the skin.

The fact that most people do not think about it is a tribute to the extraordinary accuracy and reliability of the sensory system of the skin. It is not geared so much for objective accuracy in a quantitative sense as it

is to the protection of the skin from damage. Thus a person who needs to apply a lot of force to an object such as a stiff bottle cap will keep changing the position of his hand until he is finally able to move it. A woman will pass the bottle to her husband who will succeed where she failed. In each case the failure of the early attempts is thought of as a sign of muscular weakness, whereas in fact it is usually a sign of active inhibition by the nerve endings in the skin. When the hand is moved around to get the best position, it is not seeking the position for the optimal use of the muscles; it is usually seeking the position in which the greatest area of skin can be applied so that the force per unit area of the skin is minimized. When a "stronger" man succeeds where a "weaker" woman has failed, it is often only because he has a thicker skin than she or skin that is more heavily keratinized. She has adequate strength but is inhibited by pain from her more delicate skin. There are uses for hard skin, and there are uses for soft skin. The word *tenderness* means ability to feel pain. It is really a compliment to be called tender-hearted or tender-handed, although some hands do need calloused palms.

When a person marginally fails to accomplish a manual task that requires strength, it is probably more often because of sensory inhibition at the skin level than of weakness of the muscles. This is especially true when the task requires shear stress on the skin. Small knurled knobs are often used as switches on lampstands (Fig 7–4) or to adjust some feature of a television picture. If the switch is stiff and the radius (moment arm) of the knob is small, it requires a lot of pressure and shear stress on the skin to turn the knob. We have found more direct damage to the insensitive skin of leprosy patients from such small knobs and small keys than from any heavy task in which large handles are used.

It is much easier to teach patients to use large areas of contact and large handles than it is to teach them to use less force. Occupational therapists who help patients to adapt to loss of sensation need to go around the patients' houses and work areas looking for small knobs and handles and replacing them with much larger handles. Ornamental metal knobs on drawers and cupboards are often a serious hazard. In our own office there is a metal filing cabinet with pull-handles 15 mm in diameter, flat on the inner surface, and with square edges 2 mm thick (Fig 7–5,A). If a drawer were to stick, a 10-lb pull on the knob (which can be held only by its edges) would result in pressures of 250 lb/in^2 on the fingers (about 20 kg/cm^2). People with insensitive hands might pull with 50 lb and bruise or tear their skin. The use of larger, more rounded knobs is not only safer for insensitive hands, it is more comfortable and more rational for normal hands (Fig 7–5,B). For hands with rheumatoid arthritis the problem is different. They have good sensation, so will not accept excessive stresses on their skin, but the musculoskeletal stress necessary to apply pressure to a small knurled knob may be too much for their weakened joints and ligaments.

If a small adjustment spindle or switch has to be turned by means of a rounded knurled knob, the stress on the skin of the hand is 64 times as high for a knob 5 mm in diameter as for a knob 2 cm in diameter, given that the needed torque is the same for each. This is because of the high pressure that is necessary to obtain enough friction on the small area of the knob to produce the tangential force that will turn the knob using

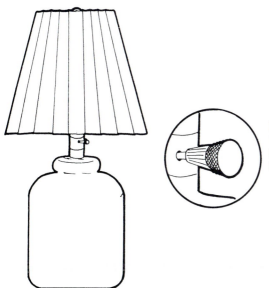

FIG 7–4.
This knurled switch is marginally acceptable. Many lamps have much smaller switches that pose a hazard to the skin. Ban All Small Handles (BASH)!

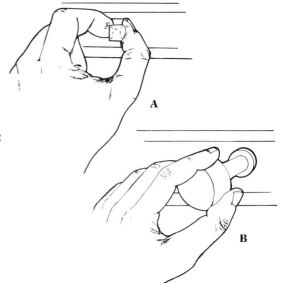

FIG 7–5.
A, this metal drawer handle could damage skin. It should be replaced with **B,** a handle that spreads the stress. BASH!

the small moment arm that is available. We call on all hand surgeons and therapists to support our campaign to *Ban All Small Handles* (BASH). Small handles simply do not make sense for anybody.

PRESSURE RESULTING IN ISCHEMIA

It is quite unusual for the hand to undergo ischemic necrosis from external pressure. Whereas hands are frequently exposed to pressures that blanch the skin and render the tissues bloodless, such pressures are almost always of short duration or are intermittent so that they allow blood to supply the tissues from time to time. Even when hands are insensitive, damage from ischemia is unusual. A finger that gets swollen from the pressure of a tight ring is one example of such damage. We have sometimes worn two pairs of surgical rubber gloves at surgery and have noted a comfortable sense of pressure on the fingers. If the operation is a long one, the comfortable ischemia of fingers becomes first uncomfortable and finally intolerable. Prolonged ischemia is agonizing.

It is not easy to use numbers to define levels of danger or safety in relation to external pressures that cause ischemia. (See Fig 7–6.)

It is relatively easy to quantify the pressures that will close a blood capillary. Many fine studies have been done on this, Daly et al.,[5] Brånemark et al.,[2, 3] and others have

shown that external pressure on the skin results, predictably, in diminution and then stoppage of bloodflow. The critical pressures are of the order of 35 mm Hg. However, the actual onset of necrosis may be delayed by many factors. It may occur only at higher pressures, depending on the stiffness of tissues that surround the small blood vessels and prevent their collapse, and it may be changed by temperature. Romanus[7] found that at a standard pressure, tissues were damaged more by 3 hours of ischemia when they were at 36°C than by 7 to 9 hours at 22°C. However, for the purpose of avoiding clinical ischemia, we find that the following numbers are a useful guide. The first number is the one we use for most short-term splinting purposes and the number in parentheses is a safer number to use when the pressure is to be maintained continuously for more than 8 hours at a time.

50 mm Hg (33 mm Hg)
75 g/cm^2 (50 g/cm^2)
1 lb/in.^{2} (12 oz in.^{2})

All numbers have been rounded off and are approximate, but then our clinical measurements are still more approximate. For short-term application (less than 2 hours), ischemia should not be a problem and the above numbers can be ignored.

In order to develop a real mental image of pressure and ischemia, we advise every therapist and surgeon to play around for a few minutes with a 50-g weight, a glass

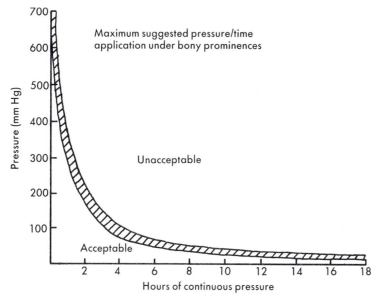

FIG 7–6.

Allowable pressure vs. time of application for tissue over bony prominences. Curve gives general guidelines and should not be taken as absolute. *(Redrawn from Reswick, M.S., and Rogers, J. In Kenedi, R.M. and Cowden, J.M., editors: Bed sore biomechanics, London, 1976, The Macmillan Press Ltd. pp. 301–310.)*

slide, and a loop of transparent plastic tape. If one hangs a 50-g weight over the finger-tips, with a glass slide between the loop of the thread and the finger, about 1.5 cm^2 of the pulp skin is flattened, and there is no blanching (Fig 7–7,A). On the little finger the area of contact is less and there is some mottled blanching. If the slide is rested on the *tip* of any finger, the area of contact is less than 1 cm^2 and there is blanching adjacent to the nail bed only.

If one uses the slide as a lever and hangs the weight on the end of the slide and supports the weight by the finger in the middle, the load is doubled and there is complete blanching with no spots of pink (100 g on 1.5 cm^2) (Fig 7–7,B). By moving the finger around one gets a good idea of how force and area interact to result in pressure and ischemia. On the back of the finger local spots of blanching over bony irregularities are seen.

If one now suspends 50 g by a loop of clear plastic tape 1 cm wide over the finger, there is no blanching. The area that is under pressure is greater, so the pressure is less. After adding more weight, ischemia commences. When the finger is tilted, blanching is seen at the edge of the sling, even with 50 g. Now apply transparent adhesive tape as a

ring around each of three finger pulps in the distal segment. The first is applied with a tightness just short of blanching the pulp. The second is applied at mottled blanching, and the third at full blanching. The earlier experiments will link each color with the pressure that causes it. Now go and have lunch or see patients and forget the tape (sensation is normal). Note the time it takes for each of the three fingertips to become uncomfortable and then painful. (See also Figs 7–12 and 7–14.)

These rather childish experiments are more educative than a dozen articles on capillary pressure that use units that are outside of our experience and practice. We have to relate usable numbers to the actual experience of the effects of pressure.

There are two conditions under which a hand may suffer ischemic damage that need comment here. One may be caused by a surgeon and the other more often by a therapist.

Pressure From Outside

The surgeon may cause pressure by circumferential elastic bandages or plaster casts. Most hand surgeons are trained to apply very bulky dressings postoperatively. They are made of fluffy gauze and are sur-

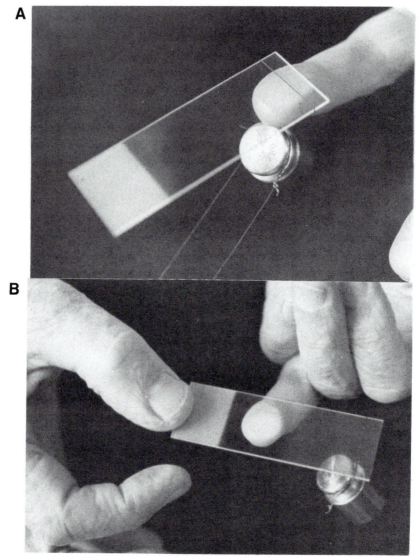

FIG 7–7.
One way to observe how much pressure it takes to blanch the skin is to play with a 50-g weight suspended by a thread from near the end of a glass slide. **A,** the fingertip supports the slide, which flattens about 1.5 cm² of the pulp. There is partial blanching. (The two threads seen at the side were only to steady the hanging weight for the photo). **B,** by moving the fingertip to the center of the slide and holding it down by one end, the 50-g weight doubles its force on the finger and the blanching is total.

rounded by turns of elastic bandages. This is a safe practice and hallowed by tradition. It is to be commended after injuries when swelling is quite likely to occur. These bulky dressings have one disadvantage. The individual finger joints are buried so deeply that the surgeon cannot control their position. In elective surgery it is often important to be sure of the exact position in which a joint is to be held. One joint may have to be flexed while the next may have to be straight. This may be accomplished by using a total-contact plaster shell (Fig 7–8). Many surgeons fear such an unpadded cast, thinking that they may cause pressure sores. We have used them for many years for all elective surgery and commend them as safe if the following mechanical considerations are kept in mind.

If a bandage is held in tension while it is applied, the pressure that results from the tension is inversely proportional to the ra-

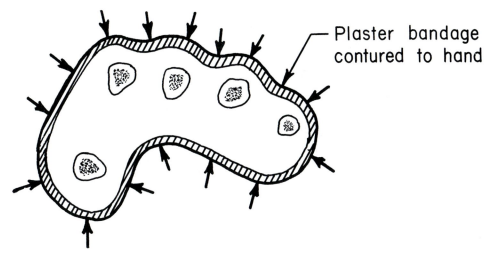

FIG 7–8.
When two to three layers of plaster bandage are laid loosely on the skin and wound dressings of a hand, it is possible to mold them around every contour and achieve total contact and avoid localized pressure.

dius of the curve or bend around which the bandage is applied. In the application of any bandage around a limb, the only force used is tension. The bandage is pulled, not pushed. When tension occurs in a flexible

fabric, all the force is within the fabric so long as the fabric is straight. There is no pressure outward from the fabric unless it is turned around a bend. In Figure 7–9 a bandage with tension T is applied around a

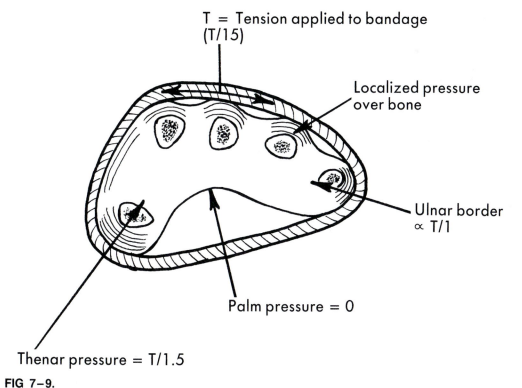

FIG 7–9.
When any type of bandage is applied with tension around a band, the pressure on the hand that results from the tension in the bandage is inversely proportional to the radius of curvature of the bandage at any point. There is no pressure when the bandage is straight. If the bandage is elastic, the pressure is summated with each extra turn of bandage.

hand. The pressure on the palm (concave) is zero. The pressure on the dorsum (convex, radius about 15 cm) is proportional to $T/15$ while the pressure around the ulnar border of the hand (radius 1 cm) is proportional to $T/1$, 15 times as great.

It is pressure that causes ischemia and it is sharp curves that result in pressure. This is one reason most hand surgeons prefer bulky dressings; they are nearly spherical, so there is no point of high localized pressure. However, in applying a circumferential plaster bandage there need be no danger of pressure from tension if one follows the rule of *no tension at all*. Any circumferential plaster bandage must be *laid on* the hand, never *pulled around it*. We often fold back part of each turn of plaster just to make sure there is no tension in it.

Quite a good routine at the end of an operation is to place a slab of plaster of Paris on the palmar side of the hand and forearm, over the immediate wound dressings, and mold it into the shape of the palm of the hand while it sets (Fig 7–10). Then a single plaster bandage is applied circumferentially around the whole hand and arm which is then rubbed into every hollow and around each bump so that each finger is clearly outlined while the plaster sets. Only about two thicknesses of plaster fabric are used on the back of the hand. If the surgeon's hands are kept constantly moving while the plaster is setting, there will never be a pressure point caused by indentation of the plaster shell by a supporting finger.

The strength of the plaster cast is in the slab. The surrounding eggshell-thin plaster may be readily slit open with bandage scissors if ever there should be a suspicion of swelling or pain from pressure. Such should not occur after elective surgery.

If the patient is to be ambulant and active, the eggshell plaster may be reinforced

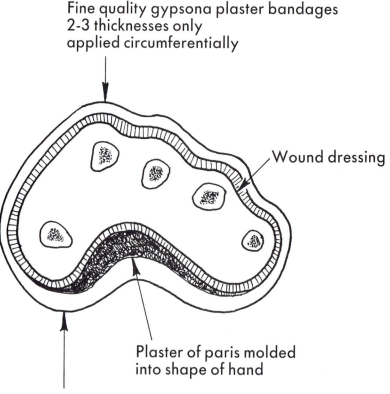

Fine quality gypsona plaster bandages 2-3 thicknesses only applied circumferentially

Wound dressing

Plaster of paris molded into shape of hand

Plaster slab 5-6 thicknesses

FIG 7–10.
A plaster slab is placed on the palmar side of a hand and an eggshell-thin plaster bandage is molded circumferentially around the hand.

after 24 hours by additional turns or slabs of plaster bandage.

Pressure From Inside

Swelling of a hand inside a cast may occur from postoperative bleeding. This should not be a problem if the tourniquet has been removed and the bleeding controlled before wound closure.

When crushing injuries have occurred and some swelling seems inevitable, we either use a bulky dressing or apply the plaster cast as separate dorsal and palmar slabs.

Since a swollen hand should decrease in volume with immobilization and elevation, and since continued positive pressure is a help in this process, it is sometimes a good idea to use a dorsal plaster and have a balloon in the concave palmar side of the hand (Fig 7–11), with dressings and a circular plaster to enclose it all. The mouth of the balloon may project out of the dressings and be inflated and reinflated with mouth pressure only, perhaps by the patient, to keep pace with the shrinking volume of the hand.

Tourniquet Pressure

Enough has been written about the value and about the dangers of a tourniquet applied on the upper arm.[4] We must just reemphasize the need to (1) monitor the level of pressure and to calibrate any dial pressure gauge against a mercury manometer, (2) remember that prolonged ischemia is harmful even though it may not result in overt necrosis of tissue or paralysis, and (3) remember that in the absence of circulating blood, tissues are much more susceptible to drying out and to burns from an overhead light. This is especially significant in the thin flaps of dorsal skin that may be lifted in a rheumatoid hand during joint replacement and in tendons that may be left exposed in the operative field.

Tourniquets should usually be released as soon as all exploratory dissection is complete, so that hemostasis may then be accomplished and the final attachments of tendons and suture of the skin are done without the tourniquet.

Finger Tourniquets.—Some surgeons use a rubber Penrose drain as a tourniquet around the base of a finger for minor distal digital surgery. The rubber band is stretched manually and clamped with a hemostat. This procedure may be safe if the surgeon is sure that it is for a brief operation (20 minutes) and if he knows how much pressure is used.

We have asked many surgeons how far or how strongly they pull on the ends of a Penrose drain before they clamp it and find that few have thought about it and none know how much pressure they produced.

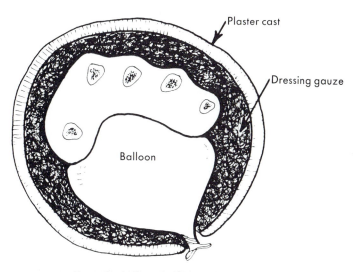

FIG 7–11.
By incorporating a balloon inside a circular dressing or plaster cast, it is possible to adjust the pressure each day while a swollen hand is shrinking. Only mouth pressure should be used.

J. A. Shaw et al.[8] have published an excellent study of digital tourniquets made of Penrose drains and of rolled rubber glove fingers. They have measured actual pressures inside fresh cadaver fingers, using different methods for stretching and wrapping both 0.25-in. and 0.5-in. Penrose drains, and also various sizes of rolled rubber glove fingers. They also checked the actual pressures produced by many surgeons who regularly use these methods. This article should be read in full by any surgeon who intends to use digital tourniquets. They conclude with the following guidelines:

1. The stretch of a 0.25-in. Penrose drain should be limited to a strain of 100% (stretched length = twice original length) or less to maintain closing pressures of less than 300 mm Hg. A 50% strain will provide adequate hemostasis in most instances while avoiding excessively high tissue pressures.
2. The stretch of a 0.5-in. Penrose drain should be limited to no more than 50% strain to maintain closing pressures of less than 300 mm Hg. A 33% strain will generate hemostatic pressures in most patients.
3. A rolled glove finger produces hemostatic pressures while avoiding excessively high tissue pressures if it is from a glove of the appropriate size to fit the hand.
4. Exsanguinating wraps using a Penrose drain should probably be avoided. Excessively high pressures can easily be generated if more than one wrap is left at the base of the digit.

We personally feel that even with these guidelines digital tourniquets should not be used for long operations. There is also the danger that an operation that is planned for 20 minutes may take much longer (e.g., trying to find a foreign body in a finger pulp). A pneumatic tourniquet on the arm is so much safer that it should usually be preferred even for small operations.

Pressure by the Therapist

It is through the application of splints to the hand that a therapist may cause ischemia. Actual pressure necrosis from splinting is likely only in insensitive hands. However, very commonly indeed a patient with normal sensation will discard a splint or wear it only intermittently because it is "uncomfortable." In such cases the discomfort is often caused by ischemia. A patient who discards a splint may be called uncooperative. It would be better to go back and check pressures under slings or under a reaction bar or to look for a flush of hyperemia when the splint is removed after 1 or 2 hours. Then we shall learn the importance of monitoring pressures, and then most of our patients will benefit from our splints and will keep them on.

Slings.—Almost everybody uses a sling as a convenient way to apply force to a digit. All manufacturers of hand orthoses provide slings as part of their prefabricated dynamic splints.

Therefore it takes some courage on our part to come out in print and condemn the typical sling and suggest an alternative system.

Our chief objection to slings is that any change in the posture of a finger results in tilting of the sling and localization of pressure to one or other edge of it (Fig 7–12). When the splint is first fitted, the position of the outrigger is usually fixed so that the spring or rubber band runs at right angles to the finger (or segment of finger). In that position the sling may have an area of 4 cm² in contact with the finger and a force of 200 g will result in a pressure of

$$200/4 = 50 \text{ g/cm}^2$$

This is satisfactory. The portrait of stresses caused by the force F in Figure 7–13 is easily explained. In the region labeled B, the elements which are originally square are compressed vertically and distended laterally. These compressive stresses are not as damaging to soft tissues as shear shown as shaded regions on the right half of the figure. The region with the darkest shading (A) shows that the square elements are deformed by shear into diamond shapes. Thus, the most damaging stresses are indicated by the shading on the right-hand side of Figure 7–13, with the most significant shear in an

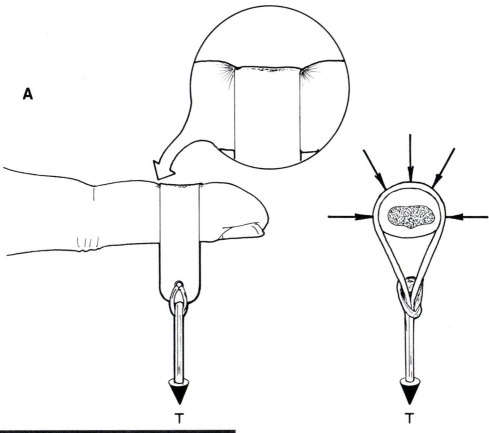

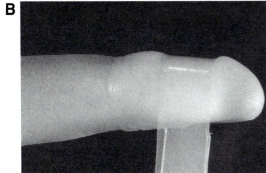

FIG 7–12.
A and **B,** a flexible sling at right angles to a digit results in (1) pressure under the sling in the line of the pull, (2) pressure from side to side because of the convergence of the two halves of the sling, and (3) some shear stress at both edges of the sling. *T* = tension.

area immediate to the edge of the applied stress. One can demonstrate this effect by experimentation on a fingertip pulp using a pencil eraser, holding a moderate stress for a period of 1 minute, and then observing the ring of inflammatory response, which will appear as slightly larger in diameter than the eraser. This edge effect is observed in the region just outside the edge and internal to the surface; it has the greatest shear. This discovery led to the practice of rounding edges of orthotic devices to minimize the sudden, sharp edge which produces the

shear. Finally, the upturned lines in region *C* demonstrate that the surrounding tissues are helping to support the external compressive load *F* by internal tensile forces.

When the patient goes home and the finger begins to respond to the tension, the joint opens out a little so that the spring no longer pulls at right angles to the finger. Now one edge of the sling rests on the finger and takes all the force and the other edge lifts off (Fig 7–14). Typically, when the line of pull is 70 degrees to the finger, a plastic sling takes all the force on less than a

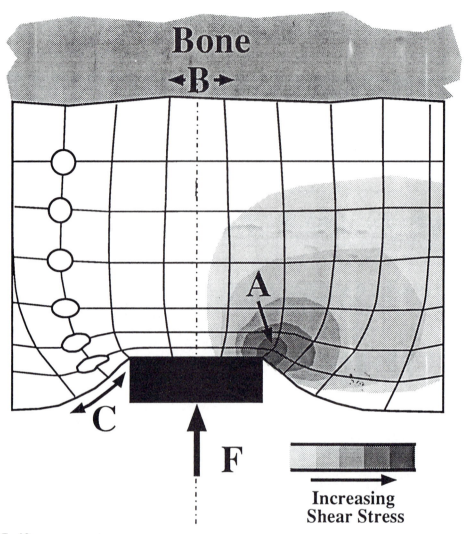

FIG 7–13.
The local representation of the stress portrait exhibited by the skin when a load *(F)* is applied. Before force was applied to this block of compliant soft tissue, its cross section was marked in equal squares. When the load is applied, pressure causes narrowing of the central squares. The squares around the edge are distorted by shear stress. This is where damage is caused. See text.

third of the available sling surface. Thus, with unchanged tension of 200 g, the pressure rises to 150 g/cm^2. This is quite unacceptable and could cause necrosis in a few hours. It will in any case be painful after a while and the patient will pull the sling off and lose confidence in the therapist.

Slings are more satisfactory if made of leather or other compliant material, but the basic problem remains the same, i.e., when the apex of the sling is pulled at an angle, the tension is transmitted to one edge of the sling.

Our solution to the problem is to form a rigid plaster shell to lie on a finger (Fig 7–14,B) and to pass a nylon thread around the shell. The thread or wire is prevented from sliding along the shell by laying a small strip of plaster across the wire at the apex. This permits easy rotation of the thread but no sliding. In such an arrangement, the angle of pull may change quite freely, and the sling will not tilt but continue to transmit the pressure evenly over its whole surface (Fig 7–15,B).

The second problem with slings is more difficult to correct. It is that there is not enough soft tissue on the dorsum of the

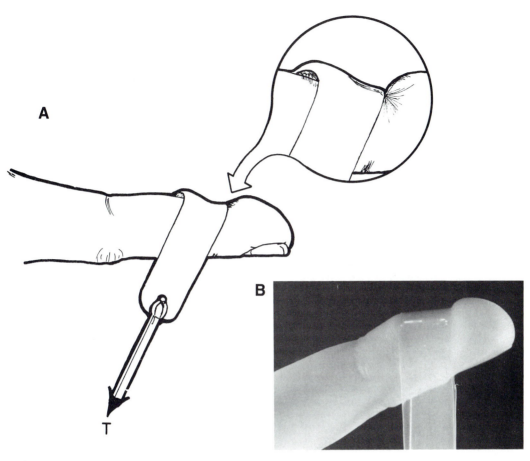

FIG 7–14.
A and **B,** when a sling is tilted because the angle of pull has changed, the pressure and shear stress is localized to one edge of the sling and is therefore greatly increased. *T* = tension.

hand and fingers to allow any standard or off-the-shelf device to apply pressure evenly on the back of a digit. This applies to slings and to reaction bars as well. When a transparent plastic sling is applied to the palmar side of a finger, it is interesting to watch color changes as increasing tension is applied. A little tension results in low pressure, producing a change of shape of the palmar pulp but no blanching. A little more tension results in a mottled blanching. More tension again and the whole area under the sling becomes evenly blanched. The same sling applied to the dorsum of the finger offers a different picture. Where it crosses a knuckle or bony prominence, there is early localized blanching over the bone. In the hollow between joints the skin may remain pink and without pressure. If the sling is made of compliant leather or soft polyethyl-

ene, some changes of shape may take place, making the pressure more even. A rigid shell is good on the dorsum but only if it is formed on the finger, not preformed, or it must have a compliant lining. The very best shell may be made of three thicknesses of fine plaster of Paris bandage (Gypsona* is the best) cut to size, wetted, and laid on the finger segment. When the plaster has set, a nylon thread is laid across the center of it and held there by a narrow strip of plaster. The position of the edges of the plaster shell are marked on the finger in case it needs to be removed and reapplied. A similar sling can be made of a square of thin heat-labile plastic splint material like Aquaplast.† It is

*Gypsona, Johnson & Johnson, New Brunswick, NJ.

†Aquaplast, WFR Corporation, Ramsey, NJ.

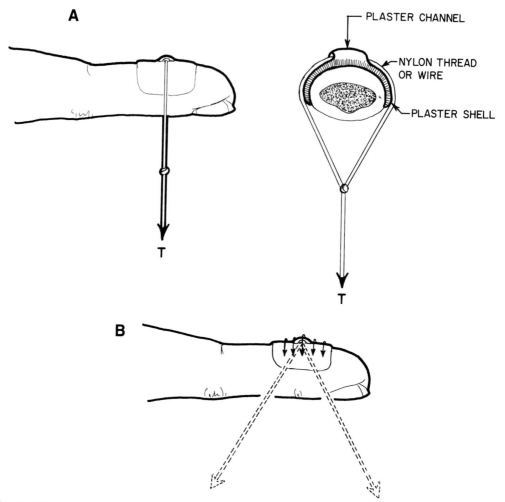

FIG 7–15.
A, by using a small plaster shell, formed on the finger, most of the dangers of a sling are avoided. A loop of thread is passed around the rigid shell, and there is no side-to-side compression. *T* = tension. **B,** if the angle of pull on the thread is changed, there is no tilting of the shell because the change of direction occurs at the top of the convexity of the plaster shell. Only the thread changes direction.

warmed and laid on the finger to conform and set.

Reaction Bars.—Whenever a force is applied to a digit in a splint or orthosis, it creates the need for an opposing or restraining force proximal to the joint that is the object of the attempt at correction.

There is a very common and very dangerous misconception about the forces involved and therefore about the pressures they produce. The misconception stems from the true statement that "to every action there is an equal and opposite reaction" and to the false concept that where a joint is in equilib-

rium the forces acting on it must be equal and opposite. The truth is that when forces act on a joint it is the *moments* that must balance, not the forces. Moments are forces multiplied by *moment arms*. So in equilibrium the force nearest to the axis must be greater than the balancing force far from the axis. Thus most commonly the reaction bar of a dynamic splint has to accept much more force than the sling or spring that is the active element. Unfortunately, standard preformed reaction bars often run transversely across several digits and are straight metal bars padded with felt. For four reasons the reaction bar is much more likely to cause ischemic pain and damage than the sling, which is more distal.

1. The sling is more often on the palmar side of a digit where the tissues are compliant. The reaction bar is often on the bony dorsum.

2. The sling is individual to a digit and encircles half the circumference, whereas the bar may be straight and have a small area of contact, and only on the dorsum.

3. The sling is often twice as far as the reaction bar from the axis of the joint that needs to be stabilized (it is the joint proximal to the reaction bar that needs to be stabilized to allow the force from the sling to work on a more distal joint). Therefore the bar has to transmit twice the force often on a fraction of the area.

4. Finally, if the force of the sling results in tilting the whole finger, this results in the proximal segment pressing on only the edge of the reaction bar, which is usually rigid (Fig 7–16).

For these four reasons gross and often painful pressures occur. This is a common and serious situation. It can be prevented by the following rather simple considerations:

1. Always calculate, at least approximately, the relationship between the active force and reactive force that will be needed for equilibrium.

2. Provide for adequate area under the reaction bar by molding it around the digits. It is much better to have a molded bar even if it is hard (plaster of Paris) than a padded bar that is straight.

3. Have the proximal part of the whole splint so snug that the hand cannot move in the splint. Tilting of a finger in relation to its restraining surface will thus be avoided. Alternatively, and especially if some movement is to be expected, the whole reaction bar or pad may be hinged on a transverse axle behind its center. Then if the proximal segment of the hand tilts, the bar will tilt with it[1] (Fig 7–17).

4. Remove the splint after the first hour (or release the tension) and look at the skin.

If there has been ischemia, there will now be a flush.

REPETITIVE PRESSURE AND SHEAR STRESS

This is a much bigger problem than most of us realize. It relates to stresses on the surface of the hand that are of the order of 1 to 5 kg/cm^2 (i.e., 13–65 $lb/in.^2$). These are much greater than the pressures we considered in our discussion of ischemia, but they are the kind of pressures we all use every day and accept as normal and comfortable. They cause ischemia for a few seconds or minutes at a time, but they cause no harm and no pain because the stress is of short duration. It is when the stresses are repeated frequently and constantly on the same tissue that they begin to cause problems. Familiar examples occur in everyday life: the hobby gardener, in early spring, who digs to prepare his soil for planting and develops a blister on his palm by the end of the weekend; the amateur builder of his own kayak (one of the authors) who tried to attach the deck fabric all in 2 days. The fabric was attached by screws every 4 cm all around the boat. After perhaps 80 screws, P.W.B.'s hand became red where the screwdriver twisted in his palm. He changed to a larger-handled screwdriver and then, later, covered that with a handkerchief. He then tried using his left hand and finally gave up. The following day he could not even start work because his hand was still sore. Later he developed an area of callus at the site.

This story identifies all the elements of the problem. First, the stress is quite acceptable and harmless for a number of repetitions. Second, it is shear stress (as with a screwdriver) that causes the problem most of the time. Third, in a sensitive hand, pain finally comes to dominate the whole work situation. In the case of an insensitive hand, this factor of pain is absent and real breakdown and ulceration are the result. Finally, the long-term response of the skin to repetitive stress is to produce a protective callus.

We have done quantitative experiments on rat foot pads and found, e.g., that 10,000

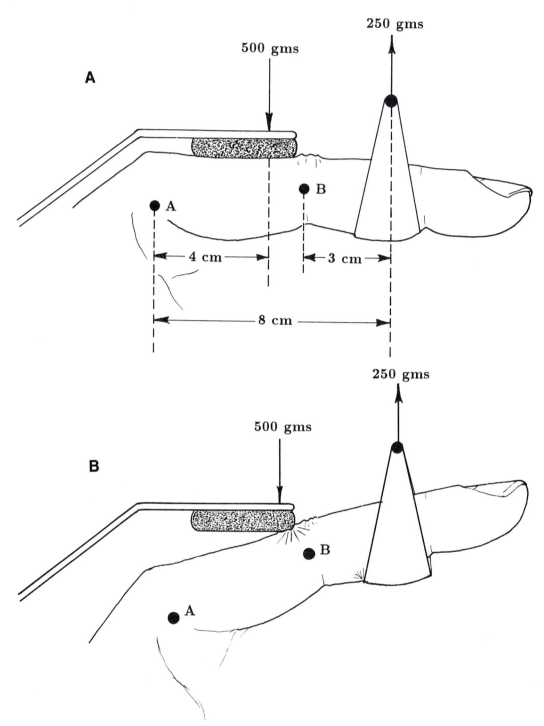

FIG 7–16.
A, diagram of a finger which has a stiff interphalangeal joint *(B)* and a normal metacarpophalangeal joint *(A).* The dynamic orthosis is being used to apply constant extensile torque to *B,* using a force of 250 g through a sling 3 cm distal to the axis *B.* To prevent the force acting on the mobile joint *A,* a padded reaction bar is placed 4 cm distal to axis *A,* to hold that joint in moderate flexion. But note that to hold the finger in equilibrium, the reaction bar, at 4 cm from axis *A,* has to accept a thrust of 500 g to balance the 250 g from the sling which is at 8 cm from axis *A:* 250 × 8 = 500 × 4. **B,** same finger as in **A** but because the proximal part of the hand was not firmly supported the whole finger has tilted. Now only the edge of the padded reaction bar presses on the finger, creating unacceptable pressure.

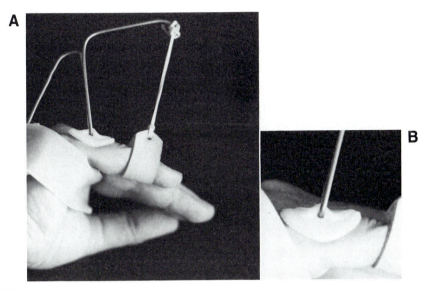

FIG 7–17.
Dynamic splinting of a proximal interphalangeal (PIP) joint flexion contracture requires control of the metacarpophalangeal (MCP) joint and proximal phalangeal segment so as to concentrate the tension (rubber band) force at the PIP joint level. Splints for this purpose require a dorsal mechanical block to control the proximal phalanx and the MCP joint. To ensure an acceptable distribution of force (pressure) over the dorsal side of the proximal phalanx, we utilize a proximal counterforce in the form of a freely hinged "foot." It is fabricated by sandwiching the outrigger wire between layers of the custom-molded plastic device, as shown. As the splint device shifts on the hand, our movable "foot" rotates to ensure even distribution of pressure over the dorsum of the proximal phalanx. *(From Agee, J.M., and Kerfoot, C.: Person communication, April, 1984.)*

applications per day of 1.5 kg/cm^2 (20 lb/in^2) pressure resulted in nothing but a little redness and swelling on the first and second days, but 10,000 repetitions daily by the seventh day caused open ulceration.

Histologic examination of those rat foot pads showed that the early change, which in the human hand we would call soreness, was a true inflammation of the dermis and subcutaneous tissue with edema and incursion of masses of inflammatory cells. The epidermis became hyperplastic. If the whole process of mechanical stimulation was halted or even diminished after 2 or 3 days, the tissues went on to resolution and hyperplasia of the skin with keratinization. If it was continued day after day, the tissues underwent blistering and necrosis.

The implications for hand specialists, both in surgery and in rehabilitation, are threefold.

First, patients with insensitive hands need to realize that there is special danger in repetitive tasks. Most such patients are aware of the danger of gross injury, but few understand that ordinary work, even with

good tools, may result in injury simply because the early warning system of soreness is missing. These patients need to change tools and handles frequently (and use large handles) (Fig 7–18). Also, they need to look at their hands to observe areas of redness and to feel their hands with a sensitive part of their body (our leprosy patients use their lips) to note if a finger or thumb has become hot.

Second, patients with deformity or paralytic limitations of their hands, even with normal sensation, are especially at risk from this problem of repetitive stress. The reason is that such patients often are limited to only one way to hold an object or tool.

The first instinct of a person doing a repetitive task is to change his or her grip from time to time. This is not so much to rest a muscle as to rest a piece of skin. People with partial paralysis, amputation, or stiffness of some fingers, or an unopposable thumb often find that they cannot change their grip. Their only way to hold a tool may involve using the side of the thumb, where there is no palmar pulp padding, or some

FIG 7–18.
By enclosing the handle of this key in a leather casing, it becomes safe for the insensitive hand that uses a lateral squeeze pinch.

scarred area of skin. These patients are often proud, in the best sense. They are proud to show that they can do things. They accept a job that they believe they can do. They can indeed do it. It is halfway through their workday that their overworked skin rebels and becomes blistered. The patient either gives up or persists in going on until severe damage results.

The answer is to recognize the problem ahead of time. If a patient is limited to just one type of use of his or her hand, the surgeon must make sure that the best available pulp surfaces are in a position to be used, even if it means that an extra operation is needed to achieve it. In the evaluation of a limited hand for the workplace, every effort must be made to fit the handles of tools to the hand of the worker. We do not ordinarily like handles that are too closely fitted to the contours of a hand, because it makes it impossible to shift the handle around. However, for a severely deformed or limited hand it may be good to make special handles from a mold of the hand and cover them with leather to maximize the area of contact and minimize shear stress (see Fig 7–18).

The third implication of repetitive stress applies to normal people in the workplace, especially at an assembly line where repetitive stress is the order of the day. Many in-

dustries have adjusted the workplace to avoid postural problems for the workers and have adjusted to proper angles of vision and appropriate back positions. The same attention has not always been given to stress on interface skin. It would be beneficial if a radiometer or thermograph machine were used to screen the hands of all workers at the end of the day, especially when new machines or new routines have been installed. We use thermography for insensitive hands in our sheltered workshop and find it very useful. If one finger or a thumb or one part of the palm of a hand shows hot at the end of the day (Fig 7–19), this means it has been subjected to undue repetitive stress. Either the worker is using the hand wrongly or the instrument is poorly shaped or positioned.

Even with normal sensation a worker may not always realize what the problem is, but feels a sense of fatigue or a tendency to slow down or quit. The correction of the problem results in a cool hand and a contented worker producing more and better work. Surgeons tend to be interested in producing good hands. Employers provide tools and machines to maximize the output of goods. Somebody must be concerned with the interface between the hand and the tool—the universal interface between man and his environment—the skin.

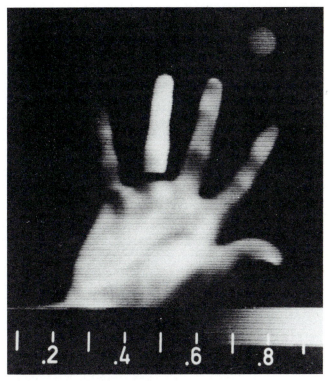

FIG 7–19.
This thermograph, or infrared photograph, of a normal hand shows typical distribution of warm and cool areas, except that the ring finger has been subjected to repetitive blunt stress. This finger showed no sign of damage, but the 5° C temperature contrast was obvious to touch and remained palpable for hours.

MECHANICAL EFFECT OF SKIN MOISTURE

The outer layer of all skin is composed of keratin. On the palmar surface of the hand and on the sole of the foot the keratin layer is thick. In response to regular stress and friction the keratin becomes especially thick and hard, as in the calloused hand of a manual worker. However, all normally calloused hands are still able to sweat, since the openings of sweat glands traverse the keratin layer. Sweat serves not only to moisten the surface of the skin but also to hydrate the keratin itself. Dry keratin and hydrated keratin, from a mechanical point of view, are totally different materials. It is instructive to shave a thin layer of keratin from a thick callus on the sole of the foot. This flake of material is very compliant and can be rolled up and pulled out almost like a piece of rubber cut from a surgical glove. If it is left in the sun to dry, it changes completely. It be-

comes hard and brittle. It cannot be stretched, and it will break if it is bent. Its flexibility and elasticity can be immediately restored if it is dipped in water.

If keratin is dry, it may limit the range of mobility at joints (Fig 7–20). Fritschi,[6] in India trying to mobilize stiff fingers, compared the effect of various modalities such as hot paraffin wax, oil massage, and simple soaking in water. He found that the water soaks were the most beneficial. These tests were on denervated fingers with dry skin and probably reflected the improved mobility of the moistened keratin layer. Another mechanical feature of dry keratin is that it is slippery. A little moisture in the skin helps a hand to grip securely and to pinch firmly. A really wet hand may slip because of a film of water, but absorbed moisture improves the texture of the keratin.

In P.W.B's hand clinic in India he had many keen young therapy aides and large numbers of patients with nerve damage of various sorts. As part of a battery of tests

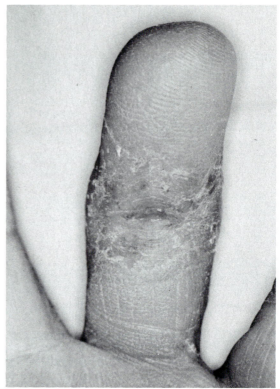

FIG 7–20.
Dry skin and dry keratin do not bend and stretch easily. This type of crack at a flexion crease may lead to serious infection. Nonsweating skin must be kept hydrated.

that all patients had to undergo was one that involved an attempt to grasp smooth cylinders of various diameters. These cylinders stood upright in a row, each on a flat base. Patients had to grasp and lift them, starting at the smallest, until they failed, usually because their clawhand would not open far enough or reach out to surround the cylinder. It was a good test and recorded the improvement that occurred in joint mobility and the dramatic improvement that occurred after surgery to restore the intrinsic action of the hand.

Several months of our records were spoiled, however, because word got out among the patients and among the aides that very good levels of achievement would occur if the patients soaked their hands and then quickly dried them just before the test. The improvement was not because the hands could open more widely, it was because the fingers were not slippery and the cylinders would not slide away when incompletely grasped. The patients were lifting cylinders two and three grades larger than before and bringing credit on themselves and their therapists.

We learned from this not only to standardize moisture content before testing but also to carry this knowledge to the rehabilitation industry and get all patients with insensitive hands to soak them every morning and keep their keratin hydrated in the workplace. They said that they felt that their hands were stronger after soaking. What they meant was that they could hold things more securely.

All denervated limbs should be soaked in water daily and then rubbed with petroleum jelly to prevent evaporation. This keeps the skin in good condition, prevents cracks, and maintains mobility.[9]

REFERENCES

1. Agee, J.M. and Kerfoot, C.: Personal communication, April, 1984.
2. Brånemark, P-I; Microvascular function at reduced flow rates. In Kenedi, R.M., Cowden, J.M., and Scales, J.T., editors: Bed sore biomechanics, Strathclyde Bioengineering Seminars, New York, 1976, Macmillan Press Ltd., pp. 63–68.
3. Brånemark, P-I, Halijamäe, H., and Lund-

borg, G.: Pathophysiology of microvascular and interstitial compartments in low flow states. In Les solutes de substitution ré-équilibration métabolique, Paris, 1971, Librairie Arnette.

4. Bruner, J.M.: Safety factors in the use of the pneumatic tourniquet for hemostasis in surgery of the hand, J. Bone Joint Surg. 33A:221–224, 1951.

5. Daly, C.H., et al.: The effect of pressure loading on the blood flow rate in human skin. In Kenedi, R.M., Cowden, J.M., and Scales, J.T., editors: Bed sore biomechanics, Strathclyde Bioengineering Seminars, New York, 1976, Macmillan Press Ltd., pp. 69–77.

6. Fritschi, E.P.: Hydrotherapy as a method of treatment for contracted fingers, Lepr. Rev. 40:117–120, 1969.

7. Romanus, E.M.: Microcirculatory reactions to controlled tissue ischaemia and temperature: a vital microscopic study on the hamster's cheek pouch. In Kenedi, R.M., Cowden, J.M., and Scales, J.T., editors: bed sore biomechanics, Strathclyde Bioengineering Seminars, New York, 1976, Macmillan Press Ltd., pp. 79–82.

8. Shaw, J.A., DeMuth, W.W., and Gillespy, A.W.: Guidelines for the use of digital tourniquets based on physiological pressure measurements. J. Bone Joint Surg. 67A:1086–1090, 1985.

9. Stenstrom, S.J.: A study on skin humidity in leprosy patients using a new type of humidity meter, Int. J. Lepr. 52(1):10-18, 1984.

10. Yamada, H.: Strength of biological materials. In Evans, F.G., editor: Strength of biological materials, Baltimore, 1970, Williams & Wilkins Co., p. 226.

11. Reswick, J.B., and Rogers, J.: Experiences at Rancho Los Amigos Hospital with devices and techniques to prevent pressure sores. In Kenedi, R.M., and Cowden, J.M., editors: Bed sore biomechanics, London, 1976, The Macmillan Press Ltd., pp.301–310.

CHAPTER 8

External Stress: Forces That Affect Joint Action

There are three major mechanical factors in the interaction of the musculoskeletal system inside the body with external environmental systems. First are the active forces of voluntary muscles that exert tension through tendons to produce torque (moments) around the axes of joints, causing movement of segments of limbs. Second is the internal passive resistance that limits the application of the force of the muscles and limits the range of motion of the joints. Third is the system of forces between the surface of a limb and the external environment. These external forces may be initiated by the body and resisted by the environment or may be initiated outside the body to affect the musculoskeletal system, as when one person fights another or helps another (e.g., by orthotic devices).

Whether force is originating from muscles to affect the environment or originating outside to move the body, the passive soft tissues serve to modify and limit the effect of the force. When the degree of soft tissue limitation is severe and unacceptable (stiff joints, stiff hand), the problem may be approached from inside or from outside. From inside, the muscles may work to stretch and loosen the adhesions. From outside a therapist may use passive movement for the same object or may apply a splint or orthotic device to improve the range of motion of a joint. This last activity is what this chapter is all about.

Just as the production of torque or moment at a joint is the object of muscle contraction, so the production of torque or moment at a joint is the objective of the external force applied by a therapist or by a splint. The ultimate goal may be to achieve free mobility or to lengthen an adhesion, but the means of achieving that goal is to produce torque. Torque from external force, as from internal muscles, is the product of the force multiplied by the perpendicular distance between the axis of the joint and the line of action of the force (Fig 8–1).

When external force is prescribed as a therapeutic modality, it requires great care and precision to avoid the harm of too much or too little. External force is most commonly applied by means of a dynamic splint or orthotic device.

TORQUE: HOW MUCH FOR HOW LONG?

We always feel a certain reluctance about using a splint on the hand. A hand is always at its best when it is being actively used in purposeful activity. A splint or orthotic device tends to inhibit the ordinary free use of the hand, and to this extent it must be bad. We use splints only when free activity may be harmful (as in a recently wounded or infected hand) or when a certain important ob-

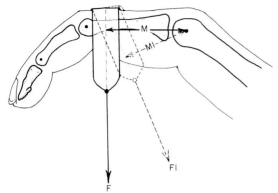

FIG 8–1.
A tension of *F* at right angles to the phalanx has a moment arm of *M* at the joint. The same tension, *F1*, through the same sling at the same point on the digit has a much smaller moment arm, *M1* (and therefore a smaller moment or torque), when the line of pull is not at a right angle.

jective cannot be achieved as well in any other way. In the case of a joint that is limited in range of motion by contracted soft tissue or scar, an external force may be necessary when the muscles cannot produce either enough torque or sustained torque for a long enough time to lengthen the contracted tissues. In other cases an external force may be useful as a stabilizer for a proximal joint in such a position that the patient's own muscles may be better able to mobilize a distal joint.

When we try to determine how much torque is needed to overcome a contracture, we should first have some idea of the nature and length and density of the structures we are trying to mobilize or lengthen. We should also have some idea of how close or how far they are from the axis of the joint (moment arm). I know of no published work that gives any information on how much force is needed to lengthen a given cross-sectional area of any living tissue. The length-tension curves that I have reproduced (see Fig 6–15) are all of dead tissue. If the tensions at the upper end of these curves were applied to living tissue they might well result in traumatic inflammation and ultimately increase the density of the contracture. The tensions at the lower end would perhaps result in no benefit. Clinical experience and intuition tell us that we should aim to produce enough torque to hold the soft tissues in the *b* section of the length-tension curve or in the lower part of the *c* section (Fig 8–2). This is the tension that holds the collagen fibers in tension (see Chapter 6).

In many stiff fingers; normal muscle ac-

tion might produce enough torque for the purpose. However, if the tissue that needs to be lengthened is dense and thick and if it bowstrings far from the axis of the joint, it may need a torque more than the muscles can supply to keep it on the stretch. Even in the common cases in which the limiting tissues are reasonably springy and close to the axis and in which the muscles can easily stretch them to the optimal extent to stimulate lengthening, the problem is time. To stimulate the active cellular action that results in the reorientation of scar and growth of skin, the contracted tissues must be held in the slightly stretched position for a long time, probably for many hours a day. This is where a splint is often superior to a muscle and may be necessary. Muscles soon tire and patients forget to contract them, so the function of the splint is not so much to pull harder than the muscles, but to maintain the tension longer.

INTERFACE TISSUES

Now comes the problem. It is one thing to calculate how much torque is needed and for how long. It is quite another thing to be able to apply an external force to produce that torque without causing ischemia of the skin and pain. In most cases in which continuous external force is applied to a digit with the idea of increasing range of motion of the joint, the limitation in the amount of force that can be used is entirely a result of the limited ability of skin and soft tissue to accept pressure without ischemia and pain.

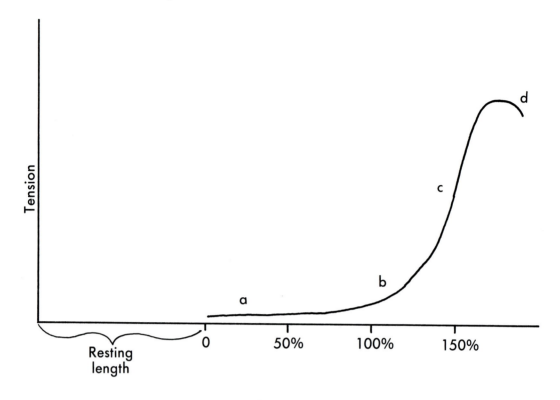

Typical biological tissue (e.g., skin)

FIG 8–2.
Length-tension curve for typical biologic tissue (e.g., skin).

The joint and the contracted tissue might safely and usefully accept much more force if it could be applied. Therefore the therapist should usually apply as much force as the surface tissues will accept without ischemia and apply the force as far as possible from the axis of the joint so that the torque is maximized by a long lever arm. This is right so long as it is remembered that a long lever has dangers as well as benefits. In some cases of rheumatoid arthritis, where lax ligaments allow some subluxation of the joint and where there may be a block to free gliding, there is a danger that external force may further damage or dislocate the joint (see Fig 6–25). In any such case it is usually best for external forces to be applied close to the joint so that the lever is short enough to encourage joints to glide, because a long lever encourages angulation and tilting of joint surfaces (Fig 8–3). Short moment arms make it even more difficult to produce sig-

nificant torque at a joint without exceeding the threshold of tolerance of the skin.

More force can be applied by skeletal traction with pins through the bone, but this is rarely justified in fingers, where the inflammation or even infection around a pin tract will produce the very stiffness that the traction is designed to overcome. Thus the objectives remain: to minimize the pressure and shear stress while maximizing the torque. To accomplish these objectives we have to spread the force. This is discussed in Chapter 7.

The problem is the same as that at the seat interface of a wheelchair patient or under the back of a paralyzed person in bed. The diagram by Kosiak[1] (Fig 8–4) was one of the earliest to define the relationship between time and pressure. It was based on sustained pressure on the foot pads of dogs. The curved line marks the boundary between the animals that were unharmed and

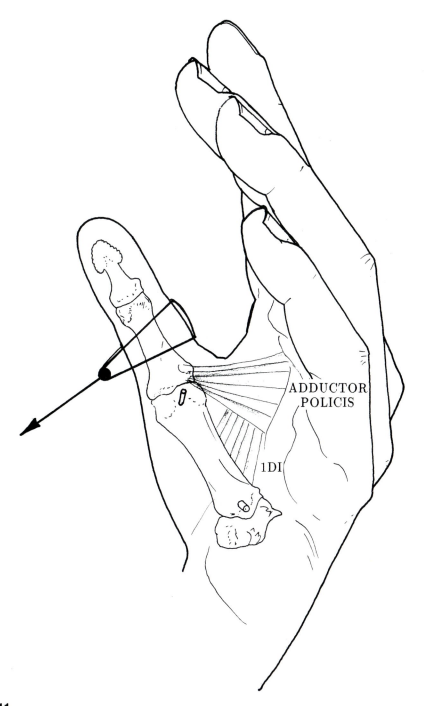

FIG 8-11.
The effect of the force of the sling applied across the MCP and CMC joints varies with the moment arm ratio of the sling's lever to the moment arm of the muscle. The 1DI (first dorsal interosseous) crosses only the CMC joint, as does most of the skin web. Only the ulnar collateral ligament resists abduction at the MCP joint and is preferentially lengthened by this splint.

Surgical correction (1) is preferable in all severe cases, especially if there is scarring.

REFERENCES

1. Kosiak, M.: Etiology and pathology of ischaemic ulcers, Arch. Phys. Med. Rehabil. 40:62–69, 1959.
2. Lister, G.D., et al.: Primary flexor tendon repair followed by immediate controlled mobilization, J. Hand Surg. Surg. 2:441–451, 1977.
3. Reswick, J.B., and Rogers, J.: Experience at Rancho Los Amigos Hospital with devices and techniques to prevent pressure sores. In Kenedi, R.M., Cowden, J.M., and Scales, J.T., editors: bed sore biomechanics, London, 1976, The Macmillan Press Ltd., pp 301–310.

CHAPTER 9

Postoperative Stiffness and Adhesions

CLINICAL MANAGEMENT OF ADHESIONS

To restore balance and beauty and power to a disabled hand is an adventure. The stakes are high. The rewards are exciting. The penalties of failure are grievous. Even today we are sometimes haunted by the specter of hands long past that had a small disability before we worked on them and had a large disability afterward. In most such cases it was not that we made mistakes in the plan of reconstruction, it was just while we were doing some good things, we allowed the overwhelming bad thing to happen—the hand became stiff. Adhesions developed, tendons became adherent, the hand became swollen and painful. Somewhere out there is somebody who does not use his hand, and who hides it from view because of what we did or failed to do.

We should not forget such patients. They should stand beside us while we plan treatment for others. They should look over our shoulders at surgery and be with us at therapy sessions and whisper reminders to be gentle and warnings to stop and think.

PREOPERATIVE PLANNING: SURGEON, THERAPIST, AND PATIENT TOGETHER

Looking back over the history of a hand that has become stiff after surgery it is easy for one member of the team to blame another. The surgeon and therapist say the patient was not cooperative. The therapist thinks the surgeon made too much scar and the surgeon is sure the therapist used too much force or else did not really try. The patient wants to sue everybody. The real problem, perhaps, was that there never had been teamwork at all or that an attempt at it started too late.

The critical moment of truth may come soon after the postoperative cast is removed and the patient, the surgeon, and the therapist observe the first attempts at active mobilizaton. On that day and in the following week they may become aware that the patient has no idea how to identify and contract the muscle that has been transferred; the joint has an inadequate range of passive motion, so active motion is frustrated from the start; the wound is infected; or the patient is already antagonistic because he or she had not realized how uncomfortable and perhaps painful the postoperative position in the cast would be for an unrelenting 3 weeks.

Any of these things may delay the effective movement of transferred tendons long enough to allow adhesions to mature and toughen until the result becomes a failure.

Preoperative preparation must include (1) attainment of full assisted-active, or at least passive, range of motion of all affected joints, (2) identification by the therapist and by the patient of the tendons to be used, and (3) training of the patient to contract the muscle to order. Above all the patient must understand how it is all going to work and that it will take time, discipline, and discomfort. If he or she cannot get excited and hopeful about it, maybe it is better not to do it. The Apollo missions to the moon were successful because in the countdown everything had been foreseen and checked. They did not wait until landing on the moon before finding out whether they had fuel for the return trip. In hand surgery, the first week after cast removal is like the moonwalk, you can only use what you had planned ahead to have ready to use.

THE TEAM

Any major surgery on the hand needs to be a joint enterprise between the surgeon, the patient, and the therapist. There may be surgeons who prefer to be therapists as well, but they need to discipline themselves to spend a lot of extra time with the patient. They will probably learn, at least when they get busier, that not only will they be able to treat more patients but also the job will be done better if the responsibilities are shared.

Every therapist is taught to respect and support the surgeon's opinion and prescriptions. Not all surgeons are taught that an experienced therapist has an opinion that is valuable and should be listened to with respect. Both surgeon and therapist should listen to the patient at every stage and demonstrate a willingness to be flexible and responsive to each other and to the patient. Every improvement of strength and mobility is a mutual triumph and a cause for mutual celebration.

The most important member of the team is the patient. Most of us professionals do not take the patient seriously enough into our confidence nor do we recognize his or her competence and value in the restoration of health and mobility. The most important variable in the patient is his or her psychologic makeup and attitude.

PSYCHOLOGY

A hand is a very personal thing. It is the interface between the patient and his or her world. It is an emblem of strength, beauty, skill, sexuality, and sensibility. When it is damaged, it becomes a symbol of the vulnerability of the whole person. The first responsibility of hand surgeons and hand therapists is not to build up their own image of competence but to encourage the patient, make him feel good about himself, and change his fear into confidence.

A family of opossums lives near our home in Louisiana. If a dog or a person frightens one of them at night, it goes at once into a rigid catatonic state. It becomes stiff and immobile from snout to tail and stays that way until we go away and it is no longer afraid. There are patients who are a little like that. We have learned to recognize the type and to be cautious about them, because their hands easily become stiff. Their eyes are wide, and they follow all our movements. Their hands are often cold, and they are reluctant to be tested. Such patients need time to gain confidence. It is good just to hold their hand gently while you talk and explore the history of their problem. It is good just to stroke their hand, and it is good to be unhurried and to ask about the family and the home and to appreciate their good qualities and the way they have faced their problems. It is necessary to emphasize very early that you, the surgeon or therapist, are not going to make decisions for the patient. It is his or her hand. Any remaining good muscles, good joints, and good sensation should be noted appreciatively.

Gradually the hand will become warmer and will relax, and early signs of confidence and hope will appear. All this time spent on developing mutual understanding and trust is not time lost. It is time saved from the postoperative period. When you have made friends with an opossum, it does not go rigid, and the same is true of a hand. We do not really understand why a whole hand may become stiff after a minor and localized injury, but we do know that it happens more often and more severely when there is fear, anger, pain, and a sense of helpless frustration. It happens less when there is hope, trust, confidence, and the patient has a sense that he or she is competent and in control.

RESPONSIBILITY OF THE SURGEON IN PREVENTION OF ADHESIONS

In Chapter 6 we discussed the nature and effect of adhesions. Now we need to think about how the surgeon may minimize the problem during the operation. We shall consider four basic principles:

1. Minimize scar.
2. Avoid anything that makes for tendon-like adhesions.
3. Ensure that all adhesions attach the tendon to movable tissues only.
4. Since adhesions are always shorter than we wish, we use their length twice, to and fro.

MINIMIZE SCAR

Before outlining specific rules about scar around tendons, we should just reemphasize the general rules about scar and atraumatic technique that Bunnell[5] laid out with such eloquence.

The great enemies of final mobility of a postoperative hand are infection, hematoma, and necrotic tissue. Every time we clamp or burn a bleeding point, we destroy tissue and leave a scar. Every time we fail to arrest bleeding, especially from a vein, we risk a hematoma.

Thus the meticulous discipline of the hand surgeon requires that bleeding points be picked up at the exact point and lifted clear of other tissues before burning them or tying them, so that the bulk of dead tissue and the probability of hematoma are both minimized. In the case of gross injuries and lacerations, one has to ask not only whether a tissue will survive but also whether in survival it will be so immobilized by scar that it will neither move nor allow other tissues to move. Excision of dysvascular tissue or even of a severely damaged digit may be a contribution to mobility.

We like to remove a tourniquet before the wounds are closed, to ensure hemostasis. If this is not done, in cases where dissection has been extensive, the use of negative-pressure drainage through very fine polyethylene tubes for 24 hours should be considered. Such tubes can be removed without disturbing the dressings and are good insurance against hematoma.

AVOID TENDON-LIKE ADHESIONS

Tendon-like adhesions form at the cut end of tendons and sometimes at the side of tendons whose "skin" or surface is injured. Therefore the most meticulous care must be taken to avoid trauma to tendon surfaces (epitenon). We never pull strongly on a tendon loop (in withdrawing a tendon) and even a gentle pull is made with a smooth rounded instrument. A large uterine dilator is a good instrument to pass under a proximal loop for traction (Fig 9–1). Even then, if there is resistance, one should not fight it but should look for an adhesion that needs to be cut even if this involves an additional incision. The extensor indicis proprius (EIP) is a good example. When it is divided distally and then identified proximally, it may sometimes be easily delivered into the proximal wound. However, quite frequently there are

FIG 9–1.
Uterine dilator

lateral connections between the tendon and other structures on the dorsum of the hand, such as the extensor digitorum. More than once we have seen a surgeon pull on the proximal loop of that tendon until the musculotendinous junction of the slender muscle began to tear. In such cases the final result has been absolute failure as a result of adhesions around the injured unit. How much better, as soon as resistance is felt, to make an additional small transverse incision, or even two, to find tendon cross-connections and divide them.

Forceps should never be used to hold a tendon except at the end that is to be discarded. Forceps bruise and crush tendon surfaces.

A tendon should never be allowed to dry or lie exposed on the skin while waiting its turn to be put into its bed. It must be left in its original bed until everything is ready for the immediate transfer to its new path and insertion. If delay is unavoidable, the tendon must be covered and kept moist by dribbling saline.

When tendons are sutured, no part of a cut end of tendon should remain exposed and unattached in a wound. If two tendon ends of equal cross section are to be united, they should be joined end to end with a tension stitch followed by the finest possible suture around the junction to get meticulous apposition.[6] If there is disparity between the size of two tendons that are to be joined, some way must be found to bury the cut end of the motor and preferably of the distal tendon too. Figure 9–2 shows some methods of interlacing and of burying a cut end and Figure 9–3 shows Brand's method of using a spread-out graft.[3] This special care is not necessary when the attachment is at the final insertion of a tendon. In such a case a tendon-like adhesion is just what is needed.

ALL ADHESIONS SHOULD BE TO MOVABLE TISSUES

The whole of every wound heals by scar. The scar of a wound is common to the whole wound. There is not one scar for the skin and one for the fascia and one for the tendon. Only a partial degree of separation of

segments of scar may be achieved by closing a wound in layers. Cut edges of fascia and interosseous membrane may appear to be at some distance from tendon grafts, but they are united by scar when the wound heals. It may be possible, as the weeks pass, to remodel and stretch some of the scar but it delays rehabilitation. It is much better to have the transferred tendon lying in contact with fat and areolar tissue only (Fig 9–4).

It is often impossible to expose the proposed pathway that a tendon is to take without cutting through some rigid immobile structures on the way. Deep fascia, palmar septa, retinaculum, tendon sheath, pulley, and bone may all become attached to a tendon graft or tendon transfer if both have an exposed cut surface in the same wound. In our early days of hand surgery we remember the frustration of using various transfers and grafts for opposition of the thumb and having them fail. On reopening the wound later we found again and again that the tendon anastomosis or tendon graft had become attached to the pulley that had been made from the tendon of flexor carpi ulnaris (FCU) or sometimes to the dense fibrous septa of the palmar fascia in the heel of the palm. Today we rarely have problems there for the following reasons. (1) We usually use a long transferred tendon rather than a graft and anastomosis. (2) We tunnel, rather than use open dissection. (3) When we use a pulley, we try to use a living fascial septum in situ such as Guyon's canal or, better, pass it through the carpal tunnel roof, as suggested by Snow and Fink,[10] where the tendon of flexor superficialis carries with it a sheath of its own living synovium that prevents direct adhesions to the cut edges of the retinaculum. (4) For pulleys elsewhere we try to use existing structures that need not be cut or even exposed. If we have to fabricate a pulley, we ensure that it is not close to a tendon anastomosis. Whenever possible, the pathway for a tendon transfer or graft should be prepared in a way to minimize any open dissection, and tendons should be passed by tunneling them through soft tissues. This can often be done almost bloodlessly and without injury to any rigid fibrous layers so that the whole tendon pathway is lined with

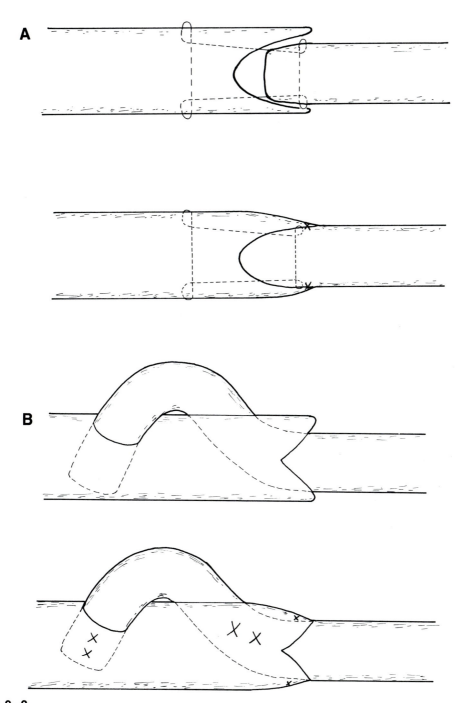

FIG 9–2.
A, for the end-to-end suture of two tendons of unequal diameter, the larger one may be made into a fish mouth of hollow end before they are sutured by a modified Kessler[6] stitch. Finally the fish mouth is closed with a fine running suture. **B,** Pulvertaft's[8] method of interlacing two tendons of unequal diameter. The larger tendon is made into a fish mouth, the smaller tendon is passed through, across, and then again through the larger tendon. All cut ends are covered with fine sutures.

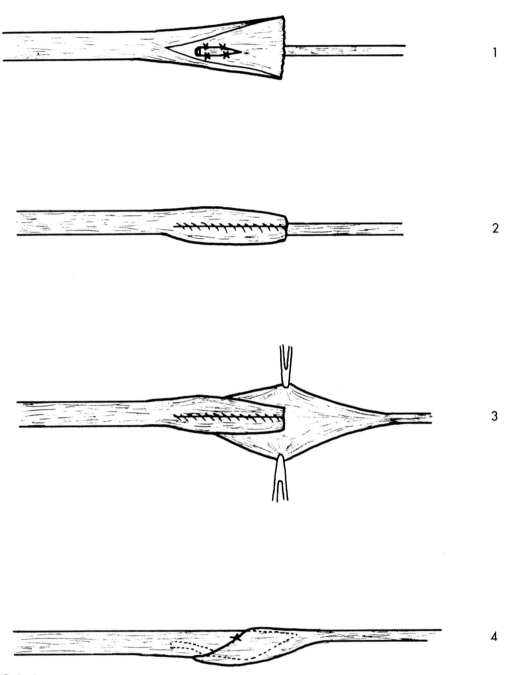

FIG 9–3.
Attachment of palmaris tendon graft to a motor that has a larger diameter. *1*, the motor tendon is split open, and the graft enters the open area through a small split. It is sutured into position. *2*, the split tendon is closed with a fine running stitch. *3*, the palmaris tendon is held by two needle-pointed forceps and firmly stretched into a membrane. *4*, the spread membrane is used to cover the cut end of the motor, which is cut obliquely at the end to avoid a sudden narrowing. (*Note:* The same technique may be used on a plantaris and on a strip of fascia lata. In the case of a plantaris it may be folded in two and sutured to the motor at the fold, leaving two exiting grafts, one of which is used to wrap the end of the motor.) *(From Andersen, J.G.:* Hand *14:339, 1962.)*

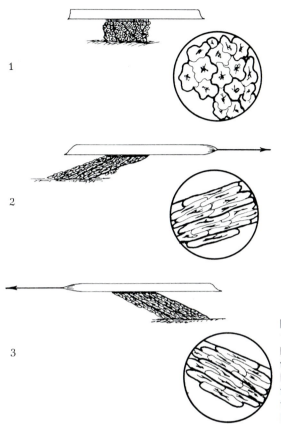

FIG 9–4.
1, tendon attached by short fibrous adhesions to a layer of fat cells lining its path. *2* and *3*, when the tendon moves, it is not necessary (or possible) to lengthen the short dense adhesions that bind the tendon to the fat. The movement is accomplished by the stretching of the surrounding fatty or areolar lining tissues.

material that will move with the tendon. (5) If it is not possible to avoid placing a tendon anastomosis or tendon free graft against raw bone or fascia, a two-stage operation may be necessary with a tendon prosthesis put in place at the first operation to create a pre-formed pathway with a lining that will permit movement.

Tendon Tunneling

Granted the superiority of tendon tunneling when compared to the use of open wounds, why is it not always used? One reason is that it is not always possible. There are times when other structures have to be inspected or repaired that require an open wound. The other reason is that tunneling is blind and surgeons are concerned that without being able to see, they may do some damage or lose their way. These are real dangers, and they have a mechanical aspect so we shall discuss them here.

First Danger: Losing Your Way in the Dark.—Blind people move confidently in their own homes but hesitantly in a strange area. A surgeon who is not familiar with anatomy should not tunnel blindly through a hand or forearm. It is not enough to know the anatomic pictures and diagrams. It is not even enough to know what the inside of a hand looks like. He or she should know how it *feels* to the tunneling forceps. Thus when an operation is planned that he or she has not done before, a surgeon should go first to a fresh cadaver hand, perhaps during an autopsy, and pass the tunneling instrument through a hand, feeling for landmarks and sensing the texture of the inner environment. Then, with the forceps lying in position, the surgeon should cut down and expose the whole length of the instrument in order to confirm its pathway and relationships to adjacent structures.

Second Danger.—There is a danger of causing damage to a nerve or blood vessel or causing adhesions to a structure that has been penetrated or injured by the tunneling instrument.

The art and science of tunneling through tissues is dependent on the sensitivity with which a surgeon can feel the quality of the tissue he or she is penetrating or pushing aside. For a surgeon to be able to calibrate the significance of what is felt, it is important that the tip of the instrument be the same every time. Surgeons who tunnel only short distances will often pick up the nearest hemostat and push it through. They learn nothing by the process.

Instruments for Tunneling.—When we first began to use tunneling seriously we were using the Bunnell tendon-passer and then the Fritschi tunneling forceps. The former (Fig 9–5) has a tapering point that penetrates tissue easily (too easily, we think) and thickens rather suddenly in the middle so that the tunnel is widened before the tendon is pulled through. The Fritschi tunneler (see Fig 9–5) has a well-rounded nose but then tapers, gradually becoming thicker

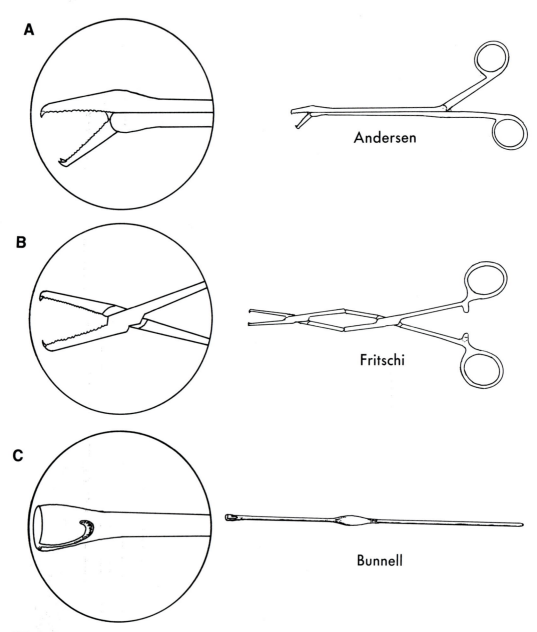

FIG 9–5.
Tendon tunnelers of Andersen,[1] Fritschi,[4] and Bunnell.[5] See text.

over a long distance. Even in good yielding tissue a slowly widening taper creates a lot of drag. It requires so much force to push the thickening shaft through the tissues that a new obstruction at the tip may not be noticed. The same is true of most hemostats, where the handles widen proximally. In contrast, the Andersen tunneler design has a correct rounded nose and a viper head (see Fig 9–5).[1] This ensures (1) that the blunt rounded nose will not perforate or penetrate critical structures, and (2) that the texture and elasticity of the tissue is evaluated by the steeply widening head within millimeters of the tip. Once the viper head has passed, the rest of the shaft is narrow and cylindric so very little further drag is experienced. Thus the surgeon is able to concentrate his or her attention and sensations on the advancing tip and head and have a good idea of texture. Because tunneling instruments need to be of different sizes, the geometry of the head should be constant as to shape and slope. The size of the handles should be in proportion to the size of the tip. In this way the surgeon gets to translate the sense of hand pressure into pressure at the tip (e.g., a big handle with a small tip would give the surgeon a false idea of the texture; it would seem easy to push through what might turn out to be a tough fascia).

We do not insist on the Andersen design. We only insist on a rounded tip and maximum thickness near the advancing end. Also it should not be necessary to open the shaft of the forceps to open the jaws; this results in tearing open the tissues that line the tunnel and causing unnecessary scar. This feature was present in the original *Scheidenschere* or tunneling forceps used by Leo Meyer[3] and illustrated in 1916. This lacked only the wide viper head.

Tunneling forceps should always be advanced empty. They carry the tendon as they retreat.

Damage to Nerves and Blood Vessels.—
It is usually possible to avoid crossing major vessels or nerves at right angles. If a blunt-nosed tunneler hits a vessel at an oblique angle, the vessel will be pushed aside. If a tunnel is to cross a nerve, care should be taken that the nerve comes to lie on the convex side of any curve the tendon may make. A curving tendon always tends to straighten under tension and may then compress any structure crossing on its concave side.

We used to worry about passing too close, or even hugging, a bone or ligament or retinacular band that could be felt with the tunneling nose. It seemed that adhesions might form. We no longer worry about that, realizing that every such structure in a normal limb is in direct living continuity with the connective tissue around it. Although the tunneler will push aside the fibers of yielding areolar tissue, it will not detach or uproot the tissue from bone or ligament or other tendon. We have never experienced crippling adhesions to rigid structures that were felt but which were not penetrated, cut, or scratched by the tunneler. When we tunnel through the carpal tunnel we deliberately feel for, and rub the nose of the tunneler, against the rough bony-ligamentous floor of the tunnel deep to all tendons. This has proved to be a beautiful path and is trouble-free. In contrast, when a strand of tendon is passed from the dorsum of the hand to the lateral band of a finger in a Fowler-type[9] procedure for the correction of intrinsic paralysis, we have found that adhesions form at the point where the interosseous membrane is perforated. This is because we have penetrated (i.e., torn) a rigid membrane. It is the wound in the membrane that makes a point of scar attachment. In the same operation we have not found adhesions between the transferred tendon and the transverse metacarpal ligament that serves as a pulley for the tendon. It is the *absence* of a scratch or wound that makes the proximity of a ligament or bone quite safe in the same bed. Even though we know that the tear in the interosseous membrane may be a point of adhesion, we still use the method because (1) an open dissection would provide even more points of adhesion and (2) those limited adhesions can be managed and even used to some advantage as a temporary tenodesis during reeducation so long as the tendon rests in its optimal position while adhesions are forming.

The Special Danger of Bifurcation.—In the normal human body there are very few bifurcated tendons. Where they do occur, as in the extensor digitorum of the hand or foot, the area contains no structure that could impact the crotch of the bifurcation as it moves distally.

In placing a new bifurcated or many-tailed tendon graft there is a real danger that the surgeon will be content to place each limb of the composite tendon in a good pathway but will forget that a few millimeters distally there is a septum that will be straddled by the two limbs of one tendon and cause an absolute block to further distal movement.

We avoid the term *bifurcation* and speak of a **V** or **Y**. When a single proximal motor divides, we call it a distal **V**. Where two muscles converge to pull on one distal tendon, we call it a proximal ∧. A distal **V** can move proximally but may be blocked distally. A proximal ∧ may move distally but be blocked proximally.

The Unforgivable Sin: Distal V.—When a downhill skier weaves around slalom posts or between small trees, it is bad to bump into them or catch on to them (adhesions), but it is unforgivable to allow one ski to go on one side of a tree and one on the other (distal **V**) (Fig 9–6). The block is absolute

and final. An example that we encounter sometimes is Brand's version of a four-tailed graft for intrinsic replacement. Here we have one motor tendon extended by two grafts, each of which is then split into two final slips. When he started using this system, Brand thought that as long as he split the tendon grafts back to the wrist level they could not be held up by any septum between them, because each slip would have plenty of separate length. Not so.

Reexploration of one or two cases showed that split tendons become united again if they remain side by side while healing. We should have guessed that in the first place. The only way to prevent split tendons from joining up to each other is to have living tissue between them to keep them apart. If that living tissue is a fibrous septum or a bone, then the resting healing position is the most distal that the tendon will ever achieve. The proximal parts of the two tendons will unite with each other and present a distal **V** to an immovable object. The downhill skier will have a tree between the legs. This is a serious cause of absolute block to further motion.

There are two possible solutions to this problem. One is always to have a distal **V** (or split tendon) heal in the most distal position that will be required of it. A distal **V** can always be pulled proximally; it is just limited

FIG 9–6.
If one motor tendon is split distally into two to serve two different insertions, both must always move the same distance at the same time. No surgeon and no skier should ever allow any immovable septum to block the movement of the two limbs at the crotch, the point where the two limbs meet.

distally. This solution may be unacceptable in many cases, because it involves holding the distal joints in an uncorrected position while healing and all adhesions will form in that position, so that the new motor will have to spend its first few weeks stretching adhesions before it can become effective.

The second solution is to make sure that the separate slips are tunneled through separate pathways in a soft tissue area that is known to be free of vertical septa or bones. In the case of a two-tailed or four-tailed graft for intrinsic replacement, the deep part of the proximal fourth of the palm (proximal to all palmar septa) is such a place. In a Fowler transfer on the dorsum of the hand, the metacarpal bones must not be exposed at surgery and the tendon tunnels should diverge from a common origin at the level of the distal edge of the extensor retinaculum of the wrist. A nonsolution is to have a wide dissection, exposing the whole area of bifurcation. This will allow beautiful movement at operation, but the healing of the wound will fill the gap of the V with scar.

Proximal Bifurcation: Proximal ∧.—

Proximal ∧ occurs every time an active motor is tapped into the side of a paralyzed muscle-tendon unit to activate it. For example, in high median palsy the extensor carpi radialis longus (ECRL) may be divided at its insertion, withdrawn proximally, tunneled forward, and brought into the sides of the flexor digitorum profundus (FDP) tendons without dividing the FDP tendons or muscles proximally. Now a proximally open ∧ is created, and free proximal movement may be blocked by any fixed structure that traverses that ∧.

In this kind of case the wide dissection that is necessary for the suture may create the kind of scar that will cripple the intended movement. Therefore we suggest bringing the ECRL out rather high in the forearm and tunneling it with a curved tunneler, so that it joins the profundus proximal to the intended suture and appears in the wound already lined up with the FDP tendons. There will still be a proximal ∧, but it will now be in an unscarred area without a septum to block it.

Proximal ∧s are less trouble than distal Vs, because they may often, and with advantage, be allowed to heal in their most proximal resting position, so that movement is not blocked by anything in the bifurcation. The dangers of a proximal ∧ should encourage surgeons, when appropriate, to cut recipient tendons and use a direct end-to-end suture rather than an end-to-side suture.

USE THE LENGTH OF ADHESION TWICE, TO AND FRO

If a tendon is attached by adhesions to a rigid object, it takes time and effort for every millimeter of length and movement that can be gained. If a goat is tethered to a stake with a 10-ft rope, it will be able to move 20 ft by using the rope first north and then south. If the same rope is tied to a fence that runs east-west, the goat will only be able to move 10 ft north or south.

If a tendon is allowed to rest at either end of its required range in the postoperative plaster cast, it will develop all of its adhesions in that position. The adhesions will ultimately have to grow to equal the length of the whole range of excursion of that tendon. If a tendon rests, while healing, in the middle of its expected range, it will be able to move twice as far, to and fro, for each millimeter that it gains in length. This is worth a lot, both for a goat and for a patient.

The Adhesion-Stretching Muscle

The surgeon should also consider which muscle will most effectively lengthen the adhesions and overcome the drag. Sometimes the transferred motor may be very strong and easy to reeducate, but in other cases it may be easier for the patient to use the opposing muscle more freely because it has not had an operation, has no problem of reeducation, and has no adhesions of its own. It may thus be easier if the transferred tendon is allowed to heal with its own muscle in the partially shortened position so that when movement is first allowed, the opposing muscle, which has not been operated on, may initiate the stretching of the adhesions by pulling the transfer distally.

AFTER OPERATION: CONSERVATIVE MANAGEMENT BY THERAPIST AND PATIENT

The effective mobilization of tendon transfers and grafts depends on the smooth coordination of several factors, each of which depends on the others:

1. Movement of uninvolved joints.
2. Passive movement of involved tendons and joints.
3. Active unresisted movement of hand, followed by active resisted movement.

The effectiveness of the mobilization is dependent on:

1. The surgeon's awareness that a tendon juncture may be only marginally secure.
2. The reliability of the patient in following orders.
3. The avoidance of strong early attempts at mobilization, which may result in violence being done to healing tissues followed by inflammation, edema, and pain.

MOVEMENT OF UNINVOLVED JOINTS

In routine elective surgery, where immobilization lasts less than a month, it is common to include the whole hand in the cast or dressing, even though the operation involves just one or two digits. If immobilization needs to be maintained for longer, it is important to limit the splints and dressings to the affected part and to insist on regular movement of the unaffected digits.

As a minimum it is good to insist on daily active movement of shoulder and elbow, and on the relief of any discomfort that may be associated with the immobilized position of any digit or of the wrist or elbow or shoulder. A potent cause of postoperative stiffness is the pain that alienates a patient from his surgeon and from his own hand. Much of this pain can often be relieved by repositioning painful joints. This is worth doing even if

it means a complete change of postoperative outer dressings and cast.

EARLY PASSIVE MOVEMENT OF INVOLVED TENDON AND JOINTS

This is a controversial subject when it refers to the first and second postoperative week. Early active movement is essential following tenolysis and is useful whenever there has been an operation that does not involve loss of continuity of a tendon. It is probable that early passive movement is beneficial following the repair of a tendon in its own natural bed. It is probably harmful in the case of a tendon free graft, where it is likely to hinder the formation and development of the new blood supply to the graft. In the case of tendon transfers it may be good to allow early (10 days or more) passive movement that moves some affected joints while the tendon is fully relaxed by positioning other joints. However, if early motion is to be permitted, the surgeon must be sure about the strength and integrity of the suture lines.

ACTIVE UNRESISTED MOVEMENT OF THE HAND, FOLLOWED BY RESISTED MOVEMENTS

Three weeks after surgery is the classic time to begin active contractions of the muscles whose tendons have been transferred.

We have tried variations from this time and currently will sometimes shorten it if we have a good patient under the close supervision of a therapist who has a lot of experience. We sometimes insist on a longer period of immobilization if a patient is going to be supervised by an unknown therapist or physician or if the patient's hand is insensitive or if he or she lacks understanding.

In the first week of active exercises the therapist and patient must have several objectives in mind that can all be pursued together or in alternating fashion.

1. General mobilization of stiff joints and restoration of fluid circulation to the hand.
2. Remodeling of tendon adhesions in-

cluding lengthening of areolar tissues to move with the tendon.

3. Encouragement of muscle fibers to grow to required length.
4. Reeducation of transferred muscle-tendon units.
5. Purposeful activity.

General Mobilization

During the postoperative weeks of immobilization, all tissues adapt themselves to the lack of movement. Skin will have lost some of the redundancy normal to the joint position and so will joint capsules.

This may gradually be restored by deliberate range-of-motion exercises, active and passive. The fluid equilibrium will have adapted to the conditions of the postoperative dressings. If the surgeon has applied a "compression" dressing of bulky fluffed gauze, the hand will have been depending on external support for 3 weeks and will look slim and handsome when the dressings are removed. Within 24 hours, however, fluid will fill the spaces in the unsupported hand, especially if it is allowed to be dependent. It takes several days of activity and elevation to get the normal pumping action going again and the normal volume of the hand restored[2] (Fig 9-7).

We strongly recommend the regular use of a hand volumometer for every major hand reconstruction case. Only if it is regularly used will the therapist recognize this routine insurge of fluid and get to know when it is excessive and whether it is being eliminated at an acceptable rate. By keeping a graph of volume it is easy for the patient to see his or her own progress and to realize how much difference it makes when the hand is kept elevated and moving.

The same volume graph gives warning when exercises have been too vigorous and are resulting in tissue strain and inflammatory edema. By comparing the postoperative volume graph with the preoperative volume of the same hand, it will be realized how much residual fluid remains even when the hand begins to appear normal.

The exercises for mobilizing the hand and getting fluids out should be simple, frequent, light, and repetitive. They do not need to involve full range of movement and need not involve the muscles that move the transferred tendons.

Remodeling of Tendon Adhesions

The exercises for remodeling scar and adhesions are most effective if they use the whole range of motion that has already been achieved. They should be planned carefully to hold tension on the restraining tissues and scar and keeping it on the stretch for a while. The tension need not be high. In the early postoperative weeks, tension must not be high because the tendon junctions may be avulsed. So the routine should be to position the hand so that the tendon adhesions are the limiting factor for the movement of a given joint and then tell the patient to pull the joint into flexion (or extension) gently but firmly and *hold it there*. Then the same thing may be done in the opposite direction to develop freedom proximally and distally. If the hand is to be splinted between exercise sessions, it may be splinted in alternating positions to hold the adhesions pulled distally one day and proximally the next.

As soon as the surgeon and therapist are confident of the surgical attachments of tendons and of basic wound healing, the adhesion-remodeling program should move into high gear. Now the patient is encouraged to pull on the muscles in really firm contraction and may even use some passive assist from the other hand so long as force is limited by (1) pain, (2) postexercise swelling as measured by volumometer, and (3) temperature profile of the hand.

At this stage the therapist needs to know what are the actual factors limiting full range of movement. Is the joint stiff because of periarticular tissues that limit joint movement or because of the failure of the tendons to move or of the muscles to contract? The use of a torque-angle curve on a finger joint in wrist flexion and then in wrist extension (see Fig 11–8) will often be a help. For example, if proximal tendon fixation is the primary cause of joint stiffness, exercises should be done with the wrist positioned to hold the tendon at its limit of movement and the finger in a position within its range of free movement. If the finger joint stiffness is pri-

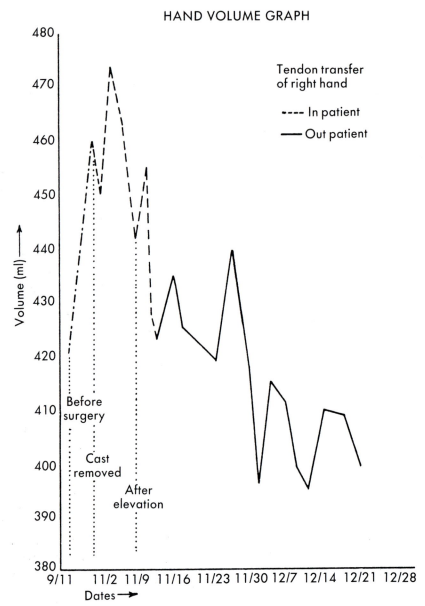

FIG 9–7.
A graph of hand volume gives a simple visible record that the patient can easily relate to and use to check his or her success in avoiding swelling.

mary, the wrist must be positioned to use the free range of tendon movement to move the finger joint to (and then beyond) its limit of range of motion. If these distinctions are not made, then after a time both joint and tendon will be limited to the range of whichever may be the primary cause of stiffness. This is one reason why careful early numeric evaluation is needed to determine primary causes, to ensure that normal joints do not become limited by restricted tendons pend-

ing mobilization of tendons and that free-moving tendons do not become restricted pending mobilization of stiff joints.

Stimulus of Muscle Fibers to Grow

The word *stimulus* here does not refer to electricity but to factors that stimulate growth of fibers in length. As far as we know the only stimulus to grow in length that muscle fibers recognize is the resting tension. If muscles are held *at rest* in the stretched or lengthened position, sarcomeres will be added until the tension is back to normal. Thus if the surgeon has attached a tendon at a tension higher than neutral, that muscle will become longer, but only if and while it is held in the lengthened position.

The implication of this is that (1) the therapist needs to know the intent of the surgeon and the approximate tension of the transfer, and (2) once the tendon junction is securely healed, the hand needs to rest in the position that keeps the muscle on the stretch to be stimulated to grow. For example, if the FCU is used to extend fingers in radial palsy, the wrist may have to be held extended while healing (to relax the tendon junction) but should be splinted in some flexion afterward, to hold the muscle fibers in a lengthened position, except during exercise sessions. The flexed position of the wrist may be alternated by splinting with the wrist straight and the fingers in flexion, to avoid joint stiffness, while the muscle is still kept somewhat on the stretch.

Reeducation of Transferred Muscle-Tendon Units

We sometimes speak of muscle reeducation as if it were the goal of muscle-balance surgery. It is not the end point but a means to an end.

When we help a patient to identify a muscle and to contract it at will, we are teaching a skill that has no real value except as a step toward the natural use of the muscle in its new situation. Once the trick has been learned and used in real-life activities, it should be forgotten and a new pathway will operate from brain to hand.

When we pick up a scalpel between thumb and fingers, we do not need to mutter to ourselves about flexor pollicis longus and adductor pollicis, we just do it. The actual neural pathway we use in real life is a complete bypass of any conscious use of named and identified muscles. If we can train a patient to do his job directly without thinking about which muscle he is using, this is a saving of time and a total gain. It is the method of choice for synergistic transfers (e.g., use of a wrist extensor for finger

flexion). The use of biofeedback for the reeducation of muscles is sometimes useful for difficult cases and when nonsynergists or antisynergists are used. However, success with biofeedback, using electromyogram, wires, lights, and buzzers, does not ensure proper use of the muscle in a real job situation. There is no reeducation that is as good as purposeful activity, and no reeducation is complete until the hand has been observed at the workbench or at needed skills, such as writing, shaving, needlework, or changing a baby's diaper.

There are many ways to build up muscle, but there are none that are as valuable to the patient as the direct use of the muscle in a task that requires strength. When a man wants to compete in body-beautiful photographic contests, he is taught to isolate individual muscles and contract them in isometric exercises in which flexors contract simultaneously with extensors. The body organizes an internal battle of muscle against muscle in which energy is expended with no external work being done. This process is sometimes called *dynamic tension* and it gets results if muscle bulk and muscle bulge is the desired goal. We find it hard to believe that a limb that is trained to behave in these antisynergistic patterns is as good as one that develops its strength in coordinated activity against external stress.

As soon as a therapist is satisfied that a transferred muscle is beginning to work in phase with others, it is time to switch to purposeful activity or at least to coordinated contraction against external loads.

Purposeful Activity

By "purposeful action" we mean that the patient thinks about what he wants to do rather than about what muscles to use. Even so the therapist will select a type of work that uses the muscles and joints that need help.

The ideal purposeful activity is work— wage-earning work. The next best is a hobby, game, or pleasurable activity. The art of the hand therapist, occupational therapist, or recreational therapist is to match the need of the hand operated on to the desire of the one who owns it. The problem is that in an average rehabilitation unit there is a limit to what is available. Woodcarving, sewing, leatherwork, clay modeling, and various active table games should be available, plus room for activities of daily living. Where possible, an open garden or yard may be used for digging, hoeing, weeding, and pruning (and even for harvesting and then chopping vegetables). Other useful garden activities include spraying with pistol-grip bottles or bulb squeezers and picking slugs off lettuce (precision pinch). The rules that Plewes[7] established for injured hands in an automobile manufacturing plant in the United Kingdom are hard to beat.

1. Whatever part of the hand needs protection is immobilized.

2. The rest of the hand, even one digit that projects out of the cast, is actively at work as early as 1 or 2 days after injury or operation.

3. Regular production machines are adapted so that the switch is high, at head level, and can be operated by the injured hand in a cast. The jig is controlled by the uninjured hand, so that the patient's focus of attention is on the good hand and on the work.

4. The patient is paid full wages for his or her work from the start.

It is much less expensive to pay a worker full wages for a period of slower work on an adapted machine than to undergo the long-term expenses of later rehabilitation of the stiff hand and the disgruntled attitude of an alienated worker. The workmen's compensation practices in the United States seem designed to produce stiff hands and unemployable workers. They cost insurance companies vast sums of money that are reflected in premiums that are paid by employees, employers, and finally by the public. Management and labor unions both act defensively and lawyers are the only ones to gain. The very worst practice of all is the one that forbids or discourages the worker from reentering the work force until every claim is settled and no further improvement can be expected.

Work is by far the best therapy for the hand and for the spirit. If it cannot be done in the workplace, the closest imitation of the workplace should be devised.

REFERENCES

1. Andersen, J.G.: The viperhead tendon tunneller, Hand 14:339, 1982.
2. Beach, R.B.: Measurement of extremity volume by water displacement, Phys. Ther. 57:286–287, 1977.
3. Biesalski, K., and Mayer, L.: Die physiologische Sehnenverpflanzung, Berlin, 1916, Julius Springer, p. 206.
4. Brand, P.W.: Paralysis of the intrinsic muscles of the hand. In Rob, C., and Smith, R., editors: Operative surgery, London, 1977, Butterworth, pp. 238–257.
5. Bunnell, S.: Surgery of the intrinsic muscles of the hand other than those producing opposition of the thumb, J. Bone Joint Surg. 24:1–31, 1942.
6. Kessler, I.: The "grasping" technique for tendon repair, Hand 5:253, 1973.
7. Plewes, L.W.: Rehabilitation in hand injuries, Proc. R. Soc. Lond.[Med.], 51:716–717, 1958.
8. Pulvertaft, R.G.: Tendon grafts for flexor tendon injuries in the fingers and thumb, J. Bone Joint Surg. 38B:175–194, 1956.
9. Riordan, D.C.: Surgery of the paralytic hand, Instr. Course Lectures 16:79–90, 1959.
10. Snow, J.W., and Fink, G.H.: Use of a transverse carpal ligament window for the pulley in tendon transfers for median nerve palsy, Plast. Reconstr. Surg. 48:238–240, 1971.

CHAPTER 10

Operations to Restore Muscle Balance to the Hand

This chapter is intended only as a general guide to the way in which surgeons should think mechanically about the restoration of muscle balance after nerve injury or disease. This is not the place to look for a catalogue of options or for a history of the development of techniques. We enter into detail only when it serves to illustrate a mechanical principle. We recognize that every hand is different and that precise mechanical data on the way in which the balance of the hand is maintained while various actions are carried out are not available even for the average hand. However, even the crudest numeric approximations are better than none at all and we hope that the attempts to put numbers to muscle tensions and moment arms will stimulate surgeons to think mechanically as they plan operations and to measure, whenever they can, preoperatively, intraoperatively, and postoperatively. Undoubtedly this will give better results and also better data than are available at present.

We have confined ourselves to figures that have been generated in our own laboratory here at Carville, Louisiana, because it has proved difficult to interpret the figures from earlier literature. It is rarely clear under exactly what conditions earlier measurements have been made. Surgeons should be cautioned about drawing conclusions from the interesting and considerable data of Von Lanz and Wachsmuth,[31] often quoted from

Weber[32, 33] and the Ficks,[16, 17] unless the reader understands German well enough to follow exactly what the measurements mean. For example the term *Arbeitsmöglichkeit*, which means work capacity (of a muscle), is a product of tension capacity and range of required excursion. It is not a simple estimate of strength. Moreover it is listed in most cases in relation to the work capacity of a given muscle *for a given vector* of joint motion, not for the total capability of the muscle. Thus Boyes,[4] quoting directly from Lanz-Wachsmuth,[31] gives an impression (not intended by the authors) that the flexor pollicis longus (FPL) has 12 times the work capacity of the extensor pollicis longus (EPL) or abductor pollicis longus (APL). Actually the APL is as strong as the FPL, maybe a little stronger, but its tension *vector* for a given movement will differ widely from that of the FPL, as will its excursion vector, which is all reckoned in with the figure for *Arbeitsmöglichkeit*.

We welcome comments and rebuttal in relation to this chapter and especially we need actual examples of preoperative and postoperative performance together with intraoperative measurements *of the capability of transferred* motors. We are grateful to Peckham, Freehafer, and Keith for examples in Chapter 13.

Much of the information in this chapter has been derived from cadaver arms and hands. These have been of good fresh-frozen quality, but the majority have been from elderly subjects and we cannot be sure that all muscles retain even their proportional relationship after a period of wasting and generalized atrophy. Gradually we shall obtain more measurements from intraoperative studies and from normal subjects following selective paralysis.

One other source of error relates to the fact that in all normal use of the hand, joints are held stable by some simultaneous contraction of muscles that oppose each other. Thus if a wrist were to be used firmly in flexion, for example, both flexors and extensors would contract, but the flexors would exert the greater moment. The extensors would exert just enough moment to keep the joints in firm congruent contact rather than allow any angular tilting at the joint that might occur if the muscles on only one side were to contract.

We have made no attempt to calculate this stabilizer function in the following studies. For example, we may state that, in order to balance finger flexors, the wrist extensors need X units of extensor moment. Perhaps we should have said that they really need $X + Y$ units, because the wrist flexors will also contract with a moment of Y units, for the sake of stability. Thus the extensors and flexors of the wrist would each provide a moment of Y for wrist *stability*, but the extensors would contract with an extra moment of X units to balance the finger flexors. So that total wrist flexion-extension moment would be:

Wrist extension moment to balance finger flexors = X units.

Wrist extension moment to balance wrist flexors = Y units.

Therefore wrist extensor moment needed for equilibrium = $X + Y$ units

In our calculations we have omitted Y and assumed simply that we need to balance active wrist flexion caused by finger flexors (X) with active wrist extension moment (X).

All of the figures we use are relative. There are no absolute figures or even average figures for strength of muscles. We use the tension fraction (see Tables 12–1 and 12–2) and multiply it by the moment arm and call that the relative moment for that muscle at that joint.

In the face of these various sources of error and speculation we have felt free to use whole numbers most of the time, both because they are more memorable and because the use of many places of decimals might give a false aura of precision.

RADIAL PALSY

As the major effect of radial palsy is at the wrist, we shall first summarize the balance of the normal wrist.

If a wrist is to be maximally stable with no need for finger movement, as when a fighter uses his clenched fist in boxing, both wrist movers and finger movers become wrist stabilizers (Fig 10–1). In such cases the total relative moment available for wrist extension is about 25 and the total relative moment for wrist flexion is 57. Thus the flexor side has about twice as much available moment as it needs for stability. The total relative moment for ulnar deviation is 32. The total relative moment for radial deviation is 32. When finger and thumb movement is primary, the wrist needs stabilization chiefly to position the hand and to hold it against the destabilization of the wrist caused by asymmetric and alternating movement of the digits. In such cases it is mainly the dedicated wrist-moving muscles that are used for wrist stability, leaving the finger-moving muscles free. The total relative moment for dedicated wrist extensors is about 13 and for wrist flexion is 22. Radial deviation is 17, and ulnar deviation is 23.

When all the fingers and the thumb are to be maximally flexed for grasp, they exert a total relative flexion moment at the wrist of about 33. This is in excess of the available (dedicated) wrist extension moment. Thus a free hand can never use its full potential for grasp, because it would force the wrist into flexion. However, the full power of grasp can be used when the hand holds the branch of a tree and supports the body weight hang-

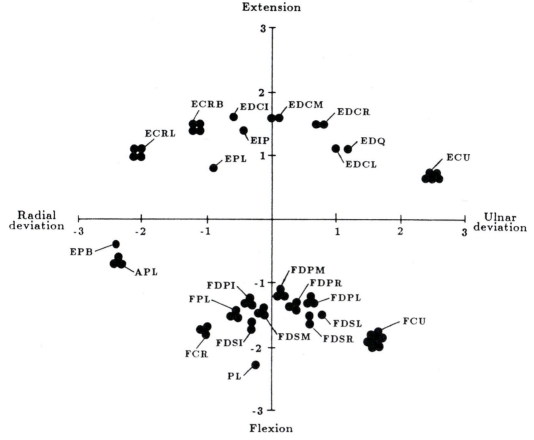

FIG 10–1.

This is not an anatomic diagram; it is a simplified mechanical statement of the capability of each muscle to affect the wrist joint. The positions of the tendons in relation to the axes of flexion-extension and of ulnar-radial deviation represent their moment arms at the wrist. The number of *circles* in each cluster is an indication of the tension capability of that muscle-tendon unit rounded off to the nearest whole number. *APL* = abductor pollicis longus; *ECRB* = extensor carpi radialis brevis; *ECRL* = extensor carpi radialis longus; *ECU* = extensor carpi ulnaris; *EDCI* = extensor digitorum communis (index); *EDCL* = extensor digitorum communis (little); *EDCM* = extensor digitorum communis (middle); *EDCR* = extensor digitorum communis (ring); *EDQ* = extensor digiti quinti; *EIP* = extensor indicis proprius; *EPB* = extensor pollicis brevis; *EPL* = extensor pollicis longus; *FCR* = flexor carpi radialis; *FCU* = flexor carpi ulnaris; *FDPI* = flexor digitorum profundus (index); *FDPL* = flexor digitorum profundus (little); *FDPM* = flexor digitorum profundus (middle); *FDPR* = flexor digitorum profundus (ring); *FDSI* = flexor digitorum superficialis (index); *FDSL* = flexor digitorum superficialis (little); *FDSM* = flexor digitorum superficialis (middle); *FDSR* = flexor digitorum superficialis (ring); *FPL* = flexor pollicis longus; *PL* = palmaris longus.

ing from the branch. The balance of the wrist is maintained, in this case, by the weight of the body suspended from it. The same is true when carrying a heavy suitcase. Gravity stabilizes the wrist so all the finger flexors may contract maximally.

If there is need to extend all digits forcefully, the long digital extensors can exert a total relative moment of 12 at the wrist, which can easily be matched by dedicated wrist flexors.

A very common action of a normal hand is when it grasps the handle of a hammer or axe and swings it downward into ulnar deviation with the forearm midprone. This action is followed by the lifting of the hammer or axe into radial deviation and then by continued alternation between ulnar and radial deviation. If the ulnar-dorsal and radiodorsal quadrants of wrist extensors were each to relax in turn when the other contracted, the available wrist extension moment would be variable, only half being used at a time. In fact the radiodorsal would be 10 and the ulnar-dorsal (in the midprone position) would be only 3, because the extensor carpi ulnaris (ECU) in the midprone position has a very small moment arm for wrist extension. Since active ulnar deviation with the fist clenched demands a strong ECU action, and since the ECU lies close to the flexor-extensor axis when the forearm is fully pronated, it follows that hammering must be difficult and weak with a pronated forearm. (It is; try it.) In fact, in many such reciprocal actions the flexor carpi ulnaris (FCU) and ECU act together and alternate with the extensor carpi radialis brevis (ECRB) and longus (ECRL). On the downstroke there is a diminished wrist support for grasp.

A glance at Figure 10–2 shows the total imbalance that results from radial palsy. No active tendon crosses the wrist dorsal to the flexor-extensor axis. Any active flexion of the fingers must result in wrist flexion, which leaves the finger flexors using the weakest part of their muscle-tension curve. The hand is almost useless.

The first priority of the surgeon is to place a strong muscle dorsal to the flexion-extension axis, so that the fingers and thumb can flex strongly again. The second priority is to consider ulnar-radial stability. If the first transfer is to be delayed for any reason, the wrist must be supported in a splint, to allow continued use of the hand and to prevent progressive shortening of the muscle fibers of all flexors.

Almost all hand surgeons have used the pronator teres as the motor for wrist exten-

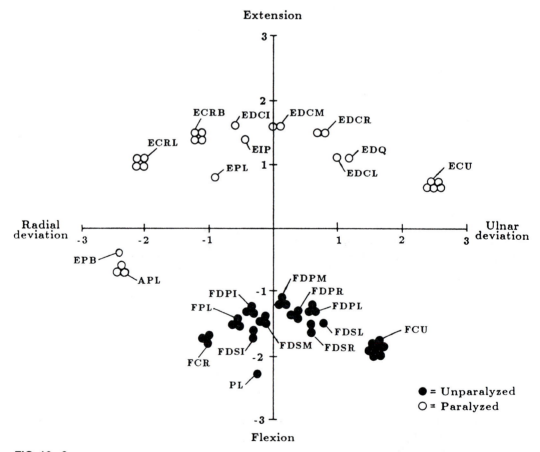

FIG 10–2.
Radial palsy: effect at the wrist. In this diagram each tendon that crosses the wrist is represented by a group of circles, each of which denotes 1.0 unit on our scale of relative tension (see Chapter 12). We have used the nearest whole number. Each tendon lies in its mechanical relationship to the axes of flexion-extension and of radial and ulnar deviation. The *black circles* are normal muscles and the *open circles* are paralyzed muscles (see Fig 10–1 legend for abbreviations).

sion in radial palsy. Such unanimity suggests that the choice has been good. The tension fraction of the pronator teres is 5.5 compared to 12 for the preparalysis total of wrist extensors. The total tension of all radial-supplied muscles that cross the wrist is 42% of the total of all muscles crossing the wrist. So the total available strength of all muscles that cross the wrist after radial palsy is only 58% of the total before paralysis. For perfect balance after radial palsy, all action would have 58% of the moment they had before radial palsy. Thus the pronator teres is not so far short of the tension needed for balance of the wrist at this reduced level. Moreover if the pronator teres is put into an insertion such as the ECRB with its large moment arm for wrist extension, its total relative *moment* will turn out to be 52% of the total moment of the normal wrist extensors, some of which have small moment arms.

However, although there is agreement that the pronator teres is the right muscle, not all surgeons agree about the insertion, and here is where a little mechanics should help. Consider some of the alternative suggestions for insertion of the pronator teres.

1. The pronator teres is inserted into both the ECRL and the ECRB.[21] This is a yoke insertion using both tendons on the radial side of the axis of deviation. This has often been recommended, perhaps because the tendon of the ECRL lies temptingly close beside the tendon of the pronator teres. It results in a strong link between extension and radial deviation. If extension without deviation is attempted, the ECRB becomes slack and only the ECRL remains effective (see Fig 5–22). This forces the wrist into radial deviation every time it is extended. When the wrist is flexed, this moves the ECRB farther distally than the longus, so the return to extension starts with the effective pull of the brevis, and then reverts to the weak, radially deviating pull of the longus as the wrist extends back to the position at which the surgeon first sutured the two insertions. We can think of only one reason why this particular double

insertion has survived so long and is still used by some reputable surgeons. It must be that the two insertions have been attached at equal tension while the wrist was held almost fully extended. In such a case the act of bringing the wrist back to neutral would slacken the ECRL, and leave the brevis as the only effective insertion for almost the whole of the range of motion of the wrist. Done in this way, such a double attachment would be, in effect, a single attachment to the brevis, and that is more reasonable, (but see the next paragraph). We hope that this fuller description of the mechanics of double insertions will help to lay the longus-brevis insertion finally to rest.

2. The ECRB is used as the only insertion. This gives a good moment arm for extension but it does have a significant radial deviation moment too, and there is no ulnar-deviating extensor moment to balance it. It is no use expecting the FCU to restore radial-ulnar balance because the FCU cannot ulnar-deviate without also exerting significant flexion moment. Thus each time the FCU is used to add ulnar-side stabilization, the wrist would lose its essential extension support. A hammer or axe might fly out of the hand each time an ulnar swing took place, because the wrist could not ulnar-deviate without flexing and this would loosen finger flexion.

   a. A possible solution to the excess radial deviation when the pronator teres is attached to the ECRB would be to link the ECU and the FCU into a yoke, with two insertions and a single motor, to make the FCU into a pure ulnar deviator (Fig 10–3). However, it would be a pity to sacrifice very useful ulnar flexion, when the problem can be solved another way.

3. A yoke attachment is made of the pronator teres to the ECRB and the ECU insertions. This would balance radial-ulnar deviation, but the ECU has a very small moment for extension and would limit the total extension moment

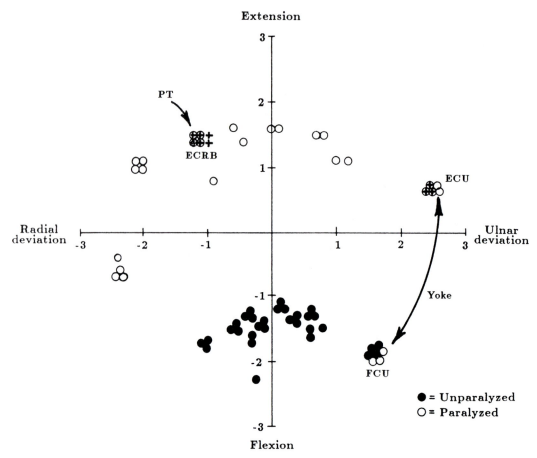

FIG 10–3.
Radial palsy: one way to maintain some radial-ulnar balance at the wrist when the pronator teres is attached to the extensor carpi radialis brevis as the only wrist extensor. (*Note:* When normal muscle *[black circle]* is transferred to the site of paralyzed muscle *[open circles]*, the *crosshatched circle* indicates activation of paralyzed tendon, while the original tendon becomes an *open circle.*)

of the pronator teres (as the ECRB did in *1* above). *This is not recommended.*

4. The pronator teres may be attached by a yoke insertion between the ECRB and a tendon graft in situ to an ulnar-side insertion that has a moment arm for extension similar to that of the ECRB. There are two ways this can be done:

   a. Attach the pronator teres to the ECRB and the ECRL together proximally. Then detach the ECRL from its insertion, pull it out proximally, and tunnel it to the base of the fourth metacarpal where it can be attached to dorsal ligaments or bone or both (Fig 10–4).

   b. ECU tendon may be proximally detached from its muscle fibers, and pulled out distally at its metacarpal

insertion. Then the proximal end may be tunneled obliquely across the forearm to join the ECRB at the point where the pronator teres is attached. It will be found that the rerouted ECU tendon will have a reduced ulnar deviation moment arm and an enhanced extensor moment arm, making it comparable to the ECRB on the other side.

RECOMMENDATION

We suggest that *4a* is a good choice and makes for lateral stability; *2a* is acceptable and has the advantage of retaining active ulnar deviation without threatening wrist extension.

If a tendon transfer for wrist extension is

done early, to restore activity to a hand in which it is still possible that some radial nerve recovery may take place, then I suggest transferring the pronator teres to the ECRB only, with the ECRB in continuity. This will give good usable wrist extension. Later, if the nerve recovers, nothing will have been lost. (The pronator teres remains as an effective pronator also, since its relationship to the axis of pronation and supination has not been altered.) If the nerve does not recover, an ulnar-side yoke (see Fig 10–4) may be added later, especially for a manual worker who recognizes the need for it.

FINGER EXTENSION IN RADIAL PALSY

A very common pattern of transfer has been that of the FCU to all finger exten-

sors,[21] and the palmaris longus to the EPL (Fig 10–5).

The FCU has a tension fraction of about 6.5 and this is more than adequate for finger extension. *However, there are two objections to this transfer.*

1. The muscle fibers of the FCU are very short. They are just adequate for wrist movement, but inadequate for finger extension plus wrist movement (see Fig 12–11,B).
2. It leaves no prime ulnar deviator for the wrist. Ulnar *stabilization* of the wrist is of great importance and ulnar *deviation* is a very significant movement and should not be sacrificed lightly.

This was dramatically illustrated for one of us (P.W.B) when he was giving the Bun-

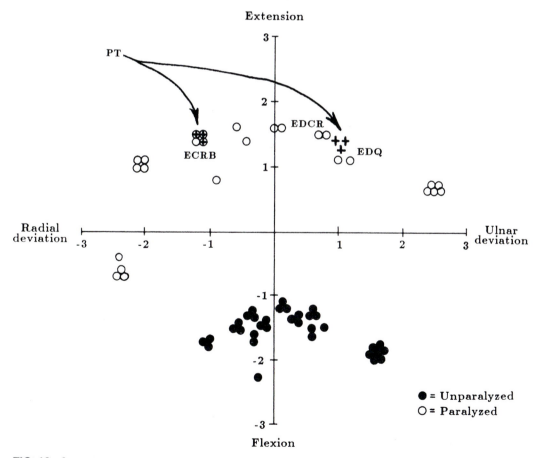

FIG 10–4.
Radial palsy. Balanced wrist extension achieved by transfer of the pronator teres *(PT)* to the extensor carpi radialis brevis and base of the fourth metacarpal (using the extensor carpi radialis longus as a graft in situ). (*Note:* When normal muscle *[black circle]* is transferred to the site of paralyzed muscle *[open circle]*, the *crosshatched circle* indicates activation of paralyzed tendon, while the original tendon has *open circles*.)

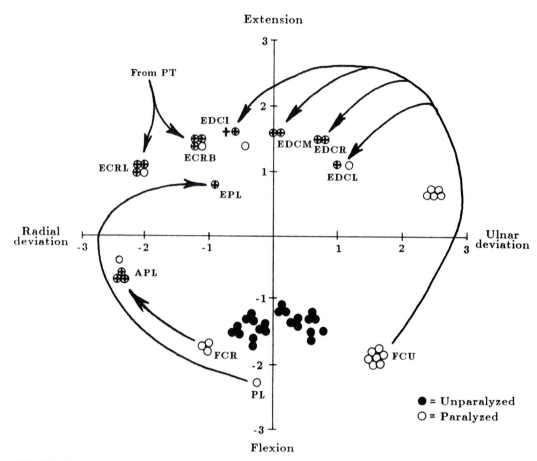

FIG 10–5.

Common pattern of tendon transfer for radial palsy. We do not recommend this. Note overwhelming radial side dominance. (*Note:* When normal muscle *[black circle]* is transferred to the site of paralyzed muscle *[open circle]*, the *crosshatched circle* indicates activation of paralyzed tendon, while the original tendon has *open circles*.)

nell memorial lecture in San Francisco some years ago. He had been speaking about the importance of keeping the FCU in situ in radial palsy and then braced himself for the clinical session in which he was supposed to comment on cases presented by local surgeons. The first case happened to be a policeman who had a radial palsy that had been successfully treated by transfer of the pronator teres for wrist extension and the FCU for finger extensors. He demonstrated it proudly to the audience.

The policeman was asked if he had any serious residual problems. Yes, he had been transferred from patrol duty, which he loved, to a desk job at the office. The reason? He could no longer use his right fist to defend himself—it would twist up into radial deviation each time he hit anybody. He

had no strength for a karate chop, and he could not shoot straight with his handgun because he could not hold the gun barrel down. The barrel jumped upward as he fired. The bullet would go above the target. The policeman graphically demonstrated his problem and then was puzzled at the gale of laughter from the audience, who may have thought I had planted this case just to prove my point about the importance of the FCU in its own place. All three of this policeman's problems would have been solved if the FCU had been left alone.

The real point was that the surgeon had never asked the critical question of his patient. The critical question is *always,* "Can you do what you want to do?" not simply, "Can you do *this* movement or *that?*" There are cases in which the FCU loss may be acceptable, if strength and stability are not so important.

Obviously this transfer works, or it would not have been done so often and for so long and by well-known surgeons. However, it cannot produce full finger extension without limiting wrist extension or else it limits finger flexion in wrist flexion. A patient will be able to extend his or her finger while the wrist is a little flexed and flex the fingers when the wrist is extended. This may be no big problem if no better alternative is available. But in fact there are two better alternatives: (1) The use of the flexor carpi radialis (FCR) (tension fraction 4.1 and with longer muscle fibers) taken around the radial border of the forearm or through the interosseous membrane to extend the fingers, or better still, (2) the use of one or two tendons of the flexor digitorum superficialis (FDS) first suggested by Codivilla[14] (1899) and now recommended by Boyes.[4] In this transfer the FDS middle and ring are taken through windows in the interosseous membrane. Boyes[3] uses one tendon to extend the three ulnar fingers and the other to extend the index finger and thumb. We prefer to use the FDS (middle) to extend four fingers and keep the thumb independent, using the palmaris longus if it is present and the FDS (ring) otherwise (Fig 10–6,A).

There would seem to be three objections to the use of finger flexor tendons to serve as finger extensors. One is that it requires a change of phase. A finger flexor becomes a finger extensor. However, this does not seem to be a severe problem although it might be in an old person. Finger muscles are highly retrainable, especially for an action such as simple extension that stands out as an all-or-nothing phenomenon.

The second objection is that finger flexion is weakened. However, in radial palsy the whole hand is 42% weaker. The finger flexors are, on balance, overpowered and cannot use their full strength when wrist extension is significantly reduced. Therefore the loss of one or two superficialis tendons is relatively inconsequential, especially with intact intrinsic muscles for control of individual finger movements.

The third objection is to the use of any flexor superficialis tendon except from the middle finger (see Chapter 12). Figures 12–62 and 12–63 show how complex the flexor superficialis is. Only the muscle to the middle finger is wholly independent. The index and little finger muscles have a common proximal muscle belly and two distal muscles that have short fibers arising from a common intermediate digastric tendon. The ring finger superficialis has a variable proportion of its fibers independent (commonly about two-thirds), but the rest are short, and arise from the intermediate tendon proximally, and will therefore severely limit independent movement of the tendon. Two thirds of its fibers will be long and independent, and one third short and moving with the flexion of the index and little fingers. Thus the ring finger superficialis should rarely be moved for a dorsal function. If it has to be, it might be good to divide its fibers from the intermediate tendon. (*Note:* This objection to the transfer of superficialis tendons does not apply when the transfer keeps to the palmar side of the wrist.)

THUMB EXTENSION BY ABDUCTOR POLLICIS LONGUS

True abduction of the thumb is a function of the median nerve via the abductor pollicis brevis (APB). The three thumb extensors are all supplied by the radial nerve, the EPL, the extensor pollicis brevis (EPB), and the APL. The last muscle should really be called the *extensor ossis metacarpi pollicis* (its previous name). Its important function is to hold the thumb metacarpal in extension as a base for the arching of the thumb in pinch. This is an important action.

In radial palsy this tendon may be powered by transfer of the FCR. This muscle is comparable to the APL in tension and fiber length. The wrist-flexing moment of the FCR is not lost in the transfer because the APL has a flexor moment at the wrist that is sometimes 50% of the FCR.

If the APL and the FCR tendons are divided at the same level and the distal APL stump is angled forward to meet the radialis end-to-end, the final balance is excellent for

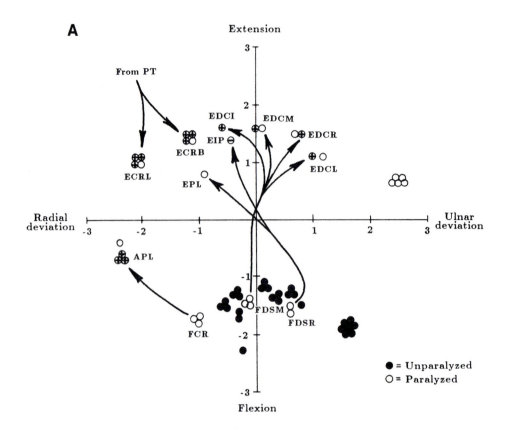

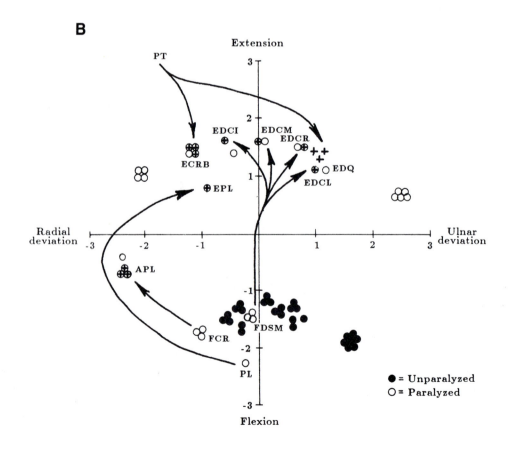

both wrist flexion and thumb abduction-extension. The FCR will have extra radial deviation moment, but this will not matter if the FCU remains in situ (Fig 10–6,B).

ULNAR PALSY

The only forearm muscles lost in high ulnar palsy are the FCU and the flexor profundus to the ulnar half of the hand (Fig 10–7,A). The division of flexor profundus between median and ulnar nerves is quite variable, and there are significant connective tissue connections between the profundus fibers to the middle, ring, and little fingers. Therefore all fingers usually flex together in ulnar palsy and most surgeons do nothing to compensate for the weakness of the ring and little fingers. This is probably justifiable in most cases. Sometimes, however, the flexor superficialis to the little finger is extremely small, and if the profundus tendons have some degree of independence between fingers, the little finger and even the ring finger may be very weak indeed. In such cases the profundus tendons should be exposed in the distal forearm and roughened and sutured side-to-side, middle finger to ring, and little fingers (Fig 10–7,B). This does not add strength, but distributes it and restores the stability of the grip by broadening its base.

The other forearm weakness is FCU. This is a great loss. It does not result in any deformity, nor does it limit the range of any free movement. However, the patient has a sense of *weakness*, especially in the common ulnar-deviating movements. The FCU is the strongest muscle that crosses the wrist and its loss is bound to be significant. Wrist flexion is weakened by more than 50% and so is ulnar deviation. Patients who have a good palmaris and who need strong ulnar deviation (carpenters) may appreciate transfer of the FCR to the FCU, leaving the FCR func-

tion to be handled by the palmaris or by a transfer of the brachioradialis.

Most surgeons do nothing about the paralysis of the FCU, and we do not disagree with this policy in most cases.

PARALYSIS OF INTRINSIC MUSCLES IN ULNAR PALSY

Because this does result in deformity, most surgeons will recommend surgical correction to balance the hand. However, because it is the deformity rather than the functional imbalance that prompts the correction, surgeons are sometimes tempted to correct the deformity without correcting the functional imbalance. The *deformity* is called clawhand and it occurs because the extensor tendons overact in an attempt to extend the fingers. Because the paralyzed intrinsic muscles are unable to stabilize the metacarpophalangeal (MCP) joints on the flexor side, the extensor muscles hyperextend the MCP joints and, in that position, are unable to extend the interphalangeal IP joints. The result is clawing (Fig 10–8).

This deformity may be corrected by anything that limits extension of the MCP joints, thus allowing the long extensors to extend the IP joints. Tenodesis, capsulodesis, bone block, or even a scar from some wound in the palm—any of these may limit MCP extension and prevent the clawing of the fingers (see Static Corrections below). Some hands are naturally so thick and stable that the MCP joints never hyperextend; such hands do not develop much clawing even though they have ulnar palsy. Others do not show clawing when first paralyzed because their palmar tissues are tight enough to hold the MCP joint stable. However, after a few months or years they develop clawing because the unopposed extensor pull has stretched the volar plate and skin and other soft tissues. Almost all cases of ulnar palsy have a *progressive* deformity. Mild clawing in the first month becomes se-

---

FIG 10–6 (facing page).
A, tendon transfer for radial palsy, as suggested by Boyes.[3] **B,** our own preferred pattern of transfer for radial palsy. (*Note:* When normal muscle *[black circle]* is transferred to the site of paralyzed muscle *[open circle],* the *crosshatched circle* indicates activation of paralyzed tendon, while the original tendon has *open circles.*)

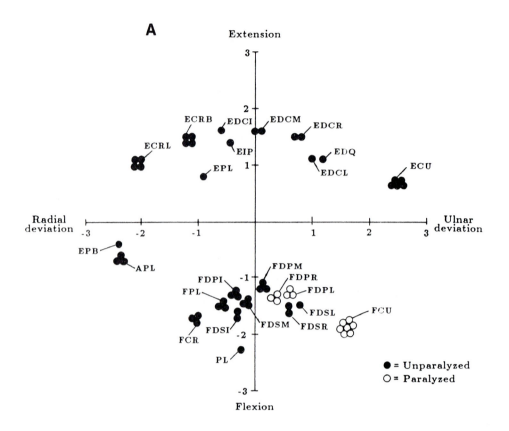

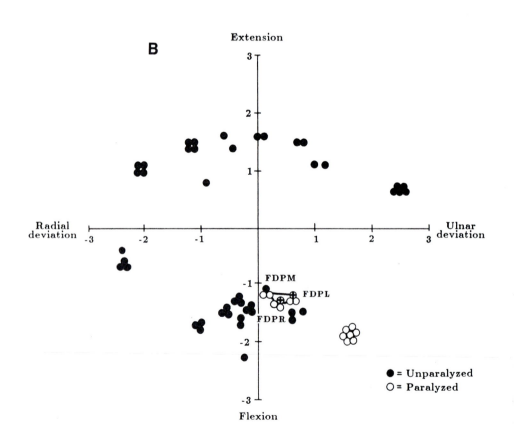

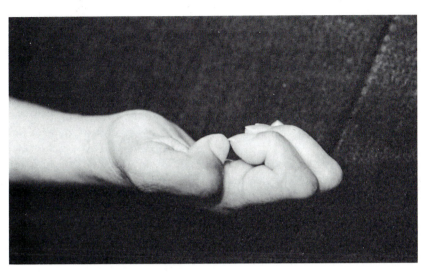

FIG 10–8.
Clawhand. Ulnar and low median palsy.

vere after several months or years simply because of the stretching of palmar soft tissues. There is often delay in correcting the clawing, because the deformity is not apparent at first.

This delay is unfortunate, because it lulls both patients and surgeons into thinking that the problem is less severe than it really is. The hand does not look bad, so nothing is done. It becomes deformed so slowly that there is no sudden moment of truth. The situation is neglected until the fingers are not only clawed but stiff in the clawed position. Correction is then much more difficult. A complicating factor in this deterioration is that the surgeon often waits for ulnar nerve recovery following nerve suture. The intrinsic muscles almost never recover adequately after repair of a high (above-elbow) ulnar nerve division, but surgeons often hope they will. Thus they wait and wait. For a nerve injury at elbow level a surgeon may wait a year or even 2 years, to be sure the muscles will not recover. Then they go in and operate but cannot get the IP joints mobilized or the stretched MCP joint tissues rebalanced.

Thus it is necessary to speak loudly and

clearly on this subject. The main defect in ulnar palsy is *functional*, not static. It is not the clawing that matters so much as the sequence of closure of the hand and the distribution of pressure on the fingers. The functional defect may be present from the beginning (though it also gets worse by waiting) and it should be corrected early in all cases of ulnar nerve injury except in distal forearm nerve lesions where the division of the nerve into definable sensory and motor bundles makes recovery of intrinsic muscles more probable. No static procedure will correct the functional defect. As soon as the fingers begin to flex, any MCP capsulodesis or single-joint tenodesis becomes slack and the finger is at once unbalanced.

THE FUNCTIONAL DEFECT

When the intrinsic muscles are paralyzed, the balance of each finger is upset. The MCP joint has lost flexors and the IP joints have lost extensors. Thus when the hand is actively closing, the MCP joint lags behind and the IP joints flex ahead. When the hand has closed on the object, it is the

FIG 10–7 (facing page).
Ulnar nerve loss. **A,** effect at the wrist. **B,** the profundus of the middle finger is attached in the forearm to the profundus of the ring and little fingers to broaden grip strength. (*Note:* When normal muscle *[black circle]* is transferred to activate paralyzed muscle *[open circle]*, the *crosshatched circle* marks the newly activated tendon while the previously normal tendon insertion is shown with *open circles.)*

fingertips that make contact and that hold it. When a cylindric object is held tightly by a normal hand, the force of grasp is exerted through the whole palmar surface of the fingers and hand (Fig 10–9). In intrinsic paralysis the same force is exerted only through the fingertips and metacarpal head areas. For a given force the pressure may be up to ten times larger because the area is often only one tenth of what is normally used. Peak pressures at the skeletal interface of the fingertip may be even larger because the *bone area* of the end of the terminal phalanx is really only about (2 mm × 5 mm) 0.1 cm². Since ulnar nerve paralysis is often sensory as well as motor, this chisel-ended bone is the point of application of force through insensitive skin. The classic picture of such a hand shows a row of ulcers or scars just in front of the fingernails on the tips of each finger (Fig 10–10).

Thus the functional correction that is needed in intrinsic palsy is for an *active flexor* for the MCP joint. This must move the proximal segment into flexion ahead of the other segments. The same tendon transfer that flexes the proximal segment may, with advantage, serve to extend the distal segments. This is not essential to straightening the fingers (this can be done by the extensor digitorum), but it is used to delay or hold back the flexion of the distal joints so that the proximal segment can move ahead (Fig 10–11).

Because of the importance of the functional balance after paralysis of the intrinsic muscles, we think that static corrections are inadequate and should not be used except when there is a severe shortage of available motors, as in multiple paralysis. In such cases an active two-joint tenodesis may be used (see below).

The intrinsic muscles, in addition to acting as primary flexors of the MCP joints and as extensors of the IP joints, have two other functions.

The first function is lateral deviation.

Control of lateral-medial deviation requires eight muscles, one on each side of each finger. Bunnell[11] tried to achieve it by splitting four superficialis tendons into eight strands and using some as flexor-adductors and some as flexor-abductors. He later abandoned this as a wasteful use of muscles, and we do not think that anybody now tries to obtain independent deviation of fingers in ulnar palsy. The index finger alone is sometimes worth providing with a separate radial deviator (first dorsal interosseus) in addition to an ulnar-deviating slip from a common tendon transfer that it shares with radial-deviating slips to all the fingers (Fig 10–12), also see below.

The second function is MCP joint stabilization. There is always a trade-off between mobility and stability in joints. The MCP joints achieve freedom of movement in two planes at the expense of a very shallow cup at the base of the proximal phalanx. The normal stability of all joints in the body depends on joint architecture plus muscle action. Capsules and ligaments are there mainly to prevent subluxation when unexpected strains catch the stabilizing muscles by surprise. The interossei assist in stabilization of the MCP joints by holding the phalangeal base firmly on the metacarpal head. If stabilizing muscles are paralyzed and if the bony configuration of the joint is shallow, the ligaments tend to become stretched by repetitive stress and partial subluxation may occur. This is seen most markedly in rheumatoid arthritis at the MCP joint when the capsules and ligaments are weak and the intrinsic muscles are also weak and wasted. We mention this stabilizer function here only to point out that there is value in having muscles that cross near the axis of a joint. Most attempts to increase moment arms by bowstringing a tendon across a joint must add to the difficulty of keeping the joint stable.

Another functional defect in ulnar palsy is the inability to flex and oppose the fifth

---

FIG 10–9 (facing page).
A, normal hand grasping a cylinder. The area of skin contact is marked in black. **B,** claw hand grasping a cylinder. The area of contact is limited to the fingertips and the metacarpal heads.

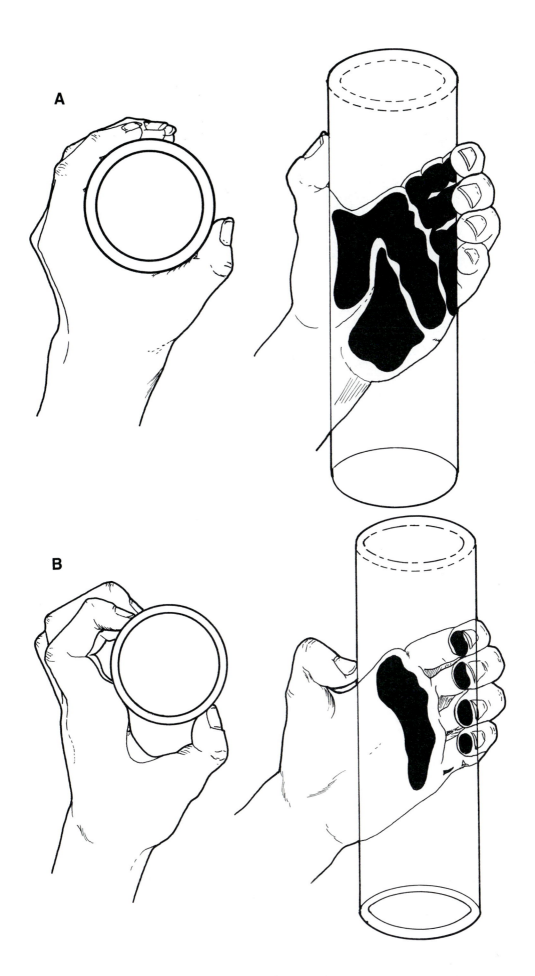

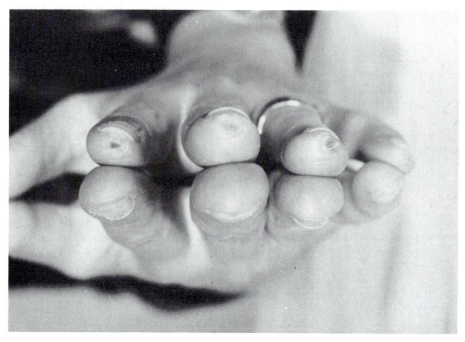

FIG 10–10.
This patient had loss of sensation from Hansen's disease. The little ulcers and scars are in the most characteristic position for a patient with functional clawing. The high stress occurs just in front of the fingernails at the end of the phalanx.

metacarpal. The opponens digiti quinti, (ODQ) is a small muscle, but its action in cupping the palm is powerfully backed up by the whole hypothenar muscle mass. In flexing and opposing the fifth metacarpal the same muscles also help the ring finger to follow the curve of the metacarpal arch and place the palm and fingers in position to grip rounded objects securely.

Patients with ulnar palsy have a flat metacarpal arch, and they feel weak in grasping any large object. There is also a weakness in

the use of the hand as a whole for exerting torque on an object such as a hammer or fishing rod, or for transmitting supination torque to a large faucet handle (Fig 10–13,A and B.) In all these actions the torque is produced by what engineers call a "couple." This is when two forces act on the same object in opposite directions. The torque is measured by the magnitude of the forces multiplied by the distance between them. In the case of the hand, major supinating torque and ulnar-deviating torque both use

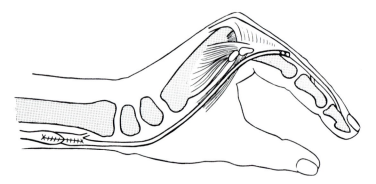

FIG 10–11.
Diagram of tendon transfer to correct intrinsic-minus hand. Note that the transferred tendon or graft passes in front of the transverse intermetacarpal ligament and into the lateral band. *(From Milford, L.: The Hand, ed. 2, St. Louis, Mosby–Year Book, Inc, p 203. Used by permission.)*

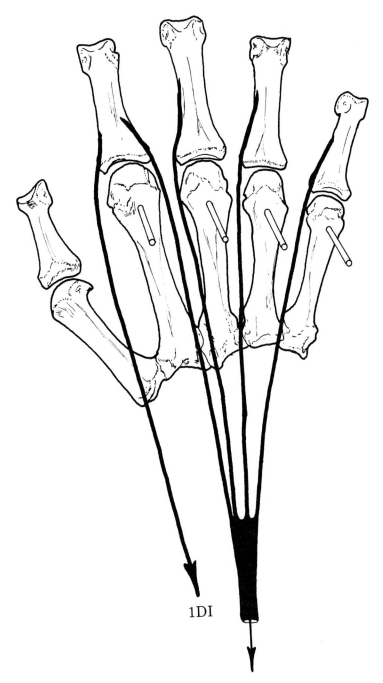

FIG 10–12.
For correction of ulnar paralysis a single motor may be used for the radial side of the middle, ring, and little fingers and the ulnar side of the index finger. A separate motor may be used for the radial side of the index finger if necessary. *IDI* = a substitute for the first dorsal interosseous.

about a 10-cm lever between the fifth metacarpal and little finger on one side, and the base of the thumb on the other. In ulnar palsy the effective lever may be only half that, because the fourth and fifth metacarpal and ring and little fingers are weak and may

be pushed backward, leaving only the stable third metacarpal to provide the other half of the couple. Unfortunately the ordinary operations for correction of intrinsic muscle palsy do not deal with this problem. Any operation (such as Fowler's)[24] that has a trans-

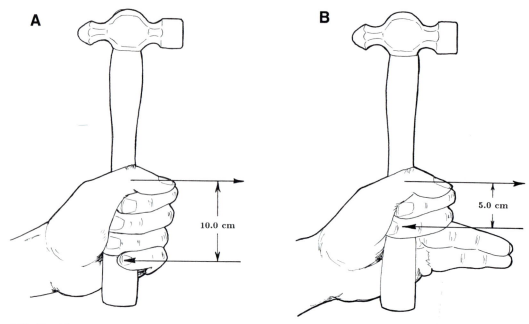

FIG 10–13.
A hand can exert torque on an object that it holds. The more common forms of this are by ulnar deviation (using a hammer or fishing rod) or by supination (screwing action on transverse handle of stopcock or large-handled screwdriver). In either case the breadth of the hand determines the leverage of the *couple*. However strong the index and middle fingers may be, the weakness of the ring and little fingers **(A)** and the mobility of their metacarpals result in the loss of 50% of the torque **(B).**

fer coming from the dorsum of the hand through the fourth intermetacarpal space actually makes the problem worse and actively reverses the metacarpal arch. Such a hand is ugly as well as weak.

Bunnell[12] advised a tendon T operation to correct this, using a transverse tendon graft across the palm from the base of the proximal phalanx of the thumb to the fifth metacarpal. A motor was attached to the center of this graft, pulling the two ends toward one another and converting the T into a Y. This is unlikely to serve thumb and little finger equally well since the thumb requires much more excursion.

If the flat or reversed metacarpal arch is recognized as a severe defect, it would justify a tendon transfer of its own, such as a metacarpal flexor for the fifth ray.[2]

ALTERNATIVE PATTERNS OF TENDON TRANSFERS FOR ULNAR PALSY

The first question relates to the number of fingers to be operated on. The ring and little fingers usually have a loss of all intrin-

sic muscles, whereas the index and middle fingers may retain median-supplied lumbricals. Thus it is common to have obvious clawing in the ring and little fingers and less obvious or absence of clawing in the index and middle fingers. A finger should be regarded as functionally intrinic minus if (1) the patient cannot hold the position of full MCP flexion with full IP extension or (2) the proximal interphalangeal joint (PIP) flexes into a claw position when a slight backward push is applied to the proximal phalanx or if the act of pinching strongly against the thumb results in PIP flexion, even when acute IP flexion of the thumb is prevented by splinting the thumb IP in extension. These two tests reveal what we call latent clawing[1] (Fig 10–14).

Depending on the results of these tests, the surgeon may decide to perform an intrinsic replacement on two or more fingers. In cases of complete ulnar palsy it is usually best to correct all four fingers even if only two are clawed, because the other two are likely to claw later.

If only two fingers are to be operated on,

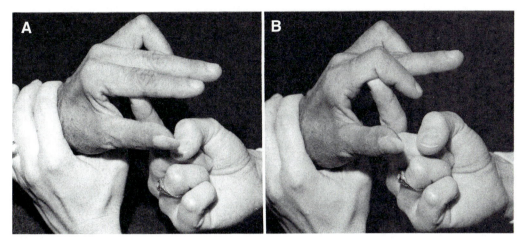

FIG 10–14.
A, a patient with ulnar palsy and obvious clawing of two fingers is still able to hold the index and middle fingers in the lumbrical position when not subjected to stress. The examiner's right index finger is not applying any pressure. **B,** the examiner now applies a little force, using the index finger to push backward on the proximal segment of the patient's index finger. At once the proximal interphalangeal joint flexes, even though the patient tries to hold it straight. This is called *latent clawing*. It suggests that all four fingers need an intrinsic replacement.

we commonly use one of the weaker motors, such as the palmaris, but in other respects the operation is the same as is now discussed for four-finger replacement. There are four levels of variability in the choice of operations.

1. Choice of motor.
2. Relationship to wrist joint.
3. Pathway in hand.
4. Method of insertion.

The only factor common to every operation for intrinsic paralysis is that the tendon transfer must lie on the palmar side of the axis of the MCP joint and on the palmar side of the intermetacarpal ligament.

Choice of Motor
The required excursion of the lumbrical tendons at the MCP joint varies with each finger but is about 15 mm for 90 degrees of flexion. An additional 5 mm may be needed for IP extension. The transferred tendon needs to move about 2.0 cm. Almost any muscle that crosses the wrist except the FCU should be able to provide for this, so the only muscle variable we need consider is strength (tension fraction).

The average tension fraction (about 2.0) of the interossei is high. However, they have a small moment for flexion of the MCP joint because they cross close to the axis. The lumbricals are much weaker (0.2) but have a larger moment arm. The flexion moment of the combined lumbrical-interosseus complex to a finger would probably be of the same order as that of the superficialis tendon. Thus a full restoration after ulnar palsy would require the transfer of very strong muscles.

However, the use of very strong muscles would also require multiple insertions around the finger including insertion into a bone[13] since normal intrinsics are multiple and have multiple insertions. Their tension is divided between bone and lateral bands and between both sides of the finger. A single insertion of a very strong muscle into one side of a phalanx or one lateral band might cause significant lateral deviation and rotation of the digit on a long-term basis and also hyperextension at the PIP joint, especially on very mobile fingers.

Fowler's[24] operation, using the extensor indicis proprius, EIP and the extensor digiti quinti EDQ uses a total tension fraction of 2.0 between four fingers. The original Bunnell[11, 12] variation of the Stiles[28] operation used four sublimus tendons with a total tension fraction of 8.3. This was probably excessive, especially as the tendons were fixed into the lateral bands where they tended to hyperextend the PIP joints, which had been deprived of their prime flexor.[5] The modified Stiles-Bunnell operation, using only one (middle finger) superficialis tendon split into four tails, has a total tension fraction of 3.4, which seems adequate.

The so-called Brand[5] procedure uses either the ECRB on the dorsum of the wrist (Fig 10–15) or the ECRL rerouted to the palmar side (see Fig 10–11). This latter was first described by Littler.[22] The tension of either of these (4.2 and 3.5, respectively) is within the range of adequacy. In recent years Fritschi,[18, 19] who has the experience of thousands of cases, has been using the palmaris longus as the motor more frequently, especially for slender and hypermobile hands. This muscle has a variable tension fraction, averaging 1.2 but often as high as 1.6, and may develop more strength with use. Although this muscle seems rather weak for the job, it has the great advantage that it is expendable enough so that nobody would feel deprived if it were to be used early after nerve injury in a case in which normal intrinsic action was recovered later.

We have no hesitation in saying that if the choice of palmaris longus allows a surgeon the confidence to operate early, it is a better choice than any stronger muscle that would only be used after a year's delay.

In any hypermobile hand an excessively strong muscle may result in intrinsic-plus, and this is worse than intrinsic-minus, because it is harder to correct.

Relationship to Wrist

A motor from the forearm extended to the fingers must cross the wrist and therefore it will affect the wrist and will be affected by the wrist.

In high ulnar paralysis the wrist is already overpowered on the extensor side in relation to the flexor side and on the radial side in relation to the ulnar side. Therefore balance would not be harmed by removal of a radial wrist extensor (ECRL), but it might be if the ECRB were removed to cross the wrist on the flexor side.

If the ECRB is to be used as a motor, it must stay on the dorsum of the wrist and be extended by grafts that pass through the intermetacarpal spaces. If the ECRL is used on the palmar side, it may pass through the dorsal part of the carpal tunnel thus exerting minimal flexion moment at the wrist. If the ECRL uses the carpal tunnel, and if the patient normally extends the wrist 30 degrees while flexing the fingers, the tendon transfer may accomplish its effect without significant actual shortening of its fibers because the muscle is actively shortened by action at the fingers while it is passively lengthened at the wrist.

This makes for very efficient action. If the ECRB is used on the dorsum of the wrist, it will continue to be used as a wrist extensor and must develop extra excursion to serve the fingers as well. If the ECRB is used on the dorsum, it *must* be kept in its sheath (or returned to its sheath after being withdrawn proximally for attachment to the graft). We

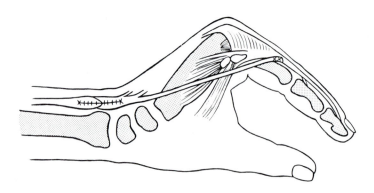

FIG 10–15.
Intrinsic-minus hand corrected by free tendon grafts from a wrist extensor muscle taken through intermetacarpal spaces to the lateral bands of each finger. *(From Milford, L: The Hand, ed. 2, St. Louis, 1982, Mosby–Year Book, Inc., p. 202. Used by permission.)*

have personally used it through a subcutaneous tunnel but abandoned that route after patients later developed bowstringing across the wrist, giving an uncontrollable wrist hyperextension.

Pathway

The pathway of any tendon graft should be considered for possible adhesions. The problem is especially urgent when there is bifurcation of tendons or, worse, bifurcation of bifurcated tendons as in four-finger clawing corrected by one motor. The problem is of a distal opening V (see Chapter 9).

In the case of either a sublimis transfer or a flexor many-tailed graft using the ECRL (extensor carpi radialis longus) or the palmaris as the motor, the bifurcation of the graft must be in the proximal palm at about the junction of the proximal third and the middle third of the distance from the wrist crease to the finger webs. If it is any more distal, vertical septa in the palm will block distal movement. We use a longitudinal incision only just distal to the carpal tunnel as a rerouting point (Fig 10–16,A). The palmar fascia is split, and the grafts, bunched together in one tunneler, are brought from the forearm into the wound. Four separate tunnels must then be made, one for each finger, from the distal insertion to the proximal palmar wound (Fig 10–16,D and E). The nose of the tunneler must appear in the proximal wound at the same place from each tunnel, or at least no vertical or transverse structure should intervene to be straddled by the grafts. When all grafts are in place they should be pulled distally while being observed proximally to ensure freedom (Fig 10–16,E). If the pulling of one graft distally results in another being pulled back into the wound, this means that the junction of the two is riding around an obstruction.

Dorsal Pathway.—When two strands of one motor have to pass from the dorsal to the palmar side of the hand, as in Fowler's operation[24] and in the extensor many-tailed graft, there are two or three fascial layers that have to be penetrated or avoided. The tunneller must be passed empty from the palmar to the dorsal side and should not be forced. The surgeon must feel for a natural opening in the fascia covering the palmar side of the interossei. This is not a continuous fascia and has many fenestrations. The dorsal interosseous fascia may have to be penetrated, but it is not thick. Once the nose of the tunneller is in the dorsal compartment, containing all the extensor tendons, it must slide proximally deep to the dorsal fascia until it appears in the proximal incision. The dorsal fascia is tough, and is in continuity with the extensor retinaculum and it should not be penetrated by any tendon or graft at any time.

The split tendons should bifurcate at least 2 cm proximal to the point at which they have to straddle a metacarpal bone (see discussion of distal V in Chapter 9).

This is one reason that it is better, in Fowler's operation, for the EIP and the EDQ each to pass through only one metacarpal space and do not have to straddle any metacarpal. The EIP passes through the second metacarpal space and then splits to the adjacent sides of the index and middle fingers. The EDQ passes through the third metacarpal space and then the two halves separate, one to the radial side of the ring finger and the other across the palmar side of the fourth metacarpal to the radial side of the little finger.

Insertion

Of the three insertions, (1) to the bone on the proximal phalanx,[13] (2) to the lateral band, and (3) to the A2 pulley,[9] the moment arms for the MCP flexion are all about the same. However, in the first two the moment arms are small when the MCP joint is straight and become larger when it is flexed, because of bowstringing. The Brooks insertion (which is on the A2 pulley)[9] has a larger moment arm than the other two when flexion begins from the straight position, but remains more nearly constant during flexion.

Where the grafts approach the MCP joints from the dorsum of the hand through intermetacarpal spaces, there is no bowstringing at the MCP joint, because they hug the intermetacarpal ligament as a pulley even when the finger is flexed (Fig 10–17). This means that the average moment arm

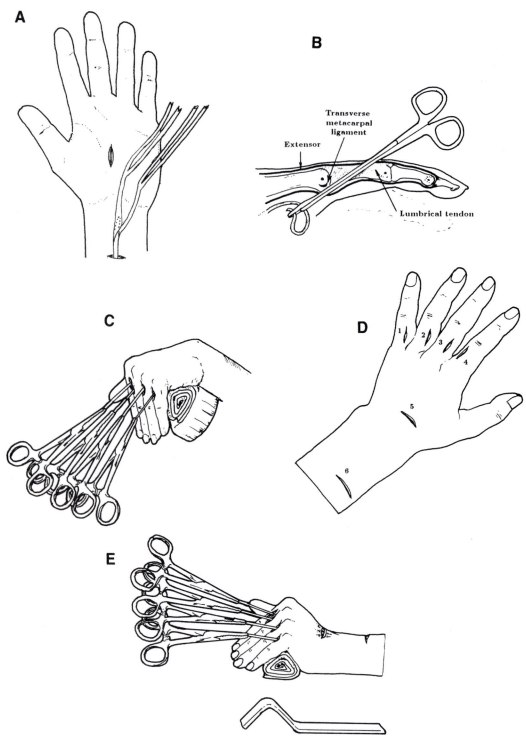

FIG 10–16.
A, the palmaris longus or extensor carpi radialis longus extended by a four-tailed free graft. The palmar incision is used for entry of tunneling forceps, which pass through the carpal tunnel, deep to all structures, and brings them into the proximal palm. From there they pass individually to each finger. **B,** the tendon tunneler enters the finger beside the tendon of the lumbrical in front of the metacarpal ligament, to pick up the tendon in the proximal palm. **C,** when each tendon graft is in place, their tension should be equal and matched while the arm is supported by a splint or bulky folded towel holding the wrist 30 degrees flexed, the MCP joints 80 degrees flexed, and the IP

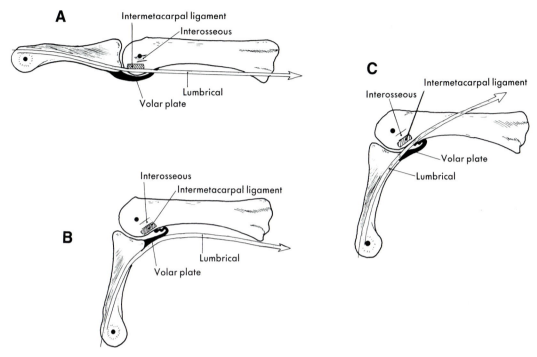

FIG 10–17.
A and **B**, with a palmar approach of grafts the new transfer develops a larger moment arm when the MCP joint is flexed than when it is extended. **C**, when the graft approaches from the dorsal side, it has no chance to bowstring even when the MCP joints are flexed.

for MCP flexion is somewhat less in Fowler's operation[24] than in the Bunnell[11, 12] or other palmar approach. This is not a defect. Normal interosseous muscles have small moment arms, probably the replacements should as well.

Note that the insertion to the lateral band has the advantage that it also extends the PIP and (DIP) joints. The insertions to bone and to tendon sheath require the action of the extensor digitorum to complete the straightening of the finger. That is acceptable, except that the extensor digitorum cannot extend the IP joints without exerting the same tension at the MCP joint where it has a larger moment arm for extension. Thus a transfer for intrinsic action must exert enough tension to overcome the moment of the extensor at the MCP joint before it can begin to flex the MCP joint. More total

muscular effort is needed to produce a straight finger.

Index Finger in Pinch

First dorsal interosseous paralysis is a special case in ulnar palsy. It is an abductor and flexor of the index finger MCP joint.

To the extent that an active MCP flexor is needed to prevent clawing of the finger, the many-tailed operations already described each include the index finger and correct the sequence of finger flexion. The index finger has also to participate in pinch, against the thrust of the thumb, and that is where the first dorsal interosseus is of unique significance.

If the thumb pinches against the side of the index finger (key pinch) (see Fig 10–28,C), it takes a very strong first dorsal interosseus to oppose it. The normal first

---

joints straight. **D**, if the tendon is to cross the wrist dorsally, the extensor carpi radialis brevis *(ECRB)* is divided through an incision *(5)* and withdrawn in an incision *(6)* and attached to a four-tailed graft. It is returned through the ECRB sheath to *(5)* and then each graft goes to one finger. **E**, with all the grafts in position the hand is positioned on the splint to relax the tendon at every joint while the grafts are attached to the lateral band.

dorsal interosseus has a tension of 3.2, stronger than either the profundus or sublimus of the index finger. To replace it, the commonly used EIP[10] has less than 30% of the needed strength and the EPB[22] is still weaker. These muscles may make it possible to spread the fingers, but not to give abduction power against the thumb for a strong key pinch. If that is needed, one might have to consider use of the ECRL[20] at 3.5 tension fraction. This, however, carries a real likelihood of an abducted index finger that cannot be adducted back beside the other fingers. This is ugly and is resented by the patient. Most people can support a key pinch simply by flexing all fingers side by side, so that the thumb pinches against a fist full of fingers.

In an ordinary pulp-to-pulp pinch the tip of the index finger lies at the axis of abduction (see circumduction Fig 4–14) of the MCP joint of the index finger. Thus the pinch from the thumb does not have much moment for adduction of the index joint. The lack of the first dorsal interosseus is not so serious as one might think. Our own solution to this problem is to encourage the *ulnar-side* attachment of an intrinsic replacement to *flex*, *adduct*, and *rotate* the index finger at the MCP joint. This results in an index finger that faces the pulp of the thumb

with its own pulp in a square pinch (Fig 10–18). Now the long flexor tendons of the index, rotating with the bone, act as *flexor-abductors* of the index to match the *extensor-adductor* thrust from the thumb. Thus the index finger supinates to compensate for incomplete pronation of the thumb. My nickname for the first palmar interosseus is the *opponens indicis*.

This makes a good pinch and it makes it unnecessary to provide a transfer to replace the first dorsal interosseus. If the patient needs a key pinch, a strong first dorsal interosseus is needed or else all fingers must be kept side by side to resist the thrust of the thumb.

Static Corrections

Although we prefer tendon transfers for intrinsic paralysis (see above), there are cases of multiple paralysis where there may be no muscle that can be spared for transfer for this task. In such cases a capsulodesis[34] or tenodesis is valuable. Better than a passive block to extension is the concept of an active-passive tenodesis in which a tendon graft is placed to cross two joints that may act to assist each other. Riordan,[24] Fowler,[24] and Tsuge[30] have all used this principle for intrinsic palsy by taking strands of tendon

FIG 10–18.
This is the kind of pinch to cultivate when a person has ulnar palsy. An ulnar-side intrinsic transfer to the index finger acts like a first palmer interosseus and circumducts the index MCP joint so as to oppose the thumb pulp. This results in a true square pinch that needs no other intrinsic muscle support.

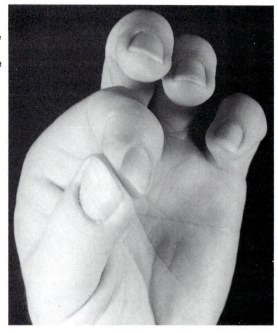

across the wrist and in front of the intermetacarpal ligaments. When the wrist moves, it either tightens or loosens the tenodesis across the MCP joint. Thus the wrist motors may be trained to modify MCP position.

Overcorrection

Overcorrection of a paralytic clawhand turns it into an "intrinsic-plus" deformity, the reverse of the previous "intrinsic-minus." It may then progress to the so-called swan-neck deformity, where flexion of the DIP joint is added to the hyperextension of the PIP joint and flexion of the MCP joint. This ugly deformity occurs most commonly in women who have slim, hypermobile hands. It is all the more distressing because such persons often take special pride in their fingers. In India and Southeast Asia traditional dances focus on hand movements, many of which involve hyperextension of all finger joints, so that the extended fingers sweep backward in an unbroken curve. If that lovely curve is changed to an unnatural zigzag of sequential joints, the hands become an object of shame, and are hidden, out of sight. The deformity occurs also in thicker, more stable hands, but takes longer to develop (Fig 10–19).

Because hyperextension of the PIP joint is often the earliest and most obvious sign of the progressive deformity, it is thought of as due to excessive extension torque at that joint, produced by the attachment of too strong a muscle to the lateral band as an intrinsic replacement. This may indeed be true, but there are other causes which must be considered. The total deformity is a *failure of balance* in the whole sequence of joints. Here are some of the factors that may encourage intrinsic-plus:

1. Anything that limits extension of the MCP joints, active or passive. This includes capsulodesis, rheumatoid arthritic joint collapse with partial subluxation, even a strong tendon transfer to the A1 pulley. The patient is constantly frustrated by his or her inability to straighten the finger, pulls hard on the long extensor tendons, which pull across the flexed MCP joints and hyperextend the PIP joints.

2. The loss of a flexor superficialis from a finger, even if it is transferred as part of the correction of some other digit. The PIP joint loses its prime flexor, and is unbalanced. Its palmar plate gradually stretches, and becomes lengthened. Flexion of the digit is always initiated by the profundus, which tends to become constantly in tension to compensate for the hyperextension. This results in flexion of the distal joint. The attempt to extend the distal joint results in

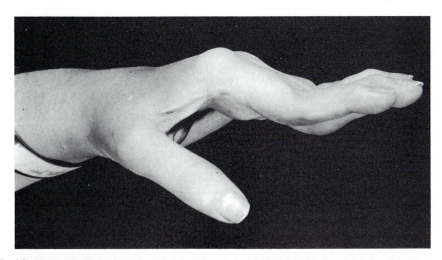

FIG 10–19.
This hypermobile hand has developed an intrinsic-plus deformity with PIP hyperextension and DIP flexion, also known as swan-neck deformity.

tension of the long extensors which now bowstring across the hyperextended PIP joint and fail to have any effect on the distal joint. The extensor fibers become stretched and lengthened at the distal joint. Once this process starts it slowly and inevitably progresses.

3. Transfer of too strong a tendon to the lateral band to correct clawing in a mobile digit, particularly if it is a superficialis tendon from the same finger. A previous flexor becomes an extensor of the same joint, and every instinctive effort to flex the joint results in more extension.

4. Arthrodesis of the PIP joint in a straight or nearly straight position. This results in a chronic tension in the profundus in an attempt to get the finger flexed into a normal resting position. The attempt to grasp makes for acute tension in a futile attempt to flex the PIP joint and only hyperflexes the DIP joint. In this case also, the extensor tendon to the distal joint becomes lengthened. If PIP arthrodesis is necessary, it should be in a normal resting position. Acute flexion posture of the distal joint may still occur, but is less likely, and the deformity is less obvious because the finger as a whole rests in partial flexion.

Prevention and correction of an intrinsic-plus deformity demands careful consideration of available mobility of the joints, and early intervention if a progressive deformity begins to develop. If the original clawhand deformity has resulted in flexion contracture of the PIP joints, and if there is difficulty in restoring full range of movement before surgery, there is no need to fear intrinsic-plus afterward. In fact, this may be an indication for using a transfer of flexor superficialis, using the tendon from the ring or middle finger, whichever is the most contracted. By contrast, if the finger is very mobile at the PIP joint, and removal of a superficialis tendon seems indicated for other reasons, it may be good to divide the tendon proximally, leaving one or both of its distal stumps long enough to attach to the A1 pulley to serve as a tenodesis to hold the joint just short of full extension. Sozen[27] has shown that this proximal division of the su-

perficialis and the tenodesis still leaves enough length for opponensplasty of the thumb. Perhaps the simplest operation for the hypermobile clawed hands would be to leave the tendon of the superficialis in continuity, but attach it to the A1 pulley in passing, while the PIP joint is in extension, so that the flexor superficialis becomes a prime flexor of the MCP joint and the tenodesis of the PIP joint at the same time. If the swan neck deformity occurs following removal of a superficialis tendon, the tenodesis shown in Fig 10–20 is recommended.

THUMB IN ULNAR PALSY

The defect in ulnar paralysis of the thumb that needs correction is a severe weakness of flexion at the MCP joint. The flexor moment at the IP joint is normal and the extensor moment is close to normal at each joint. There is a gross weakness in the flexion-adduction moment at the carpometacarpal (CMC) joint, but this need not always be corrected if the skin and fascia of the thumb web are normal (see Chapter 12).

To compensate for the weakness at the MCP joint it is necessary to use a muscle that is comparable with what has been lost. The combined tension fraction of the adductor pollicis (3.0) and the (FPB) (1.3) is 4.3. A number of surgeons have used the convenient extensor indicis, which has a tension fraction of 1.0, less than a quarter of what has been lost. We suggest instead using either a flexor superficialis (FDS) [middle], 3.4) or the ECRB (4.2) extended with a tendon graft.[25] The tendon should approach the thumb from a point on the thenar crease just distal to the midpoint from wrist crease to the index finger web. If the FDS (middle) is used as a transfer, it may use a split in the palmar fascia as a pulley or if the ECRB is used it may come through the intermetacarpal space. We have even taken an FDS (middle) tendon back through the interosseus membrane in the forearm, then across the dorsum of the wrist, and back through the second intermetacarpal space to the thumb. It worked very well.

Whichever tendon by whichever route, the final insertion must cross the MCP joint

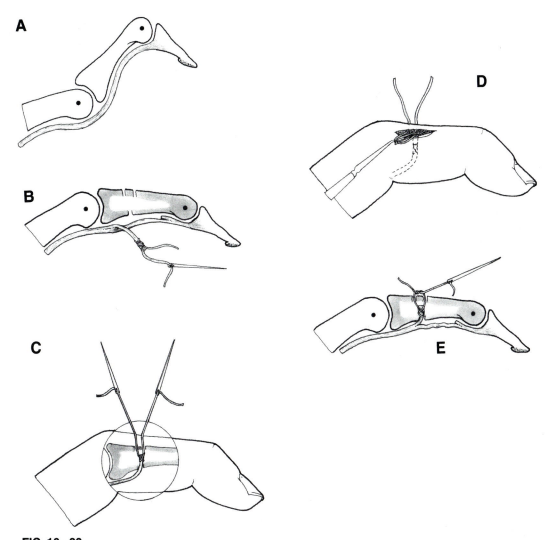

FIG 10–20.
Profundus tenodesis for swan-neck, or intrinsic-plus deformity. **A,** swan-neck deformity with flexor profundus as the only flexor for the PIP and DIP joints (the superficialis tendon having been removed for transfer). **B,** profundus tendon split. One slip is left in continuity and the other prepared by a zigzag silk stitch with both ends left long. A hole is drilled through the anterior cortex of the base of the middle phalanx, through which two fine holes are drilled through the dorsal cortex. **C,** the tendon end is drawn into the bone through the anterior drill hole by passing sutures on straight needles dorsally through holes in the bone and out through the skin. **D,** silk sutures are drawn out by a blunt hook passed between the bone and the extensor expansion. **E,** the sutures are tied together while checking the tension of the tenodesis, which should keep the DIP joint just short of full extension.

of the thumb at about 1.5 cm distal to the axis of flexion-extension. In the case of pure ulnar palsy the attachment may be made to the tendon of the adductor. If, however, the paralysis is ulnar and low-median, the same muscle-tendon unit may serve to assist pronation of the thumb by approaching it on the radial side and being attached to the extensor tendon about halfway along the proximal phalanx or it may be attached to the

tendon of the APB at its insertion. Thus the transfer will still result in MCP flexion but will also be a medial rotator (pronator) of the thumb and assist in opposition.

If it should be determined that the patient will not easily learn to control a tendon transfer or if the MCP is very hypermobile and unstable, an arthrodesis of that joint in slight flexion, abduction, and medial rotation, will usually allow a very satisfactory pinch, controlled at the terminal joint by the FPL and at the CMC joint by the skin and fascia of the thumb web. In such a case the index finger must be the moving element of pinch and the thumb will just lean back in its cradle of skin and fascia to receive it. To understand the role of the thumb web in ulnar palsy see Chapter 12.

LOW MEDIAN PARALYSIS: LOWER FOREARM AND WRIST LEVEL

In low median paralysis all the forearm muscles are normal. There is paralysis of the APB, the opponens pollicis, part of the FPB, and the lumbricals to the index and middle fingers.

Mechanics of Balance

Since the interossei are intact in the fingers that suffer lumbrical paralysis, and since the adductor is adequate to compensate for any partial paralysis of the FPB, it is unusual to have any observed defects in low median palsy except for those resulting from paralysis of the APB and the opponens pollicis.

The sensory deficit in the thumb and index finger is more likely to be serious and may result in failure to use the pinch even after motor balance has been restored.

The rebalancing of the thumb after low median palsy is not difficult if it is done soon after the paralysis. It needs a moderately strong motor (APB, 1.4; opponens pollicis, 1.9; total, 3.0). The direction of pull determines the ratio of effectiveness as between abduction and rotation. When there is no ulnar weakness a flexor brevis replacement should not be necessary. In compensating for the loss of the opponens pollicis it is more important to restore its pronating effect than its effect as a metacarpal flexor, because the latter may be handled by muscles supplied by the ulnar nerve.

Bunnell[11, 12] recommended a tendon transfer to pull toward the pisiform bone. Littler[22] is quite happy with a pull toward the center of the wrist (palmaris longus insertion), which gives excellent abduction and fair pronation. Thompson[29] used a more transverse direction that is very suitable for

low ulnar-median palsy but unnecessarily limits wide abduction in simple low median palsy.

If the corrective surgery is done late, it may have to work against some shortening of the dorsal skin and fascia of the web and a functional shortening of the powerful adductor muscles, which also tend to supinate (externally rotate) the thumb.

This is more serious if the patient has developed a habit of a lateral squeeze pinch while waiting for recovery of the thumb opposition. In all such cases the surgeon must make sure that the range of the dorsal skin and fascia is corrected to allow full play (see Fig 10–23) and may consider transferring part of the insertion of the adductor from the ulnar to the radial side of the thumb to reverse its rotatory influence while retaining its other actions.[15]

Attachment of a transfer for opposition may be to the tendon of the APB or it may be taken on to the extensor tendon on the dorsum of the proximal phalanx. This latter probably enhances pronation.

The motor may be a superficialis tendon or the palmaris longus or the EIP. The last is taken around the ulnar side of the wrist or through the interosseus space of the forearm to the ulnar side of all the flexor tendons except the FCU. A good operation for low median palsy is to swing the insertion of the (ADQ) around, and make it an abductor-opponens of the thumb.[15, 22] This has the advantage of giving some muscle bulk to restore the shape of the thenar eminence.

Just as important as the provision of a motor to abduct and oppose the thumb is the removal of any passive restriction to opposition that may be caused by the thumb web.

The key to the understanding of the thumb web is to realize that it is the *dorsal* skin and fascia that is subject to alternating folding and stretching with varying positions of the thumb. The palmar skin and fascia may be regarded as a more or less constant-length passive support from the thenar crease to the dorsum of the metacarpal. It is a little tighter in full extension, a little, but not much, looser in full pronation of the thumb. The dorsal skin, by contrast, falls into loose folds in extension and adduction and becomes tight in full abduction and pronation. The same is true of the dorsal fascia.

It is a good exercise to get a strip of adhesive tape folded on itself so that only the last 1 cm^2 is adhesive. This sticky end is laid on the back of the MCP joint of a thumb. Now the other end is pulled across the back of the middle of the second metacarpal, and the thumb is moved through full circumduction while the tape is free to move (Fig 10–21). The excursion of this dorsal skin is about 4 cm and its passive *moment arm* to the CMC joint is also about 4 cm. This is a big strong passive restraint if the skin is tight; it is too strong to be overcome by a tendon transfer on the opposite side, which is supposed to oppose (pronate) the thumb. A problem about this skin and fascial re-

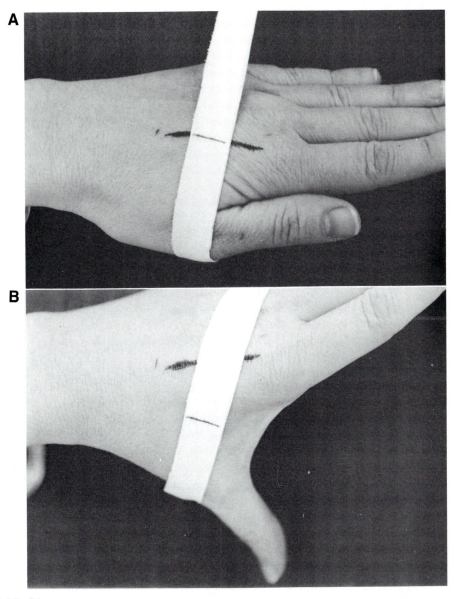

FIG 10–21.
A, a strip of adhesive tape is folded on itself to leave only 1 cm^2 of adhesive surface at one end. This is attached to the dorsum of the thumb just proximal to the MCP joint. Full circumduction of the thumb demonstrates how much the dorsal skin of the web (represented by the tape) has to be free to move, and **B,** how strongly it can limit pronation if it is tight.

straint is that it is dorsal around the thumb. This means that it has a strong passive supinating moment. The more the thumb is abducted, the more the skin becomes tight. The more the skin becomes tight, the more it supinates (or externally rotates) the metacarpal.

If a tendon is transferred to restore opposition of the thumb while there is still residual tightness of dorsal skin in the web, the skin will hold the dorsum of the thumb back so that the final position of the thumb is in limited abduction and only partial pronation (opposition). Then, in pinching, the index finger will not meet the pulp of the thumb but the side of the thumb (Fig 10–22). This can result in the problem of crank action (see Chapter 5) and then in progressive loss of metacarpal abduction and rotation.

Therefore, before any operation for the correction of intrinsic palsy, the thumb must be pulled passively into full opposition to make sure that the dorsal web will allow it. If not, a few weeks of passive stretching and splinting or a sliding webplasty may be needed.

The sliding webplasty[6] frees the dorsal skin and fascia of the web by an incision on the back of the hand (not in the web). The skin and fat and fascia of the web are undermined in a plane of areolar tissue and then the thumb is held in full abduction with full rotation into opposition. Any restraining strands of tissue are then identified and divided (Fig 10–23).

In some cases the distal end of this incision is continued by being angled back to form a small Z-plasty centered on the free edge of the web, but most often this is not needed. A full-thickness free graft is then placed in the skin defect over the second metacarpal.

LOW ULNAR AND MEDIAN PARALYSIS

This results in clawing of all fingers, which may be corrected in the same way as for low ulnar palsy. It also results in loss of all the intrinsic muscles of the thumb, which is often treated in the same way as a low median palsy, because the loss of opposition is the most obvious feature of the palsy. This must not be done, because the loss of the ulnar-supplied adductor and short flexor muscles is at least as significant.

We believe that the best way to treat low ulnar-median palsy is to transfer two tendons, one for abduction-pronation and the other for adduction-pronation of the thumb. The stronger of the two transfers should be on the adductor side where the power is needed and the weaker one for abduction, which is only a positioning muscle. A good combination is an FDS (middle finger) for

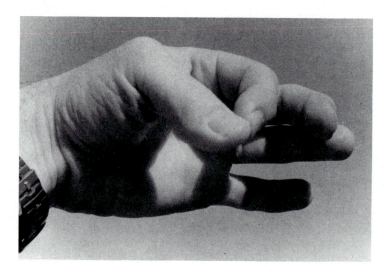

FIG 10–22.
Index finger pinching against the side of the thumb. This type of pinch, if the web is tight, gives rise to a crank action and supination of the thumb.

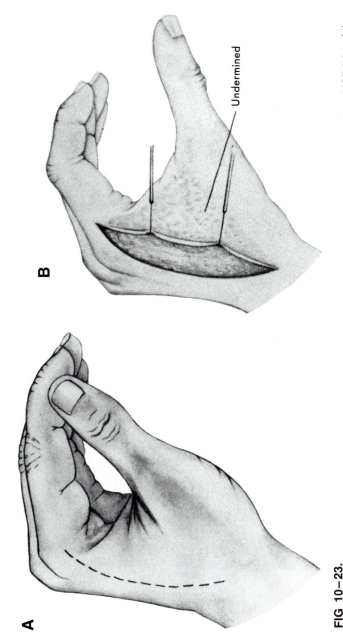

FIG 10–23.
A, an incision is made over the full length of the second metacarpal bone from a point on the radial side of the MCP joint of the index finger, curving to the dorsal aspect of the bone, and carried proximally almost to the wrist joint. **B,** the incision is deepened to the areolar plane in which the skin moves freely over the deeper tissues. The radial flap is reflected and freed back as far as the thumb, dissecting into the free edge of the thumb web and down to the tissues over the CMC joint of the thumb. While an assistant pulls the thumb into abduction and full opposition, the surgeon divides the doral fascia and any tight bands that seem to hinder full opposition. Usually it will not be necessary to divide paralyzed muscles. *(From Rob, D., and Smith, R., editors: Operative surgery, ed. 3., Woburn, Mass., 1977, Butterworth Publisher, Inc., p. 144. Used by permission.)*

adduction, as for ulnar palsy (see p. 208), and an EIP, around the ulnar side of the wrist, for abduction.

If only one tendon is to be used, it should be given the prime task of opposing the thumb as a whole (CMC joint) and a secondary task of stabilizing the MCP joint in a good position and *restricting its movement.* Alternatively, the MCP joint may be fused in slight flexion-abduction and rotation toward pronation.

The reason for this is that it is not possible to give independent control of the MCP and IP joints when there is only one muscle to provide flexion and adduction (FPL).

Our own technique for a single tendon transfer (flexor superficialis) is to take the tendon from the direction of the pisiform bone toward the axis of the MCP joint and to split it before it reaches the axis. From there we pass half of it to the dorsal and ulnar side of the MCP joint and the other half to the palmar and radial side. So half of the tendon is an extensor-adductor of the MCP joint and the other half of the tendon is a flexor-abductor of the *same joint*. The net effect is that the MCP joint is stabilized—*not mobilized*—in a position determined by the relative tensions of the two halves of the tendon (Fig 10–24).

Low ulnar-median palsy of the thumb can be treated as above, provided the thumb is of the thick and stable variety. However, if it is slender and hypermobile it is very difficult to balance the three joints with so few muscles, even when the intrinsic replacement is split to stabilize the MCP joint by an insertion on both sides. In such cases, if the thumb begins to develop chronic MCP flexion, hyperextension, or abduction, it is better to reoperate soon, because the deformities tend to be progressive. Whatever the secondary problem may be, the answer is always arthrodesis of the MCP joint in about 15 degrees of flexion, 10 degrees of abduction and 15 degrees of medial rotation. Experienced surgeons, operating on hypermobile thumbs, will often choose to perform arthrodesis of the MCP joint at the same operation as the tendon transfer.

FIG 10–24.
Brand transfer to restore opposition; a newer modification in which a different pulley is used is described in the text. *(From White, W.L.: Surg. Clin. North Am. 40:427, 1960. Used by permission.)*

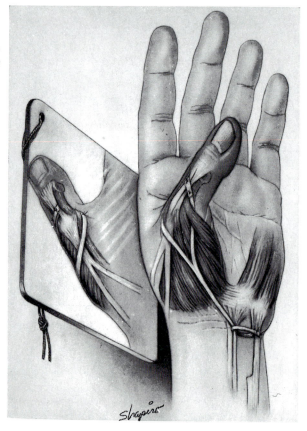

The transferred tendon has to change direction near the heel of the palm and needs some kind of sling or pulley. We used to use Guyon's canal as a pulley for such a transfer, but since Snow and Fink[26] published their use of a hole in the carpal tunnel roof, we have used that and prefer it. They realized that the superficialis tendon carries a sheath of its own synovium at that level that prevents adhesions from forming between the tendon and the tough fibrous roof of the carpal tunnel (Fig 10–25). Thompson[29] brings out the tendon from the ulnar side of the proximal palm and uses the ulnar border of the palmar fascia as a pulley. This is a good pathway as it causes opposition and some abduction of the thumb without hindering adduction. Others have used a sling made from the FCU tendon. This results in a pure abduction-pronation, which is acceptable for simple median palsy or for the abductor component of a two-tendon transfer, but is not suitable for a single-tendon transfer for low ulnar-median palsy.

Extensor Pollicis Longus: The Spoiler

The old proverb says that "the good may be the enemy of the best." In low-ulnar median palsy the EPL is useful, because it has a vector for adduction of the CMC joint. If it is contracted along with the FPL, it results in adduction of the thumb. It enables a patient to hold things between the side of his or her thumb and the side of the hand (see Chapter 4). However, the very usefulness of that trick movement may make it very difficult to restore true opposition by tendon transfer. The patient quickly comes to depend on the EPL every time he or she wants to pinch or hold anything. After the operation, while the patient is trying to re-educate the transferred tendon to work in phase, he or she finds it much easier and quicker to use the EPL and pick up objects by lateral squeeze. This totally defeats the

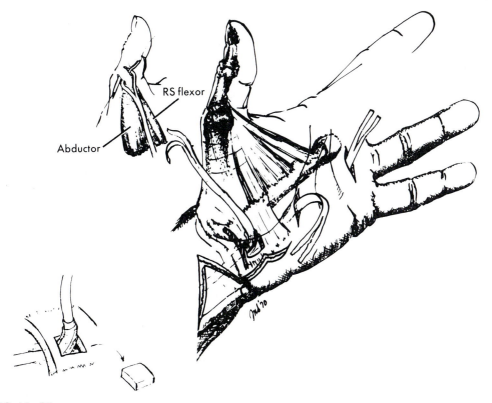

FIG 10–25.
Diagram showing use of an opening in the roof of the carpal tunnel as a pulley for the flexor superficialis tendon (with sheath) for opposition of the thumb. *(From Snow, J.W., and Fink, G.H.: Plast. Reconstr. Surg. 48:238, 1971. Used by permission.)*

new tendon transfer. This is a very common cause of failure of surgery for ulnar-median palsy. It may not be recognized by surgeons who are content to observe their patients only in the office, where they can prove to their own satisfaction that the transfer works. They can take a photograph to prove it. They fail to notice that when the patient goes to pick something up, the old lateral squeeze is used—not the new opposition. A year later the patient may be unable to identify or use the transferred tendon and the dorsal web is again contracted.

All this can be avoided by one of the following:

1. Operate soon—before the patient has a chance to develop the trick movement as a habit pattern.
2. Reeducate thoroughly and at a work bench— not only by exercises.
3. If the patient is older and has a habit pattern established, consider rerouting the EPL tendon to lie over the APL at the lateral aspect of the wrist. (It must be freed from its retinaculum.)
4. In really established cases and in older patients with predicted failure, use the EPL as the abductor-motor. The tendon is divided near the MCP joint; pulled out on the dorsum of the wrist; and taken through to the anterior forearm via an opening in the interosseus space between radius and ulna and to the ulnar side of the flexor tendons. It is then tunneled back *to its own stump* and sutured end to end. It thus remains an extensor of the MCP and IP joints of the thumb but becomes an abductor rather than an adductor (Fig 10–26) of the CMC joint.

Note that this will only work if the thumb is free to abduct and oppose. It may require a webplasty to be done at the same operation to permit it. The only defect of this excellent operation is that the patient may have difficulty in pulling the thumb back to make a flat palm. If this proves a problem, a small adjustment to the angle of approach of the APL or EPB should correct it.

Crank Action

This is another spoiler that results in late failure of operations to restore opposition of the thumb and pinch after low ulnar and median paralysis.

The mechanism is explained in Chapter 5. It begins with a failure to fully correct the contracted dorsal skin and fascia of a tight web. Now any attempt to abduct and pronate the thumb ends up with limited rotation, so that the index finger pinches against the thumb on the ulnar side rather than on the pulp. This may result in stretching the ulnar collateral ligament of the MCP joint if the MCP joint is held straight. If the MCP joint tends to be flexed and especially if the IP joint is straight, the segment of thumb distal to the MCP joint is used as a crank by the index finger to rotate the metacarpal into supination. *This is a progressive deformity*. If it is unchecked the patient may finish up, years later, with his first metacarpal lying beside the second metacarpal, externally rotated and with the two phalanges pointing forward. The MCP joint flexes for this action because the metacarpal is so rotated that MCP flexion brings the distal half of the thumb at right angles to the palm.

It is hard to salvage the thumb from this disaster, because by now the dorsal skin and fascia are tight and the tendon transfer for opposition is lost to consciousness.

An alert surgeon or therapist should pick up on the earliest stage of crank action and recognize the danger of a postoperative thumb that attempts pinch with (a) a flexed MCP joint, (b) a straight IP joint, (c) incomplete rotation, and (d) the index finger pinching against the side of the thumb.

Any such thumb needs reeducation directed to a pulp-to-pulp pinch (a night splint may be needed to lengthen the web). If quick improvement is not seen, it is probably best to reoperate and lengthen the web and fuse the MCP joint in 10-degree flexion, 10-degree abduction, and 15-degree rotation. Before finishing the operation, the surgeon should check to see that pulp-to-pulp pinch can be achieved without strain.

Alternatively, attempts to restore pulp pinch should be abandoned, and a key pinch should become the new objective.

Joints of the Thumb

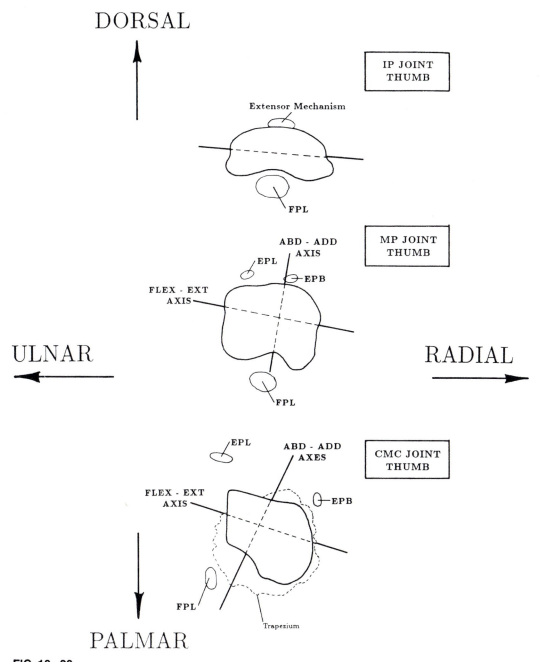

FIG 10–26.
Note that the extensor pollicis longus (EPL) tendon is more of an adductor than an extensor of the carpometacarpal joint of the thumb. If the tendon is rerouted to approach the thumb from the front of the wrist, it retains its extensor moment for the MCP and IP joints, but becomes an abductor-pronator of the CMC joint.

HIGH MEDIAN PALSY

In high median palsy there is loss of pronation of the forearm and of flexion of the thumb and all finger flexors are lost except for the ulnar-supplied profundus (Fig 10–27).

The loss of forearm pronation is quite a disability and may become almost irreversible while waiting for recovery of muscles. Thus it is important to maintain the range of pronation after nerve suture. If the median palsy is irreversible, one should consider removing the strong supinating force of the biceps by rerouting the biceps tendon so that it pronates rather than supinates.[35] The ECU (tension fraction, 4.5) can be a useful pronator by being brought across the front of the forearm to the lower radial side.

Flexion of the fingers may be accomplished by side-to-side suture of all the profundus tendons in the forearm. This will allow the ulnar-supplied profundus to move all fingers. With exercise this may become quite a strong grasp. The long flexor of the thumb may be activated by the ECRL, or the thumb and index finger may be made to flex together by making the ECRL move both. However, the required excursion for the FPL is less than that of the profundus to index; therefore the index finger would have only limited flexion.

The thumb retains good short flexors and adductors. Abduction can be provided by transfer of the ADQ or by the EIP (tension fraction, 1.0) taken around the ulnar border of the wrist, or by the ECU, which can be directed across the wrist to the MCP joint of the thumb using a free graft. Goldner[20] has suggested using a flexor superficialis tendon as a graft in situ for this purpose and this is a

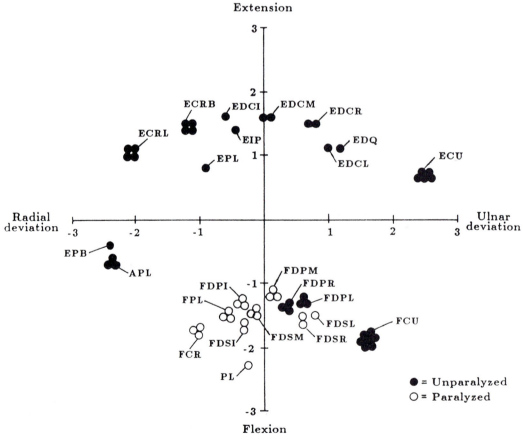

FIG 10–27.
Balance at the wrist following high median palsy.

good idea, but his suggested use of the FCU (tension fraction, 6.7) provides too strong a motor with too limited a range for this task.

The brachioradialis (tension fraction, 2.4) is an obvious resource muscle in high median palsy but it is difficult in a technical sense to mobilize. It is also difficult to reeducate as part of a multiple transfer. We prefer to leave it in situ to use at a second operation if it is found that the primary transfer needs to be supplemented or balanced.

In high median palsy, finger flexion is so weak, even after tendon transfers, that just one wrist extensor (ECRB) in situ is adequate to stabilize the wrist during grasp.

Because high median nerve injury will also affect sensation in both the thumb and index fingers, most patients will use the hand only for gross grasp and leave fine finger discriminatory movements to the other hand. Therefore, as Moberg[23] has said, it may be better to aim for a good firm *key pinch* rather than a pulp-to-pulp pinch that the patient may not use. Key pinch is performed best in midopposition rather than full opposition.

HIGH MEDIAN AND ULNAR PALSY

This gross disaster leaves a hand very weak and often clumsy because of sensory deficit. However, because of numerous radial-supplied muscles that cross the wrist, a fairly useful balanced hand may be obtained after transfer.

The most obvious losses are finger flexion and thumb flexion. The ECRL and the ECU are the muscles most often used to power the FDP and FPL.

An intrinsic replacement may be provided either by Fowler's[24] operation, using the EID and the EDQ or, better, by the Riordan[27] modification, using only the EIP. This tendon is detached at the MCP joint, withdrawn at the wrist, split, and directed to the fourth and fifth finger lateral bands, passing in front of the transverse intermetacarpal ligament. A two-tailed tendon graft is added proximally to the EIP tendon to serve the index and middle fingers[25] or the brachioradialis may be attached in the forearm

to the superficialis of the middle finger, which is then divided distally and withdrawn in the palm, where it is split into four tails, one for the lateral band of each finger.

Abduction of the thumb may be by the EIP taken around the ulnar side of the wrist or through the interosseous space at the lower end of the radius and ulna.

A muscle that has not had much attention is the extensor digitorum to the middle finger (tension fraction, 1.9). It has a discrete muscle belly that is proximal in the forearm and a long tendon that can be traced along the back of the hand and freed from its junctura and other tendon slips that converge on the middle finger from other parts of the extensor digitorum complex.

This long tendon can be used for abduction of the thumb, around the ulnar side of the wrist, or for adduction and flexion of the thumb, by taking it through the second interosseous space between the second and the third metacarpals to attach to the base of the proximal phalanx of the thumb. Here it will have a moment arm of about 4.0 cm for the CMC joint and should add considerably to the effectiveness of the thumb if it is properly reeducated. (Middle finger extension can be managed by a slip from the extensor to the ring finger.)

Another approach to the thumb in high median and ulnar palsy is to use the technique suggested for key pinch in tetraplegia (see below) in which the FPL tendon is rerouted behind the flexor tendons in the palm and around the hook of the hamate. In the case of ulnar-median palsy the tendon may be powered by a transfer rather than by tenodesis as in tetraplegia.

TETRAPLEGIA AND OTHER SEVERE PARALYSIS

In planning hand surgery, the two main differences between tetraplegia and various very severe lower motor neuron paralyses is that in the former there is gross paralysis in all the other limbs. There is also a very significant muscle tone in the paralyzed muscles throughout the arm.

In the flaccid paralysis of severe lower

motor neuron palsy, there need be little hesitation in using tendon transfer and tenodesis combinations to bring some use to an unbalanced and useless limb. In tetraplegia, no surgeon should rush in to transfer tendons or do tenodeses until every implication of the action has been thought out and discussed with the patient and therapist. The situation is often dominated by the need for ability to transfer from bed to chair and to avoid limiting any action or trick motion the patient already has that might be lost following tenodesis.

For example, if a patient has elbow extension and is able to help in transferring in and out of a wheelchair, then any tenodesis on the palmar side of the wrist and fingers might prevent the flat hand–extended wrist posture used in transfer, or attempted use of the hand to lift the body in this way could destroy the tenodesis. If the patient can learn another way to transfer, then the situation may change.

The other factor in tetraplegia is the good muscle tone of the finger flexors. Once a patient has learned to use this by controlling the active wrist extensors, he or she may be reluctant to risk losing it in exchange for a more precise and stronger grip or pinch, which is linked by tenodesis to loss of range of motion at the wrist. Time spent in artificial simulation of the effect of proposed tendon transfers by dynamic splints and simulation of proposed arthrodesis by plaster casts is always worthwhile.

To obtain a simple useful prehension without the use of tenodesis across the wrist, there must be a minimum of two good muscles below the elbow. These are likely to be radial wrist extensors and the brachioradialis.

One muscle is used to extend the wrist and the other to flex or oppose the digits. If there is a third muscle, it probably should not be used for another primary or synergistic action, but to oppose one or both of the primary movements above. This is because it is very difficult to obtain good range of motion if a muscle has only gravity to extend or stretch it.

Zancolli[35] has a good way to use one muscle to extend the wrist and flex a digit (or

digits) by attaching a flexor tendon from the thumb to the proximal part of the tendon (Fig 10–28) of the wrist extensor. A similar technique may be used to have one muscle flex the wrist and extend the thumb, thereby giving a full wide range of key pinch that does not depend on gravity and therefore may be used in any position.

The most consistent forearm muscle to survive in tetraplegia is the brachioradialis, and it may be the only one. Moberg[23] was not the first to use it but has been the pioneer in its effective use as a wrist extensor, linked to a tenodesis-controlled key pinch. He has also emphasized that it is necessary to have a functional elbow extensor to prevent the brachioradialis from flexing the elbow instead of (or as well as) doing its intended work at the wrist and below. Moberg uses a transfer of part of the deltoid as an elbow extensor.

We describe the anatomy of the brachioradialis insertion in Chapter 12 and suggest tubing the flat part of the tendon, where it has to be detached from its lateral aponeurotic insertion. We also suggest a possible slide of its linear origin partway down the humerus toward the epicondyle to lessen its moment arm at the elbow and free more excursion for the wrist.

The brachioradialis is about as strong as the FPL or FDS to the index finger and has the largest potential excursion of any forearm muscle. However, most surgeons regard it as a muscle of last resort as far as transfer is concerned. This reluctance must be a result of its dependence on elbow position and the tendency of the tendon to become adherent (as a result of lateral adhesions to its raw edges) and perhaps to the fact that it is often used in situations where there is no active muscle to oppose it.

If the brachioradialis is the only remaining muscle below the elbow, it has to depend on gravity to flex the wrist. This may work well provided that the range of motion of the wrist is maintained in the postoperative weeks by the therapist, using passive movements if needed. The same problem exists for the tenodesis for key pinch. The action of the brachioradialis will pull the thumb toward the index finger, but what

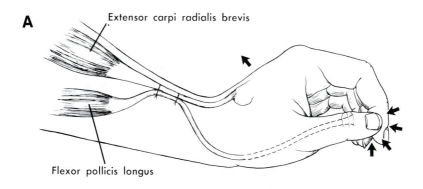

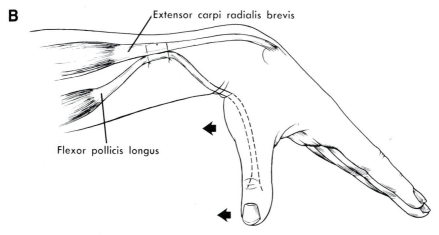

FIG 10–28.
A, key pinch is obtained with wrist extension by an active extensor carpi radialis brevis (ECRB) with tenodesis to the flexor pollicis longus (FPL). To achieve correct tension, with the wrist in complete passive extension, the FPL tendon is sutured to the ECRB when pinching is produced. **B,** with passive wrist flexion, pinch is released. *(From Zancolli, E: Clin. Orthop. 112:101, 1975. Used by permission.)*

will open the pinch wide enough to be useful? This may need a tenodesis of the EPL.

Since there is to be only one motor for the thumb, it is necessary to decide about the joints that are to be moved. If we expect the tenodesis of the FPL to control three joints, it is certain that one of the three will hyperextend while another will hyperflex and that the weakest joint will determine the final strength of pinch. What most surgeons actually do (though they may not realize it) is to rely entirely on the skin and fascia of the web to hold the metacarpal against extension because the flexion moment of the FPL at the CMC joint is hopelessly inadequate (Figs 10–29 and 10–30). This means that the thumb metacarpal is held in a constant position during pinch, and the MCP joint does all the active flexing (see Fig 10–32,B).

This is satisfactory if only a pulp-to-pulp pinch is desired, and if the fingers have a motor that can produce good MCP flexion. In pulp-to-pulp pinch the thumb need move very little but the fingers must have active flexion (see Fig 10–30).

If only one or two motors are available, Moberg[23] has demonstrated the value of using only the key pinch. This is successful because it requires no active movement of the fingers, only positioning in flexion to provide a passive anvil for the hammer of the thumb. The problem of key pinch is that it requires movement of the whole thumb, including the metacarpal (see Fig 10–29). Therefore one cannot rely on help from the dermal fascial cradle of the thumb web. The metacarpal has to flex, because without it the MCP joint must flex 60 degrees, and this is not practical (Fig 10–31).

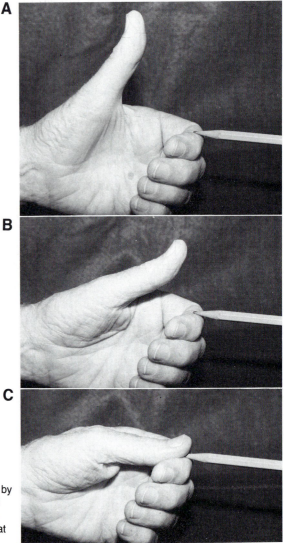

FIG 10–29.
A–C, in normal key pinch all the movement is by
the thumb (in these photographs the pencil did
not move). Note that the CMC and MCP joints
both flex a little, and the IP joint does not flex at
all.

To power the thumb for key pinch, Moberg[23] frees the pulley of the FPL at the MCP joint to give the FPL a better moment arm there (this has no effect on the moment at the CMC joint), and he fixes the IP joint with a Steinmann type of pin. He really makes the tenodesis of the FPL into an MCP flexor of the thumb. This is satisfactory if the MCP joint has enough mobility to allow the thumb pulp to reach the index finger without the need for flexion of the metacarpal at the CMC joint (see Fig 10–31). If it cannot quite reach, the metacarpal will have to flex, the skin and fascial support will become slack, and the FPL would have to control the MCP and CMC joints, resulting

in a very weak pinch. Zancolli[35] sometimes fuses the CMC joint, thus concentrating all the available power on the MCP and IP joint. Some patients are unhappy with a fixed fused thumb and we prefer to change the course of the tendon so that an effective moment may be delivered to the CMC joint. This may be done by routing a tendon across the palm obliquely from an insertion distal to the MCP joint of the thumb across and behind all the flexor tendons proximal to the vertical septa of the palm. This tendon may cross the wrist just to the radial side of the ulnar artery, using as a pulley either the ulnar fibers of the palmar fascia or the muscle origins from the hamulus of the

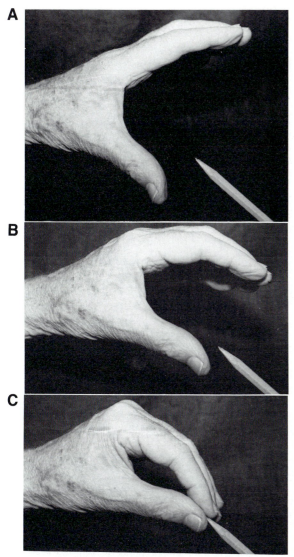

A

B

C

FIG 10–30.
A–C, in pulp pinch most of the movement is not thumb, it is fingers (and mostly the MCP joints of fingers). In these photographs the pencil is stationary. One can pinch comfortably with no movement of any thumb joint or even of any IP joint of a finger. The thumb may therefore rest back in the dermal fascial cradle to avoid the need for muscle action or the flexor pollicis longus may be used to flex the IP joint while the MCP joint is fused.

hamate. Guyon's canal is a useful pathway. The tendon may then be tenodesed in the forearm. The transferred tendon has a 3.0 cm moment arm for CMC flexion, compared with the normal 1.5 cm of the FPL.

Now the FPL has a 1.5-cm moment arm at the MCP joint and a new 3.0-cm moment arm at the CMC joint. Since the MCP joint is about halfway from the thumb pulp to the base of the thumb, there should be a good chance for the tendon of the FPL to control both joints when opposed by the thrust of a pinch force on the pulp. The backward moment from the thrust of pinch would be twice as much at the CMC as at the MCP joint, while the flexion moment from the

FPL would also be double at the CMC as at the MCP joint (Table 10–1).

If, in any individual case, the MCP joint was found to hyperextend during pinch, one could always perform an MCP arthrodesis and allow pinch to be powered by the 3.0-cm moment arm at the CMC joint (Fig 10–32,C). A full key pinch, moving the MCP joint 15 degrees (Fig 10–32,A), will use about 1.5- to 2.0-cm excursion of the tendon, which could result from about 60 degrees of a wrist-extension motor (brachioradialis) at the insertion of the ECRB. Such devious tendon pathways require a rather aggressive postoperative mobilization to give a full range of movement of the thumb.

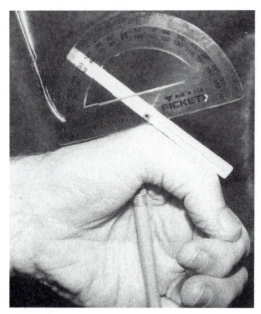

FIG 10–31.
One cannot do a key pinch without some flexion of
the CMC joint. In this photograph the metacarpal neck
is forcibly supported with a pencil eraser to keep the
web extended. One has to flex the MCP joint 60 degrees to pinch. This is very uncomfortable (see also
Fig 10–28, B).

It may be an advantage to provide an extensor for the thumb by tenodesis of the
EPL proximal to the wrist. If this is to be
done, the excursion of the EPL for thumb
extension must be carefully tested at operation to match the excursion of the EPL tenodesis released by wrist flexion.

For example, if the brachioradialis is capable of extending the wrist 60 degrees
(from 40-degree flexion to 20-degree extension) and if the transferred FPL has a 2.0-cm
moment arm at the wrist, transfer and tenodesis will give the FPL a 2.0-cm excursion
to flex-adduct the thumb. However, it may
take only 1.0 cm of EPL excursion to extend
the thumb and open the pinch (because its
moment arm is smaller than the rerouted
FPL. Therefore the EPL must cross the
wrist closer to the axis of the wrist so that
the same wrist movement will give the EPL
tenodesis only half the excursion that the
FPL has. This can be done by freeing the
retinaculum of the EPL on the dorsum of
the wrist and allowing the EPL to cross the
wrist nearer to its radial border. The surgeon must test the effect of wrist flexion and

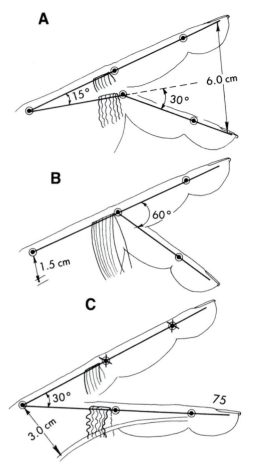

FIG 10–32.
A, normal key pinch uses about 15 degrees of flexion-
extension at the CMC joint and about 30 degrees at
the MCP joint with IP joint straight (see Fig 10–29,
C). The web gives no support during normal key
pinch. **B,** if the flexor pollicis longus (FPL) is the only
thumb flexor, it will fail at the CMC joint, and the metacarpal will rest in the dermal-fascial sling of the web.
Now the MCP joint has to flex 60 degrees for the tip of
the thumb to reach the side of the index finger. If the
MCP joint cannot flex so much, the metacarpal will
have to move and the whole pinch becomes weak,
because the sling is loose. **C,** if it is determined to use
CMC flexion for key pinch, the FPL must be rerouted
to give a large moment arm at the CMC axis (or another muscle may be used) and the dermal-fascial
sling becomes unimportant. The MCP joint may be
fused or it may move with the CMC if the moment
arms are adjusted so that the CMC moment arm is
just about double the MCP moment arm and the IP
joint is fused.

extension on the FPL and EPL before the
EPL tenodesis.

This double tenodesis should result in
quite a firm key pinch and a 5-cm width of
opening when the wrist is flexed. If the

TABLE 10–1.

Mechanics of Pulp-to-Pulp Pinch and Key Pinch in Severe Paralysis*

Joint	Pulp-to-Pulp Pinch	Key Pinch
CMC	May do without muscle power by resting in dermofascial sling of thumb web	Cannot use dermofascial sling of web; needs flexor muscle power with good moment arm (suggest 3.0 cm)
MCP	May undergo arthrodesis or have pulley advancement	May remain intact or be fused
IP	Should be free to move if MCP joint is controlled as above	Better without any movement

CMC = carpometacarpal; MCP = metacarpophalangeal; IP = interphalangeal.

pinch proves to be firm, it will be necessary to fuse the IP joint of the thumb. Moberg's[23] success with a simple pin stabilization only proves that there was not much strength in the pinch. A pin cannot stabilize a joint against significant repetitive stress.

This modification of Moberg's original key pinch tenodesis for tetraplegia was worked out by us together when he visited our hand biomechanics laboratory at Carville. Eric Moberg reports to us that his first cases with enhanced moment arms for CMC flexion have achieved a strong pinch.

REFERENCES

1. Andersen, J.G., and Brandsma, J.W.: Management of paralytic conditions in leprosy, All African Leprosy and Rehabilitation Training Centre, P.O. Box 165, Addis Ababa, Ethiopia, 1984.
2. Beine, A.: Combined opponens replacement of thumb and little finger: a preliminary report, Int. J. Lepr. 42:303-306, 1974.
3. Boyes, J.H.: Tendon transfers for radial palsy. (Bull. Hosp. Jt. Dis. Orthop. Inst.): 97–105, 1960.
4. Boyes, J.H.: Bunnell's surgery of the hand, ed. 5, Philadelphia. 1970. J.B. Lippincott Co., pp. 13–15.
5. Brand, P.W.: Paralysis claw hand, J. Bone Joint Surg. 40B:618–632, 1958.
6. Brand, P.W.: Orthopaedic principles and practical methods of relief. In Cochrane, R.G., editor: Leprosy in theory and practice, Bristol, 1958, John Wright & Sons, pp. 255–265.
7. Brand, P.W.: Tendon grafting. J. Bone Joint Surg. 43B:444–453, 1961.
8. Brand, P.W.: Paralysis of the intrinsic muscles of the hand. In Rob, C., and Smith, R., editors: Operative surgery, Vol. VII, ed. 3, Woburn, Mass., 1977, Butterworth Publishers, Inc.
9. Brooks, A.L.: A new intrinsic tendon transfer for the paralytic hand, J. Bone Joint Surg. 57A:730, 1975.
10. Brown, P.W.: Reconstruction for pinch in ulnar intrinsic palsy, Orthop. Clin. North Am. 2:323, 1974.
11. Bunnell. S.: Surgery of the intrinsic muscles of the hand other than those producing opposition of the thumb, J. Bone Joint Surg. 1:24, 1942.
12. Bunnell, S.: Surgery of the hand, New York, 1948, J.B. Lippincott Co.
13. Burkhalter, W.E.: Restoration of power grip in ulnar nerve paralysis, Orthop. Clin. North Am. 2:289, 1974.
14. Codivilla, A.: Sui trapianti tendinei nella practica ortopedia, Archivio d'Ortopedia, 16:225, 1989.
15. De Veechi, J.: Opposition of the thumb: physiopathology; a new operation: adductor muscle transplant, Bol. Soc. Cir. 10 Urug., 32:423–436, 1961.
16. Fick, A.: Statische Betrachtung der Muskulature des Oberschenkels, Z. Rationelle Med. 9:94–106, 1850.
17. Fick, R.: Handbuch der Anatomie und Mechanik der Gelenke unter Berücksichtigung der bewegenden Muskeln, 1904–11. Vol. 3, Spezielle Gelenk und Muskelmechanik, Jena, 1911, Gustav Fischer.
18. Fritschi, E.P.: Reconstructive Surgery in Leprosy, Bristol, 1971, John Wright & Sons Limited. p. 64.
19. Fritschi, E.P.: Surgical reconstruction and rehabilitation in leprosy, ed. 2. New Delhi, 1984. The director for Southern Asia. The Leprosy Mission.

20. Goldner J.L.: Tendon transfers from irreparable peripheral nerve injuries of the upper extremity, Orthop. Clin. North Am. 5:351, 1974.

21. Jones, Sir R.: Tendon transplantation in cases of musculospiral injuries not amenable to suture, Am. J. Surg. 35:33-341, 1921.

22. Littler, J.W.: Tendon transfers and arthrodeses in combined median and ulnar nerve paralysis, J. Bone Joint Surg. 31A:225, 1949.

23. Moberg, E.: Surgical treatment for absent single-hand grip and elbow extension in quadriplegia, J. Bone Joint Surg, 57A:196, 1975.

24. Riordan, D.C.: Tendon transplantations in median-nerve and ulnar-nerve paralysis, J. Bone Joint Surg. 35A:312, 1953.

25. Smith, R.J.: Original communications. Extensor carpalradialis brevis tendon transfer for thumb adduction: a study of power pinch. J. Hand Surg. 8:4–15, 1983.

26. Snow, J.W., and Fink, G.H.: Use of a transverse carpal ligament window for the pulley in tendon transfers for median nerve palsy. Plast. Reconstr. Surg. 48:238, 1971.

27. Sozen, S.: Opponens plasty using transverse carpal ligament window and tenodesis of the distal end of the superficialis tendon of the ring finger to the A1 pulley. Presented at 42nd Annual Meeting, American Society for Surgery of the Hand; videotape, September 1987.

28. Stiles, Sir H., and Forrester-Browne, M.F.: Treatment of injuries of the peripheral spinal nerves. London, 1922, H. Frowde and Hodder & Stoughton.

29. Thompson, T.C.: Modified operation for opponens paralysis, J. Bone Joint Surg. 24:623, 1942.

30. Tsuge, K.: Tendon transfers in median and ulnar nerve paralysis, Hiroshima J. Med. Sci. 16:29-48, 1967.

31. Von Lanz, T. and Wachsmuth, W.: Praktische Anatomie, Vol. 1, Pt. 3, Arm. Berlin, 1935, J Springer, pp. 154–243.

32. Weber, E.: Ueber die Längenverhältnisse der Fleischfasern der Muskeln im Allgemeinen. Berishteüber die Verhandlungen der kgl. Sachs, Gesellschaft der Wissenschaften in Leipzig, Sitzung, August 16, 1851.

33. Weber, W., and Weber, E.: Mechanik der menschlichen, Gehwerkzeuge, Göttingen, 1836, Dietrich.

34. Zancolli, E.A.: Claw-hand caused by paralysis of the intrinsic muscles: a simple surgical procedure for its correction, J. Bone Joint Surg. 39A:1076, 1957.

35. Zancolli, E.A.: Structural and dynamic bases of hand surgery, ed. 2, Philadelphia, 1979, J.B. Lippincott Co., pp. 236 and 244.

CHAPTER 11

Methods of Clinical Measurement in the Hand

Measurement is useless if it is not precise and repeatable. However, precision requires some expenditure of time and sometimes requires costly apparatus. In this essentially practical handbook, we have tried to select methods of measurement that are as simple as possible for obtaining the information that is needed. Even so we do not suggest that busy surgeons and therapists spend time making a whole range of evaluations that will not be of practical use to them or to the patient.

We have observed the "eager beaver" syndrome in young hand specialists. The symptoms are brought on by a laudable desire to do clinical research and publish papers. Because the specialist is not sure just what is going to be important, the next stage is marked by the printing of an extensive hand evaluation sheet on which every muscle is named and graded; every joint has a space for range of motion; total sensory maps are routine; and standardized functional tests, strength tests, pinch tests, work aptitude tests, psychologic tests, and photographs fill the second and later sheets. The sad part of this syndrome is not the early stage of excessive time spent on evaluation and paper work. It is the late stage in which the eager beaver, snowed under by a mountain of meaningless data and burdened with the cost of it, finally says, Why bother with all this? I can get good results by instinct, observation, and guesswork, and I can make

more money that way. So he has a bonfire of unused forms and becomes a run-of-the-mill hack surgeon.

It is for fear of the final stage of reaction that we include this warning against the early stage of excess. Measure what is relevant and measure it with precision. Record what is relevant and record it in a way that makes it easy to see and review at a glance. There must be flexibility in the setting and later modification of goals. There must not be flexibility in the disciplines of accurate measurement. Better not to measure than to record figures that are meaningless. Measurements and tests must always be servants to the educated, alert, and sensitive mind.

OBJECTIVES

MONITORING INDIVIDUAL TREATMENT

However well accepted a treatment method may be, there is no certainty that it will be of benefit to any one individual patient. If it is not going to be helpful, it should be abandoned quickly, before it wastes more time. This is especially true of splinting; many splints continue to be prescribed long after they have ceased to have any good effect, at which time they are a hindrance to active use of the hand.

Objectivity is essential. Everybody wants

223

to see improvement and that wish is re-flected in extra effort to obtain good figures for the record. When the surgeon is expecting recovery after nerve suture, it is almost impossible to avoid applying extra force when testing pinprick or extra pressure on the touch applicator for sensory evaluation. When the therapist is expecting improvement in range of motion, he or she cannot avoid pulling or pushing the joint a little harder during later measurement. When patients want to believe they are gaining strength, they put more effort into the dynamometer test each week. This might not be bad except that it often encourages the continuation of treatment methods that should be changed or abandoned.

Therefore, time should be saved by measuring range of motion only on significant joints and by assessing fewer areas or modalities of sensation, but that saved time should be spent in assuring precision and repeatability in the measurements and in measuring more frequently, especially if there is doubt about the choice of treatment modality.

ENCOURAGEMENT OF PATIENT

Almost all evaluations should be shared with the patient. Success depends on the patient. It is the patient's hand. This is one reason why most measurements should be recorded graphically. Few people (including surgeons) readily comprehend a page of numbers. Almost everybody can relate to a line that goes up or down on a graph. It is often a waste of time to make a list of numbers and then make a fancy graph to display them. The graph should be kept in the clinic area and measurements recorded directly on it as the examination is made. Sometimes the patient may mark the graph while the therapist calls out the numbers.

In this way the patient establishes an identification as part of the therapeutic team. He or she will soon recognize that on the days the arm is allowed to hang down the volume graph goes up and that half an hour of exercises with the arm up will bring the line down; the patient may also note how range of motion improves when volume

diminishes. When the final moment of truth arrives, and the patient knows that the imperfect hand is the best that he or she can expect, the patient is more likely to smile and say, "We did our best" than to say with a frown, "You did a poor job; you will hear from my lawyer."

EDUCATION OF SURGEON AND THERAPIST

Quite apart from helping the individual patient and without any pretensions to serious research and publications, the habit of precise measurement opens the way for the surgeon to become a far better surgeon and the therapist to become a true professional. We emphasize this aspect of self-education before we mention serious research, because we know that many surgeons in private practice and therapists who are tied to a timetable of scheduled patients feel that they cannot undertake Research (with a capital R). Research involves exploring a lot of literature and it involves statistics and it takes a lot of time. Because of those constraints, many physicians become chained to the methods they learned at school or are victims to whichever supply house spends the most money on advertising.

It need not be so. Even in a busy private practice, the selective use of repeated objective measurements can quickly reinforce the value of some methods of treatment and suggest that others be discarded. It may not be a big series and may not be worthy of publication, but it is educational and very satisfying. Even a single case, perhaps a stiff finger, may be treated alternately with one of two methods, dynamic splint or serial cylinder casts, and in a couple of weeks the graph of range of motion may show a tendency to go up on some days and down (or less up) on other days. Very interesting! But only if the angles are measured at standard torque (Fig 11–1).

RESEARCH

The need for objective measurement in research is so obvious that it needs no argu-

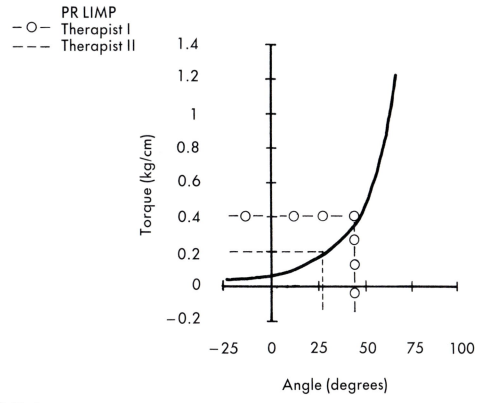

FIG 11–1.

Two therapists, measuring the angle of extension of an MCP joint, came up with 30 degrees and 50 degrees, respectively. A full torque-angle curve demonstrated that therapist II had been using just double the torque of therapist I.

ment. We will tell just one true story. It was when the first ultrasound machine was imported into India and was allocated to Physiatry. A physician was sure it would loosen up stiff joints and ran a series of cases. Six months later he had figures to prove that the ultrasound-treated patients increased their range of motion faster than the controls. But by chance a visiting electronic engineer did a check of all equipment in the hospital. He found that the ultrasound machine had been shipped with an open connection between the motor and the head. This had not been closed when the machine was installed, and no patient had ever actually received any ultrasound treatment. They had heard a humming noise from the box. The physician was transparently honest, but somebody had allowed his faith in the treatment to affect the torque he had applied to the fingers during measurement of joint angles. It is better to have double-blind methods in which the evalua-

tor does not know which cases are the controls.

LEGAL AND INSURANCE CONSIDERATIONS

There is nothing so valuable in a courtroom as a patient's record that includes a piece of graph paper with lines showing progress or with colored areas relating to sensory mapping, especially if the patient has seen them before and perhaps helped to make them. Even if the actual findings are not of great significance, the graph is evidence of constant and careful review.

DISADVANTAGES OF MEASUREMENT

These are: (1) diversion of the therapist's time and attention that could be spent on treatment, and (2) danger of losing sight of the main goal among many numeric objectives.

MODALITIES

In outlining a number of different modalities of measurement, we do not suggest that all should be used. We record them in order to comment on them.

RANGE OF MOTION OF JOINTS

Active

Active range of motion is that range that patients can achieve on their own with no external support—even from their other hand. It is best to take the measurement at the same time of day and with the hand in the same basic posture each time. The goniometer should scarcely touch the hand during measurement; otherwise it may give support or assistance to one limb of the joint.

For certain types of paralysis, additional rules or constraints may apply. For example, in evaluating a preoperative clawhand with intrinsic paralysis, we measure the best angle of extension of the proximal interphalangeal (PIP) joints that the patient can achieve while the metacarpophalangeal (MCP) joints are not hyperextended. It takes a while to explain, and the patient will say that he or she cannot extend like that. The patient has to understand that after surgery he or she will be able to do better, but for now we need to record limitations. If the patient is allowed to hyperextend the MCP joints, the PIP joints will extend further, owing to assistance by the tension in the palmar plate. But that is no longer "unassisted" by the special definition for this test.

Assisted-Active

In this measurement the patient still controls the joint but is allowed to have a proximal segment stabilized by external support to allow him or her to concentrate on the joint being measured. In these measurements the goniometer may be pressed on the hand and may even serve to stabilize it. For example, in intrinsic paralysis the goniometer may press the proximal segment of the finger into flexion, allowing the extensors of the finger to use their best tension on

the PIP joint. Another example occurs following repair or grafting for flexor tendon injury. Often MCP flexion becomes free quite early, and IP flexion is much more difficult to achieve. The assisted-active range of motion is measured with the MCP joint supported in extension to allow the best tension of the long flexors to act on the IP joints while they are being measured.

Passive Range of Motion

The advantage of this measurement is that it can be more objective. It takes no account of the patient's will or strength; it measures only the passive restraints. However, even the passive range of motion may vary with the position of other joints.

In the case of tendon repair and tendon grafting, the passive range of motion may define the limitation caused by joint structures while the active and assisted-active range of motion may be dominated by adhesions around the tendon proximal to the joint.

In intrinsic paralysis the passive range of motion should reach and be maintained at 100% of normal while awaiting surgery. The active range of motion is always poor before surgery and should approach the passive range after surgery. If the assisted-active range of motion is as good as the passive before surgery, it means that the result of tendon transfer should be good; but if the assisted-active range is much less than the passive, it means that there is a defect in the extensor mechanism (e.g., a boutonnière type of attenuation of the dorsal expansion at the PIP joint) and a regular tendon transfer will fail unless the dorsal tendon complex is restored.

A very simple method of recording range of motion was suggested by Glanville.[18] He uses short lengths of multicore soldering wire threaded through fine rubber tubes. They may be laid along the back of the hand or digits and bent to conform to the back of each segment. Soldering wire will bend at a touch and then hold its new position with no tendency to spring back. The wire is lifted off the finger and laid on paper where its shape is traced with a pencil. This is done first in flexion and then in extension with

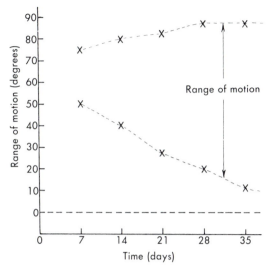

FIG 11–2.
Graph of range of motion of a stiff PIP joint at standard torque during 1 month, showing improving range caused mostly by an increase in the angle of extension.

the proximal segment of the wire model lying over the same line each time. At each subsequent test the same procedure is followed on the same record sheet with each tracing superimposed on the previous ones.

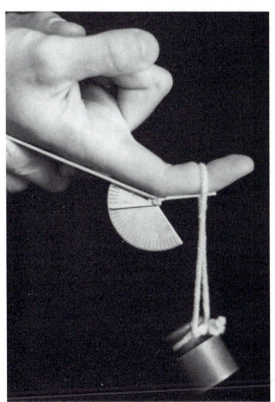

In this way any improvement is clearly seen, and when improvement stops, the failure to improve is also very obvious. If numbers are needed at a later date, a goniometer is used on the diagram. We sometimes use short lengths of solder wire to record the best angles of a series of joints and then measure each angle from the wires for the record or for a graph. The solder wire record is excellent for the patient's own use also, to follow progress at home.

Torque-Angle Measurement

In its simplest form this is a method of making the passive range of motion more objective (Fig 11–2). It may be done by choosing a standard weight or spring (e.g., 250 g) and attaching it to a loop of string that is slipped over the finger to pull at some defined anatomic level, such as at a flexion crease (Fig 11–3). The finger or hand is positioned so that the string is at right angles to the finger segment while the angle is measured. In this method the actual torque is not known, because the lever arm (moment arm) is not measured. However, the

FIG 11–3.
A simple way to obtain a torque-angle measurement is for a hand to be positioned so that a hanging weight, of, say 250 gm, is pulling at right angles to a segment of a finger and at a finger crease. Alternatively, the weight may hang over a pulley wheel, so that the string is horizontal and the hand is more easily positioned.

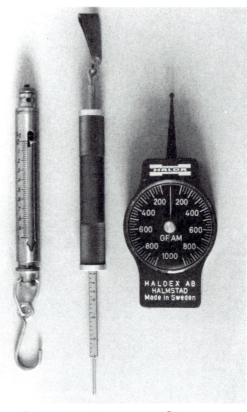

FIG 11–4.
We have used many different devices for applying torque to joints. **A,** a simple spring scale, from a hardware store, used by sport fisherman for weighing their catch. It is calibrated in grams to 1 kg. **B,** a push-pull device to which, at one end *(top),* a sling may be attached for measuring pull while a pencil eraser (not shown) is used to cap the other end, so that the narrow metal end (used sometimes as a dolorimeter) does not cause harm when used to push a digit into flexion. **C,** a Haldex lever device in which the dial records the force needed at the end of the lever to displace it. We attach a string sling to the end and pull at right angles to the lever. We also use the end of the lever to push with. In either case the reading of the dial is recorded. *(**B** and **C** available from Fred Sammons, Inc., Box 32, Brookfield, Ill.)*

A   B   C

torque is known to be the same every time, since the lever arm is the same each time (the same crease or mark is used) and the tension is the same (250 g). Thus if a change in the angle is observed after a period of treatment, it can only represent a real change in the hand (see Fig 11–1). Steve Kolumban, a physical therapist, used this method when assessing the efficacy of cylindric plaster of Paris casting vs. dynamic splinting in releasing contracted PIP joints. In order to have two comparable groups of contracted PIP joints he assessed the passive extension angle at 400 and 800 g.[20, 28]

Various forms of pull or push devices are shown in Figure 11–4. Goniometers that can give angle readout with one-handed control to free the other hand to control the torque are shown in Figure 11–5.

Torque-Angle Curves or Torque–Range-of-Motion Curves.—We have given this a separate heading because, although it is simply an extension of the torque-angle range of motion, it does take extra time and may be perceived as too much of a research method for the average surgeon and therapist to undertake. This is wrong. It is easier than it looks and is most useful in a practical sense. However, even if the torque-angle curve is not used, the simple torque-angle measurement above should still be used routinely. The torque-angle curve is a series of torque-angle measurements that are made with a different torque for each. With the goniometer in position and with some type of push-pull spring positioned at an anatomic marker (e.g., skin crease), the tension force (or the compression) is varied through perhaps five readings (e.g., 100, 200, 400, 600, and 800 g or perhaps 4, 8, 12, 16, and 20 oz). At each pull (or push) the angle is measured and marked on a graph (Fig 11–6).

The result is a curve that reflects the mechanical quality of whatever it is that prevents free motion of the joint. Normal soft tissues, such as skin and fat, often become

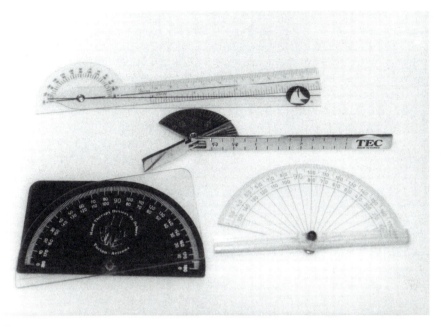

FIG 11−5.
Various goniometers for measurement of joint angles. The moving arm must be free to move at a touch. For torque-angle measurements the center model seems best because its arms are flat on the digit.

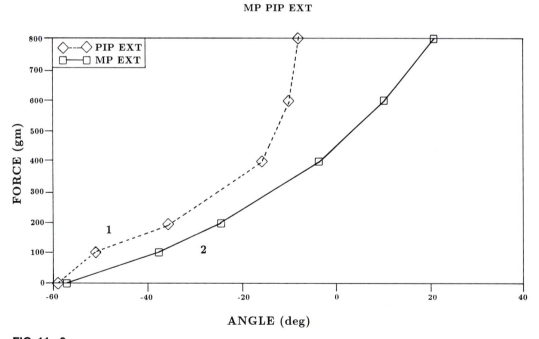

FIG 11−6.
An example of a normal hand depicting PIP extension which (1) graphically demosntrates the typical abrupt halt to further extension of the joint at just beyond the straight position. (2) The normal MCP joint has more mobility than the IP joint into hyperextension with a soft (springy) end feel *(read left to right) (From Breger-Lee, D., Bell-Krotoski, J., and Brandma, J.W.: J. Hand Ther. 3:7−13, 1990. Used by permission.)*

shortened or "take up the slack" when there has been disuse of a joint in a poor position. Such tissues will then restrict joint motion. Their torque-angle curve is a characteristic gentle curve (Fig 11–7,B). Tendon, fascia, and dense collagen scar may also restrict motion when there is tenodesis or a similar parallel-fibered collagen band. This type of tissue has a very steep curve, and the range of motion hardly changes between different torques (Fig 11–7,A).

Much of the value of a torque-angle curve may be obtained by using just two measurements for a joint, one at 300 g and one at 800 g. This two-torque differential gives good prognosis for improvement when it is wide and poor prognosis when it is small. The slope of the torque-angle curve will vary also according to the distance of the restraining tissue from the axis of the joint, e.g., even skin will give a steep curve if it crosses the joint far from the axis (bowstrings) because of the large moment arm of the restraining tissue.

If a viscous restraint is suspected, as when edema is a factor, a different tech-

nique may be used. A constant torque (say 500 g) may be applied and then a series of joint-angle measurements made at intervals of a few seconds. A progressive angle change at constant torque over just a minute or two suggests a fluid or viscous element in the stiffness of the joint (see Fig 6–20).

The influence of muscle shortness on joint limitation can often be evaluated by the repetition of a torque-angle measurement while a proximal joint position is changed (flexion of wrist to relax finger flexors and then extension to relax extensors (Fig 11–8) or, in the case of intrinsics, positioning of the MCP joint while measuring the IP joint and vice versa).

The significance of torque-angle curves may not be immediately obvious to a therapist. However, after using them on a few cases of stiff hands, both therapist and surgeon will come to realize that some types of torque-angle curves give good prognosis for improvement through therapy while others suggest that only surgery will be effective. The chief difficulty about torque-angle curves is in the matter of dexterity in trying

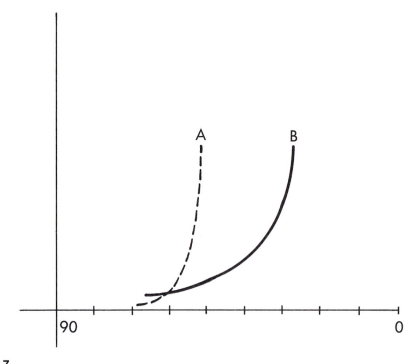

FIG 11–7.
A, steep torque-angle curve of a PIP joint, suggesting a problem with ligament, scar, or tenodesis. **B,** gentle torque-angle curve of a PIP joint, suggesting disuse stiffness.

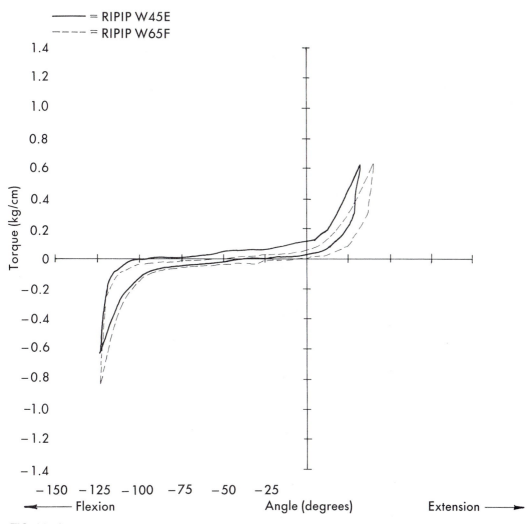

FIG 11–8.
In this torque-angle loop of the PIP joint of a right index finger, the *solid line* is measured with the right wrist extended 45 degrees *(RIPIP W45E)* and the *dashed line* is measured with the right wrist flexed 65 degrees *(RIP-IP W65F)*. Note that there is very little difference in flexion, but that wrist extension limits PIP extension by about 8 degrees.

to stabilize the patient's hand, hold the goniometer, hold the tension device, and record the angles. It takes practice and is quite a challenge. Therapists may like to try one or more of the following ideas. The two hands of the therapist are used (hand 1) for stabilizing the patient's hand, including the proximal segment of the involved joint, and (hand 2) for pulling (or pushing) the distal segment of the involved joint, using the calibrated spring device. Now the control of the goniometer and the recording of the data are the difficulties.

Control of the Goniometer

1. If the goniometer has a short flat arm and a friction-free joint, hand I may hold the flat arm on the proximal segment and nudge the free arm into position with another finger.

2. A short length of solder wire can be bent around the joint and laid down on graph paper to be recorded later.

3. A single overhead light may be used to cast a shadow on a "sunburst" protractor of 360 degrees lying flat on the table (Fig

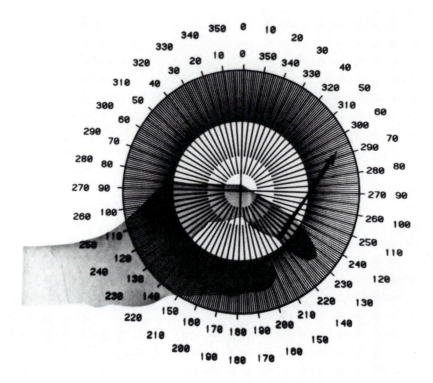

FIG 11–9.
Shadows of a finger cast by a overhead lamp on a 360-degree "sunburst" protractor lying on the table.

11–9). The hand is held midprone so the fingers flex in a plane parallel to the table and cast a shadow on the protractor. The backs of the fingers may even be lengthened by short Kirschner wires or toothpicks taped to the backs of the fingers to make it easier to read the shadow.

4. The very neat little electronic goniometer devised by Cantrell and Fisher[14] may be used, with one arm held by the thumb of hand 1 and the other arm held on the back of the distal segment by a rubber band. The absence of any hinge between the arms of this goniometer makes this system free of any constraint.

5. A similar but nonelectronic device on the same principle, made in our laboratory, gives direct readout on a dial (Fig 11–10).

Recording.—If a whole series of numbers have to be written down, it involves dropping everything between each measurement, and this vastly increases the time taken for the test. The numbers may be written down by an aide or by the patient or may be recorded on a voice-activated tape recorder or by one with a foot switch, so the figures may be marked on a graph at the end of the test.

Summary

We have suggested many methods for measuring the range of motion of joints. We would like to think that in the future all passive range-of-motion measurements would be made at a standard torque, but realistically we suggest that this refinement be used whenever a stiff joint is being subjected to continuing treatment to improve its range of motion. This should always be monitored by periodic torque-angle measurements at a single standard torque, and the results should be recorded on a graph of range of motion against time (in days or weeks of treatment).

The indication for torque-angle curves (a series of angle measurements made at one time with a different torque for each) is for an analysis of the nature of stiffness, with the idea of planning and evaluating treat-

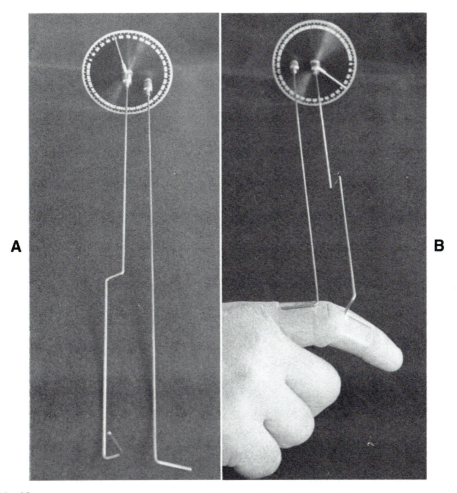

FIG 11-10.
A, homemade goniometer (affectionately called "rabbit ears") based on an original idea of Cantrell and Fischer.[14] Each of the two wires is bent to a right angle at one end. The other end attaches to a 360-degree metal protractor. One wire is fixed to the protractor and the other rotates freely at its center, carrying a pointer that defines its rotational position on the scale of the protractor. Each short right-angled limb is taped to the dorsum of one segment of a digit (or wrist or elbow). The long wires run parallel to the joint axis to the protractor. The joint may move actively, passively, or by imposed torque while the angular changes are read out on the protractor. It is often convenient to suspend the protractor by a thread from a stand. The value of this system is that the axis is unrestrained, and the two wires may be close or far apart without affecting the angle readout. An electronic version has a potentiometer at the center of the protractor. **B,** "rabbit ears" device taped to an IP joint.

ment (Figs 11–11 and 11–12). It need not be done often—perhaps once at the start of treatment and again when the graph of improvement begins to plateau. The same is true of viscosity curves (a series of angle measurements during 1 minute of constant torque); these are one-time studies or to be repeated at longer intervals after treatment. Several therapists have further developed and used torque range of movement in a clinical setting since the first edition of this book.[10-13, 17] (Figs 11–13 through 11–16).

STRENGTH

Strength is a very difficult thing to measure. Most of us do not attempt to measure the actual tension capability of individual muscles or even of a group of muscles (e.g., finger flexors). Instead we use a functional assessment of output of strength (e.g., grip strength or pinch). This is reasonable as long as we realize that it is not a measurement of actual muscle capability. It is a measure of what a patient is willing and able to ask of

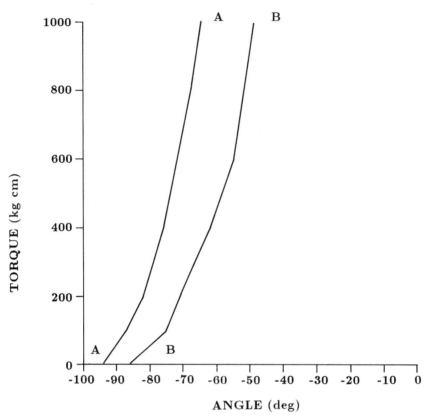

BEFORE AND AFTER EXERCISE OF STIFF PIP JOINT

FIG 11–11.
Torque-angle curve of a stiff index finger PIP joint. *A*, taken just before an exercise session and *B*, taken after the exercise session.

his muscles. A man may have a muscle capability for a 100-lb grip, but with rheumatoid arthritis he may have pain feedback that tells him that anything over 20 lbs will put too much strain on a joint or a ligament. He therefore grips with 20 lbs. He might do better after local anesthesia or steroid injection. These do not strengthen his muscles, but they make him willing to put more stress on his joints (perhaps to his own harm).

Moberg[25] has demonstrated that it is skin sensation that keeps us most accurately informed of joint position. We must add to this that it is skin sensation that is used to monitor strength and forms the most commonly used feedback for controlling the use of strength.[9] We hold the handle of a hammer with a grip that gives skin feedback that tells us it is secure. Our leprosy patients habitually use twice as much strength or more

(Fig 11–17) in this kind of task, because they do not have skin sensation. They use extra strength to try to obtain the feedback from skin that gives them the sense of a secure grip. In cases of nerve injury the opposite may be true. After nerve suture and partial recovery a patient often has showers of paresthesias ("pins and needles") whenever he or she touches an object. Such unpleasant sensations often inhibit strong use of the hand and give a low performance on the dynamometer and a false idea of weakness.

Having said all this, we still do use dynamometers and pinch meters as a functional test, but we await better methods of assessment of strength. It is important to use existing equipment in a way to make it as repeatable as possible; for example, the bar of the Jamar dynamometer should be set to the size of the hand and the same setting used at

OBLIQUE RETINACULAR LIGAMENT CONTRACTURE

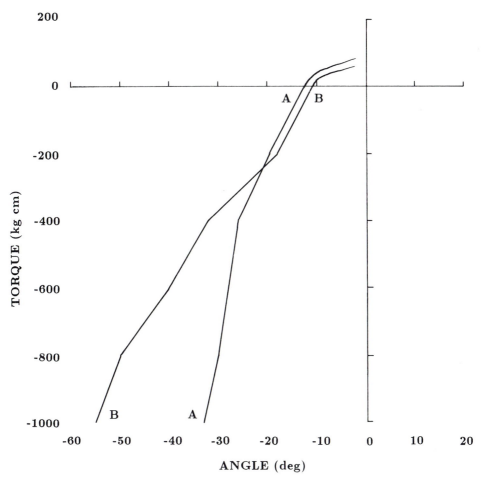

FIG 11–12.
Torque-angle curve of a distal interphalangeal joint moving into flexion (toward the left), *A,* measured with the PIP joint held extended, *B,* measured with the PIP joint held flexed. This demonstrates a degree of tightness of the oblique retinacular ligament.

subsequent tests. At each test the results of three attempts may be recorded. Mathiowetz[21, 22] has extensively researched grip strength and discusses in a review article standardization of grip strength assessment and important variables (e.g., age, hand dominance) that the surgeon and therapist need to be aware of.

The evaluation of flexion strength at individual digits poses many problems. If a spring scale is used on each finger by means of a sling around one finger segment, and if the spring is held at right angles to the segment, it is quite possible to obtain a reading. However, the coordination between fingers at the mental level and at distal anatomic levels is such that it is almost impossible for a person to exert maximum force on one finger without some force being used on other fingers. If all fingers are permitted to contract unresisted except for the one being measured, then the physical interconnectedness of the muscles and tendons will result in an unduly high reading, because the finger being measured is receiving assistance from the adjacent fingers.

Thus if the patient is told to flex only one finger, the force will be less than maximal because of neurologic inhibition, and if he or she is allowed to contract all together, the

PRE POST SERIAL CASTING
RING PIP EXTENSION

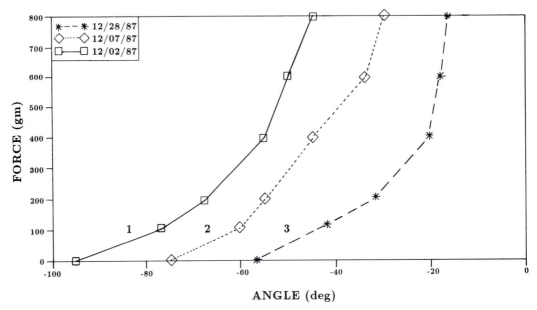

FIG 11–13.

Plaster of Paris casting of the PIP joint of the ring finger into extension. *1*, the curve created before casting. The gentle slope indicates that the joint has some mobility. *2* torque range of motion (TROM) 1 week after casting. Improvement is noted, with a slope indicating more casting may be beneficial. *3*, TROM after 3 weeks of casting. Note that the steep curve indicates a plateau has been reached in remodeling the joint. If the curve continues to be steep in the next week, consider discontinuing the casting. *(From Breger-Lee, D., Bell-Krotoski, J., and Brandsma, J.W.: J. Hand Ther. 3:7–13, 1990. Used by permission.)*

TWO TORQUE ANGLES
SLOPE

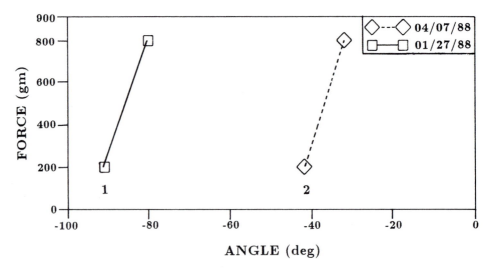

FIG 11–14.

A slope created by two torque angles is depicted with applied loads of 200 g and 800 g of force (tension) *1* before, and *2*, after casting *(read from left to right)*. Note the improved range of motion of the patient's PIP extension following serial plaster of Paris casting. 1/27/88—200 g, −93 degrees; 800 g, −80 degrees. 4/7/88—200 g, −43 degrees; 800 g, −34 degrees. *(From Breger-Lee, D., Bell-Krotoski, J., and Brandsma, J.W.: J. Hand Ther. 3:7–13, 1990. Used by permission.)*

FLEXOR TENDON TIGHTNESS

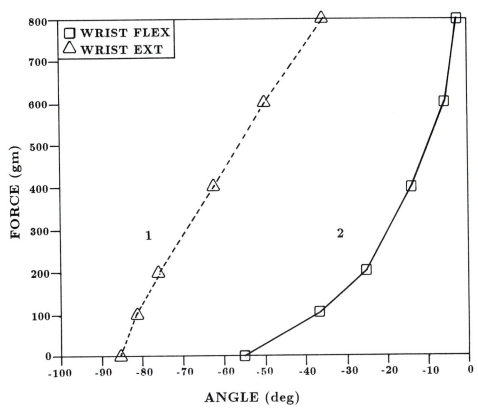

FIG 11–15.
Torque-angle curve of the PIP joint moving into extension measued *(1)* with the wrist at 45 degrees of extension and *(2)* with the wrist at 45 degrees of flexion. With the wrist extended a decrease in PIP extension is seen. This can be indicative of flexor shortening or tightness.

measured force on one may be more than maximal because of enhancement from other muscles. A device by Schneiderwind[29] (Fig 11–18), using a cylinder with four transducers, one for each finger, permits full force to be exerted by all fingers with equal resistance to each, while each may be measured separately. Such a device is not yet on the market and would be expensive and perhaps seldom used. If anybody wants to study the effect of any treatment on the flexion strength of a single digit, probably the simplest way would be to use just one force transducer embedded into a wooden cylinder so that it projects a few millimeters. The hand would then be positioned to grasp the cylinder so that the critical segment of the digit (e.g., the tip pulp) lies over the transducer. Then the patient would be told to grasp maximally with all fingers and the

force-transducer readout would give a reproducible number (except for the variables of the patient's enthusiasm, fatigue, blood sugar, sensory feedback, and the time of day). If this sounds difficult—it is.

Maximum Tension Capability of a Muscle or Group of Muscles

All the figures for muscle that we have calculated for cadavers are of *relative* tension. If these are to be translated into actual tension in any individual subject, it is necessary to get some estimate of the actual tension in at least one muscle, on the basis of which all the others can be assigned tension capabilities by the use of the relative scale.

We suggest the use of the flexor pollicis longus (FPL) for this purpose, since flexion of the IP joint of the thumb can neither be accomplished nor assisted by any other mus-

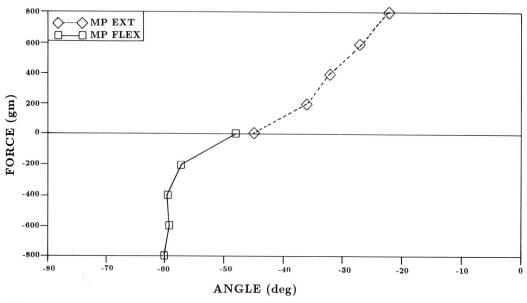

FIG 11–16.
Torque-angle curve graph of MCP flexion and extension of this patient's ring finger. Note flexion curve is steep, coming to an abrupt halt, representing the limited movement of that digit into flexion, whereas the curve for MCP extension represents more freedom of movement of the digit into extension. This patient required an MCP arthroplasty in order to allow for increased MCP flexion. *TROM* = torque range of motion. *(From Breger-Lee D., Bell-Krotoski, J., and Brandsma, J.W.: J. Hand Ther. 3:7–13, 1990. Used by permission.)*

cle. For this measurement it is necessary only to have a force transducer, such as Schneiderwind's, or even any strong calibrated spring scale with a sling. Probably the simplest method is to use the common pinch meter of the pattern shown in Figure 11–19. The subject is permitted to place his or her hand in any comfortable position such that (1) the distal segment of the thumb is horizontal and (2) the pulp of the thumb lies on one arm of the pinch meter and takes the thrust of the thumb on the distal pulp at the level of the middle of the thumbnail bed. In that position the subject is allowed to apply maximum force downward on the meter, which is resting on the edge of a table (Fig 11–20). The subject may support or reinforce the hand in any way as long as the distal joint is not supported in any way and the terminal segment remains horizontal.

In most subjects the distance from the axis of the joint to the middle of the thumbnail bed will be about three times the moment arm of the FPL at the distal interphalangeal (DIP) joint. Therefore, if the subject can exert 8 kg of flexion thrust by his or her pulp at the level of the thumbnail, it means that the FPL is contracting with a force of about 24 kg.

Pinch Strength

The standard pinch meter is an acceptable device for measuring pinch strength (see Fig 11–19) as long as it is recognized as an estimate of a total functional end point. It does not measure muscle strength as such but summates the ability of the patient to stabilize all joints, and it may actually be measuring the weakness imposed on the pinch by a single unstable or painful joint in a chain of well-supported joints. For example, arthrodesis of the MCP joint of an unstable thumb may sometimes result in a pinch meter reading that is 2 or 3 times higher than before. No new strength has been added; only stability has been added.

The pinch meter should never be held firmly by the therapist or fixed in any way while it is being used. A pinch is often weak because of instability of just one limb of the

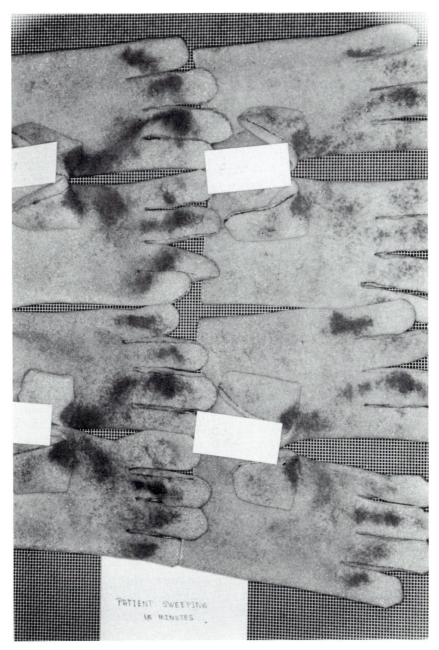

FIG 11–17.
Polyurethane gloves impregnated with microcapsules that rupture under stress and release blue dye. These gloves are used for standard tasks for a standard time by patients without sensation. They demonstrate areas that are used to excess and areas not used at all.

two digits involved (e.g., the thumb may be strong and the index finger weak or vice versa). By having the pinch meter held or stabilized by the therapist, the patient is able to use the stability of the meter device to pinch against the stronger of the two digits and produce a much higher reading than he or she could if the meter was free to move.

A very simple pinch meter is often used in India. It consists of a smooth metal disc, 2 cm in diameter, that has a hole drilled near one edge and is connected by a nylon loop to a calibrated spring. The patient pinches on the disc and the therapist pulls on the spring until the disc slips out of the pinch. The pinch strength is recorded as "can hold

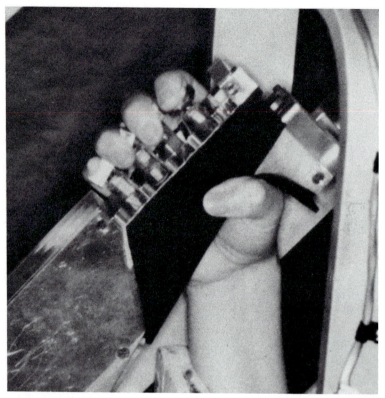

FIG 11–18.
Device for measuring grip strength of individual fingers (designed by Walter Schneiderwind[29]).

disc against *x*-g pull." The test should be made with normal finger moisture. Dry skin should be soaked for a few minutes and then dried before testing because the disc will slip out more easily between dry fingers.

SENSATION

This is a book about mechanics, and we do not plan to become involved in discussions of neurology. However, most methods of sensory evaluation involve mechanical

FIG 11–19.
The pinch meter is freely suspended while used to test a patient's strength in key pinch.

FIG 11–20.
Since the flexor pollicis longus (FPL) is the only muscle to flex the distal joint of the thumb, the actual tension capability of the muscle can be measured by exerting maximum flexion strength of the thumb IP joint against a transducer such as a pinch meter resting on the edge of a table. The proximal part of the thumb may be supported and the distal segment should be horizontal while the subject exerts maximum force down on the pinch meter with the middle of the thumbnail being directly over the terminal ridge of the pinch meter. The ratio of the moment arm of the FPL at the joint and the meter at the thumbnail is about 1:3, so that the reading of the meter may be multiplied by 3 to obtain the tendon tension.

forces, and many errors in evaluation are based on errors in mechanics. Also we are committed in this book to recommending rather simple and practical methods of evaluation of the hand, and sensory testing is one area where much time is wasted and many results are often open to question.

Objectives

For most surgeons sensory testing has as its object the answer to two questions: (1) Is the nerve recovering that was injured or compressed a few days or weeks ago, or that was repaired a few weeks ago? If not, should it be explored? (2) How well can the patient use the hand with the limited nerve supply that is present?

The first question is anatomic and the second is functional. The research-oriented surgeon or neurologist has additional questions such as, What nerve endings recover faster, pacinian corpuscles or the Merkel discs? or, are fast-adapting sensory endings, rather than slowly adapting endings, selectively recovering? These and many other questions are of profound interest and significance, but it is possible to make all the important decisions about the management of any individual case without knowing all the answers. More important, if surgeons insist on testing all modalities fully, they will use so much time that either all tests will be

skimped and hurried (therefore inaccurate and useless) *or* they will be done so seldom that they will not help in planning ongoing treatment.

Therefore we suggest that most surgeons choose just one test or two to evaluate recovery on a quantitative anatomic basis and one test or two to evaluate recovery on a functional basis. Then if more time is available, these few tests may be repeated more often to get a time-related profile of recovery or loss. Almost all tests use a mechanical (sometimes thermal) stimulus and are evaluated subjectively by the patient. We are all aware that there is a significant variability in the patient's subjective response. We may be less aware of the variability of the mechanical stimulus that we give.

Touch Threshold Testing

The Basis of Touch Threshold Measurement

1. The touch applicator never touches a nerve ending. A sharp needle may penetrate the epidermis and cause direct stimulation, but the Frey hair or Semmes-Weinstein fiber or a wisp of cotton or ballpoint pen never touches the nerve ending. The touch stimulus is activated by the *mechanical deformation of the skin.* The applicator bends

the skin and the skin deformation causes the firing of the nerve ending.

2. A single stimulus to a single nerve ending will rarely be transmitted up to the brain and perceived. The axon of a sensory nerve will fire (a) if several of its endings are stimulated at the same time—this is summation in space—or (b) if one ending is stimulated a number of times in succession——this is summation in time (or it may be both).

In our biomechanics laboratory Judy Bell-Krotoski, certified hand therapist, and William Buford, bioengineer, have explored some of the variability of testing stimuli. Using highly sensitive force transducers,[4] which are recorded on the screen of an oscilloscope, it can be shown that what a physician thinks is a single touch stimulus is usually a series of spikes of varying stress, dependent on hand tremors and pulse and magnified by the lever arm that carries the applicator. Thus what was intended as a single stimulus may result in a whole chain of firings from the nerve ending and may be perceived by the patient when a momentary touch of the same pressure would not be perceived.

To get a repeatable touch threshold measurement, it is best to (1) standardize the time element and (2) vary the area affected (and therefore the number of nerve endings). To standardize the time element, the duration of every application of the touch stimulus should be standardized and must not be repeated until the tester returns to the same site after testing other sites. Those examiners who use two or three repeated touches on the same spot must use the same technique at every spot that is tested.

The best way to vary the number of nerve endings that are stimulated at one touch is to vary the firmness of the touch. For example, if a small applicator of 1-mm^2 area is pressed on the skin with a 1-g force, it may create a dimple that is 5 mm in diameter. Thus it is probable that every nerve ending in that area will be stimulated. If the same size of applicator is pressed on the skin with a 5-g force, the dimple or crater may be 1 cm across and will be deeper, so many

more nerve endings will be affected. Normal skin, with a high density of endings, will register with 1 g and the patient will give a positive response. With partial denervation a larger area of skin will contain only the same number of endings, so the 1-g test will be negative and the 5-g test will be positive. Note that the area of skin that dimples is not dependent so much on the diameter of the applicator as on the force of application. The shape of the crater, for a given force, depends on the mechanical qualities of the skin and supporting tissues (Fig 11–21).

However, for a given force, there are a number of other variables, notably the thickness of the keratin layer of the skin. A very thin skin will dimple deeply but not widely and will simulate fewer superficial endings but more strongly. The same pressure on heavily keratinized skin may affect a wider area, but the movement (or deformation) may be so slight that nothing will register. A very small applicator, cut obliquely at the end, may even begin to penetrate a thin epidermis and elicit a pricking response rather than touch. Then again, the very same applicator using the same force on a

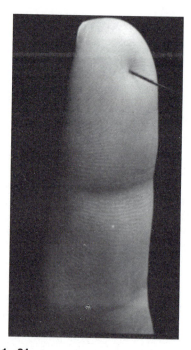

FIG 11–21.
In touch threshold testing it is the size and depth of skin deformation that determines the number of nerve endings that will be stimulated.

normal person will be recognized as touch on a fingertip, where there is a high density of endings, but may be negative on a forearm where there are fewer endings. The same test on a normal young person will yield positives at a very light level of pressure that will be negative on a much older person.[31]

With this number and variety of variables it is not possible to define absolute normal on the basis of a hand-held applicator applied by a variety of testers. However, this must not discourage us from using the test as the best that we have and as an excellent means of monitoring sensory change or recovery on the same patient, especially when tested with the same instrument by the same tester time after time.

Recommendations for Touch Threshold Testing.—For the sake of simplicity and standardization we recommend the following:

1. The tips of all touch applicators should be of standard size and shape (e.g., a ball of 1 mm diameter). At present such sets are not on the market.

2. The force of application should be varied through a range from light touch to deep pressure.

3. Not more than five forces need ordinarily be used in diagnostic tests and perhaps one or two for monitoring progress or recovery.

4. The five fibers should either be numbered simply 1 through 5 or preferably should be referred to by the load or force (e.g., 0.1 g up to 100 g).*

5. The system of grading the Semmes-Weinstein fibers by numbers that represent the logarithm of ten times the force required to bend the fiber results in figures that go to two decimal places and that have no clearly obvious relationship to actual touch.

*Research Designs, Inc., which markets the Semmes-Weinstein filaments, is now marketing, at our suggestion, a minikit of the five most used filaments, numbered 1 through 5 and color-coded. Sensory testing filaments assembled at our Gillis W. Long Hansen's Disease Center are also available through the Hansen's Disease Foundation, Inc., Filament Project, Carville, LA 70721.

Few surgeons or therapists think easily in logarithms. When one is recording the result of tests that are subject to wide variation as a result of human error and judgment, it is better to use round numbers. We remember a professor of ophthalmology who wrote a paper in which he said: "I have observed that 27.19% of lepromatous leprosy patients have moderate to severe dryness of the eyes." How is it possible to define moderate to severe dryness with enough precision and repeatability to justify two places of decimals in percentages? He would have better said ". . . about one patient in three."

The greatest single source of error in conventional touch threshold testing is that the patient is given other clues than the intended stimulus. The commonest clue is that of movement. Even a 1-g fiber will move a finger that is passive and free to move. That movement, perhaps too small to be seen, will be perceived proximally in the forearm by tendon-muscle movement even when the finger is insensitive. The patient who then points to the totally insensitive index finger and says he or she has been touched there is not cheating. Patients do not know how they know, but they know when they have been touched and are telling what they know. The patient is a good guy; it is the tester who is the dupe.

Bell-Krotoski and Buford[4] also showed that testers using two-point discrimination usually press with different total force when using two points than with one. Thus if the finger is free to move, the patient has a clue that more force means two, less force means one. For these reasons, every hand must be fully supported while being tested. A hand may conveniently be supported by a lump of putty, into which it is pressed (Fig 11–22).

Visual clues are common. If the patient can see the tester's face, he or she knows that the tester looks carefully when a part is touched, but does not look when not touching. So the patient says yes and no from facial expressions. The converse is also true. A really experienced tester uses other clues that are as valuable as the yes or no answer. The tester will observe how confidently and promptly the patient responds to a test on the normal hand and how hesitant and un-

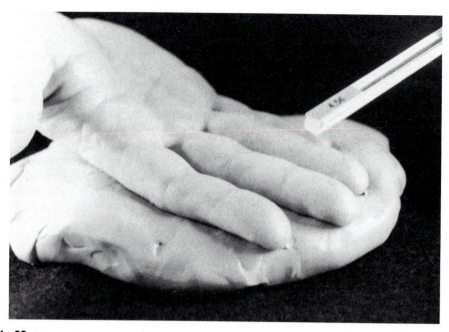

FIG 11–22.
A lump of soft putty is a convenient way to support the fingers to prevent movement while testing for touch thresholds.

sure he or she sounds when the partially denervated hand is being tested. There is really no substitute for observation and experience in testing. In spite of all the weaknesses and sources of error, the touch threshold is probably the best method of mapping a whole area of sensory loss for comparison with later similar maps.

Technique.—A neat blindfold or a pair of eyeshields should cover the patient's eyes as a matter of routine or there should be a screen to prevent visual clues (Fig 11–23). The patient is asked to point to the place where he or she has been touched as soon as it is felt. A delay in pointing, or a pointing close to but not at the point,* is given a different rating. No point is marked as sensitive until two positive responses have been

*In a progressive sensory neuropathy, a delayed response or misreference in space always indicates loss of nerves or nerve endings. In recovery after nerve repair, misreference may only mean that the brain has not yet adjusted to the changed pattern of reinnervation that may already be complete. Many studies have appeared in recent years in which nylon filaments have been used in the assessment and evaluation of nerve function, especially in leprosy and diabetic patients, and studies of which the filaments themselves have been the subject of research.[2–8, 16, 19, 23, 32]

given to stimuli, separated by touches at other sites. Successive stimuli should be scattered randomly, so that the patient cannot anticipate where the next stimulus will be. A known sensitive area should be established early, and it should be touched again if two insensitive areas have been touched in succession. A tension builds up if the patient knows he or she is being touched again and again and feels nothing. This tension makes the patient desperate to say yes or to point to where he or she might have been touched. As soon as something definite is felt and can be pointed to, the tension is relaxed and the patient becomes objective again.

The test may be speeded up by accepting a yes or no response from the patient rather than requiring him or her to point. If so, an occasional false touch movement must be made without actually touching to check the continued objectivity of the patient. A 2-second delay in response is recorded as negative. The examiner may use felt pens with colored washable ink to record areas of different touch thresholds. Finally the limb may be photographed or the map reproduced on a sheet of paper.

FIG 11–23.
The examiner's hand is supporting the patient's fingers in a comfortable position while testing for touch threshold.

When monitoring a recovering nerve after suture, it saves time to inspect the first map of sensory loss and draw a line through the middle of it from proximal to distal and define the proximal end by anatomic landmarks. At the next testing time it may not be necessary to test the whole area or make a whole new map but only to test a row of points along that line to determine whether and how fast sensation is growing distally.

Sharp or Dull

The second mapping test that is rather simple is one for sharpness and dullness sensitivity using two applicators, one of which is pointed (not very sharp) and the other rounded. The technique is similar to the touch threshold test but there is no need for avoiding movement of the digits, because the patient knows he or she is being touched. The patient only has to say if the touch is by the sharp point or the dull (blunt) point. If the sharp point is really very sharp (e.g., a hypodermic needle), the tester will be afraid to touch firmly with it for fear of penetrating the skin and drawing blood. Therefore the touch will be light with the sharp and firm with the dull. This is a clue. This is not just a test of pain, it is a discrim-

inatory test and a number of different nerve endings are involved. Patients who cannot feel pain may still distinguish sharp from rounded. Even penetration of the skin may not always be painful in a normal person; any smart mosquito knows how to penetrate the skin painlessly; it just takes a very sharp proboscis and a knowledge of where the nerve endings are likely to be. The important thing with sharpness and dullness testing is that the patient must first feel the test on normal skin and understand exactly what his or her responses mean.

Two-Point Discrimination

The two-point discrimination test is not a mapping test for an area of loss; it is a test of the quality of sensation that has returned and therefore of the ability of the digit to function as a sensory organ. Thus it is really a function test. It is related to the number of nerve endings in a given area of skin, but the level of achievement may be improved by sensory education even with no additional nerve recovery. It is a good predictor of the ability to recognize objects that are touched and held in the hand. Thus the two-point discrimination test should not usually be used for mapping areas of loss or recovery[33] but for quality of (1) nerve ending density in the skin and (2) reeducation of the area of recognition in the brain.

The two-point discrimination test is subject to most of the same errors as the threshold touch test and one or two more. The two points must touch simultaneously, or the patient will have the obvious clue that the pressure builds up in two steps in time, even though he or she cannot distinguish separation in space. Although the two-point test is not a threshold test, the pressure applied should be light, perhaps 5 g. The higher the pressure, the wider the area of skin that is deformed and stimulated. At a seminar at Carville on sensory testing, we invited a number of physicians, surgeons, neurologists, and therapists to apply two-point tests on a fingertip. The applicator was wired with strain gauges, and the force applied by each tester was recorded. One of the participants (Omer) consistently used about 4 g, others used 10 to 20 g, and some used 30 to 40 g. A 40-g application will deform a far greater area of skin than 4 g. Moberg, who uses about 5 g, points out that if the two points are applied transversely to a fingertip, the use of high pressure will often allow one point to register on the territory of the dorsal cutaneous nerve on one side of the finger and the other point may deform the skin on the dorsal territory on the other side. Thus two-point testing may give a positive response, 7 to 10 mm, in a median nerve area that is really totally anesthetic. Moberg insists that false positives, caused by intact dorsal nerves, are so common that he always blocks the radial nerve before testing median nerve areas for two-point discrimination. The same holds for ulnar or median nerves when only one is damaged in a finger (e.g., the ring finger) supplied in part by both. Moberg[24] would block the normal nerve before testing the quality of the other. Omer does not do this nor do we in routine cases, but we need to be aware of the source of error and block nerves in doubtful cases.

Pain

Pain is so difficult to test that we doubt that it should be routinely evaluated. Except in a few diseases, pain is lost and then recovers along with other modalities of sensation. The evaluation of pain independent of other modalities is important sometimes in medical diagnosis and to our understanding of the nature of disease and injury and mechanisms of recovery, but this is beyond the scope of this book. In conditions such as congenital indifference to pain and in some radicular neuropathies patients will recognize and respond to painful stimuli and call them pain even though they do not feel that they are in any way unpleasant. Such people are not protected by pain and may damage themselves as severely as patients with leprosy who have lost all skin sensation. For the routine evaluation of nerve injury, as well as of leprosy, an evaluation of sharp and dull sensation gives a picture that is likely to parallel pain. If a closer parallel is needed, then heat and cold perception or just heat perception may be tested, which may be carried to the point of threshold pain with minimum danger.

Hot and Cold

A common fallacy is to speak of temperature sense. The skin cannot perceive a scale of temperature. It senses heat flow. A wooden handle may be at the temperature of boiling water but will not be felt as very hot, nor will it cause a burn. A metal handle at the same temperature will be felt as very hot and may cause a burn. Thus in testing heat and cold perception the applicator must *deliver* a standard quantity of heat per unit of area and time, no matter what temperature it is. Heat flow depends on the conductivity of the applicator and of the skin and also on the temperature differential between the two. The perception of heat also depends on the mechanisms for dispersion of heat (i.e., the circulation). The same heat applicator at the same temperature will be felt as hotter if the circulation has just been occluded by a tourniquet.

Probably the best instrument for testing heat and pain is the Hardy-Wolff-Goodell* dolorimeter, which has a metal cone that de-

---

*Williamson Development Co., Inc., West Concord, Mass.

livers a standard beam of infrared heat to a standard area of skin, which has to be blackened with carbon black before the test. The patient has to signal the onset of feeling heat and then pain. The test is quantified by time in seconds. There is a possibility of producing a tiny burn if the patient cannot feel and if the instructions on the machine are not followed carefully. This is a research tool and we like it, but it is rather expensive for routine use. The use of test tubes with hot water in one and cold water in the other is, unavoidably, so variable that it is probably not worth using.

Protective Sensation

Many testers use their own terminology and threshold levels to denote what is often called protective sensation. For example, a level between 4.31 = 2 g and 4.56 = 4 g on the Semmes-Weinstein scale has been designated as *protective* sensation only, and a level greater than 6.65 = 300 g has been designated as *untestable*. These have been correlated with experience. There is a tendency to have repeated injury in hands that are rated as having lost protective sensation.

However, we prefer not to state that a certain level is safe and another is unprotected. It is all relative. The time-response to heat of patients with partial nerve loss follows a slope on a graph, and all that one can say is that a person with minimal nerve loss will feel pain from a standard heat source faster than a person with a more severe loss. We once watched a leprosy patient walking down a corridor swinging his arms. His knuckles hit a projecting doorknob and he walked on about two strides, then suddenly stopped and looked at his knuckles, and then angrily looked back at the doorknob. He had just a little residual pain sensation. It was slow, yet perhaps it would have protected him from being deeply burned from a moderate heat source. He would move his hand away after 1 or 2 seconds and suffer only a blister.

We need to explain these things to our patients, and few will learn it before they experience their first burn. We should use the occasion of that first injury to evaluate the sensory threshold in the injured area (af-

ter healing) and also the severity of the stimulus that caused it and reiterate and dramatize our earlier warnings about the lack of protective sensation.

Tinel Sign

This is an old and fairly reliable index of nerve recovery. It should be done gently, tapping first in the distal insensitive area and finishing at the level of nerve injury or suture. The level at which the tingling is first felt must be marked and measured relative to an anatomic landmark. Omer has demonstrated the value of the tuning fork as a method of standardizing the stimulus (of tapping) so that the Tinel test becomes more significant and repeatable.[26]

FUNCTIONAL SENSATION OR SENSIBILITY

Sensation is purposeful. The mere growth of nerve fibers and nerve endings is not meaningful to the brain until impulses are generated and return to the brain associated with movement and sight and recognition and usefulness.[15, 35] Thus sensory reeducation is an active process and is progressively achieved; it is accomplished by the patient when he or she is offered a variety of objects and textures to hold, run through the fingers, and work with. The subsequent blindfold tests, picking up and recognizing familiar objects, are arbitrary but may be standardized by timing. In real functional tests a patient is free to move his or her fingers and manipulate the objects and textures he or she is trying to recognize. Such tests are not a waste of time; they are an important part of the treatment. Nerves and nerve endings must be thought of as an active purposeful outreach from the brain, which is trying to evaluate its own environment. Sensory nerves actually grow faster and arborize more freely when their growth is sensed as meaningful by the brain. While nerves are growing they are simultaneously making proximal connections in the brain. These new synaptic channels are serving to replace old patterns of sensory recognition based on axons that have been destroyed and to establish new patterns based on what the new nerve endings are saying. This process can only develop with use. No proximal pattern recognition can take place until it has impulses to correlate. Thus a person with a recovering nerve should be encouraged to play a constant game against self, alternately closing and opening the eyes while playing with an infinite variety of shapes and textures. The formal pick-up tests with standard objects are simply interruptions in a life of constant exploring tactile play.

Various attempts have been made to develop standardized pick-up tests with standardized timing. However, so many factors, such as intelligence and dexterity, come into the picture that Omer now recommends testing a nerve-damaged hand against the other hand of the same patient and recording the speed with which a tray full of standard objects is picked up and recognized by the good hand and then by the other.[27]

In pick-up tests there is no cheating. The patient uses whatever sensation he or she has, including the parts that have never been damaged. The surgeon should watch this activity before recording the recovery as excellent based on any tests. if the patient after "recovery" from median nerve injury still uses the backs of the fingers and the ring and little fingers to recognize objects, the result is not yet excellent.*

PRESSURE

Pressure at the surface of the body is not easy to measure. The problem is that any device, such as a pressure transducer, has thickness and occupies space. If such a device is placed under a sling or bandage or plaster cast, it makes a little bulge and thus concentrates pressure on itself. Probably the simplest device is the pneumatic bubble transducer produced by New Generation.† This consists of a thin plastic bulb with a

---

*For further reading on physical evaluation of nerve injury I suggest Omer, G.: Physical diagnosis of peripheral nerve injuries, Orthop. Clin. North Am. 12:207–228, 1981, and for discussion of the management of nerve loss Parry, W.: Rehabilitation of the hand. ed. 4, 1981, Sevenoaks, Kent, England, Butterworth & Co.

†New Generation, Temecula, Calif.

copper foil electrode in each wall. The bulb is inflated through a fine tube by a sphygmomanometer bulb. When the pressure of the air in the bulb exceeds the external pressure, it lifts the two walls apart and breaks the circuit. The pressure is read from the digital pressure evaluator. This inexpensive device may be used to train oneself to estimate the relationship between the tension on a sling and the pressure that it causes on a finger. It is also useful to learn about the actual pressure environment inside a cast or pressure bandage and to learn the relationship between actual pressures and the signs and clues (such as blanching of the skin) that are often used to monitor pressure on the skin.

VOLUME

The volume of the hand is a sensitive index of edema and of inflammation. It is reliable, easy to measure and understand, and can be recognized by the patient as evidence of his or her compliance with the treatment program.[1, 23, 30, 34]

The volumometer in Figure 11–24 may be constructed of acrylic sheets cemented at the seams. So long as the same volumometer is used each time, it does not matter whether the water level of the overflow measuring jar reaches exactly to any anatomic level. The essential point is that the hand must be immersed to the same stop point every time; therefore there is a rod across the volumometer that stops the hand at the web between the middle and ring fingers. Our original design has a flat, level overflow spout sprayed with silicone water repellent. The water rushes out into the measuring jar as the hand is immersed and then stops dead without drips. There is now a commercial volumometer with the same basic specifications (Volumeters Unlimited, 1307 Sandra Way, Redlands, CA 92374). The technique is so simple patients enjoy doing it themselves. Measurements should be recorded on a graph (see Fig 9–7).

Very small volumometers may be constructed for volume measurements of individual fingers in case one needs to investigate, for example, the inflammation induced

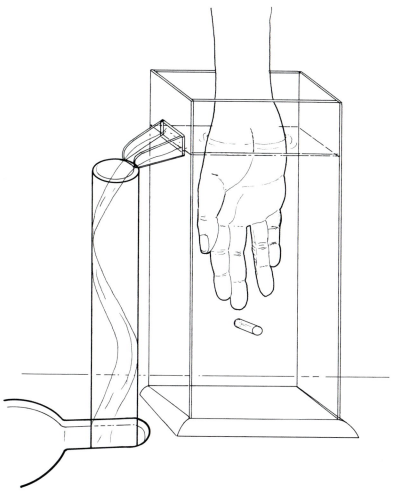

FIG 11–24.
Hand volumometer. The rim of the overflow spout is wide, level, and flat. It is sprayed with silicone water repellent. There is no dripping. The flow is fast, then stops. The transverse rod ensures immersion to the same level every time.

by the stress of different amounts of force used to overcome single joint stiffness and contractures. In the absence of a volumometer, a rough measure of swelling may be made by a flexible measuring tape placed around the same part of a finger each day. However, this is an inferior and inaccurate method. Its only advantage is that the patient can do it at home.

TEMPERATURE

Temperature is still neglected in most hand therapy centers. It is a very valuable index of inflammation. Until it is used regularly one does not realize how easily a therapist or a patient may overstress healing tissue or even normal tissue. It is also an excellent index of the inflammatory activity of rheumatoid arthritis in individual joints and of the extent and spread of the inflammation of infection. It is of special value in treating the insensitive hands of patients who cannot feel the pain of inflammation, and in the management of malingering or of hypochondriacal patients who are not good witnesses of the amount of pain they experience. In all temperature measurements the following must be recognized:

1. The actual temperature is rarely important. It is the (ΔT) delta temperature or

relative temperature that is important. What we record on the graph is the *difference* between the area at risk and some comparable area that is subject to the same ambient temperature. Whereas the temperature of the normal hand may vary through many degrees in different environments, there is rarely more than 1°C difference between one part of a hand and the same part of the other hand when both are measured in the same environment. For example, if an injured or stiff finger, or one that has been operated on, is being followed through healing and treatment and the PIP joint temperature is 35°C while the temperature of the normal joint on the other hand is 32°C, then we record +3°C on the graph. The following day the affected finger may be 35°C, but the normal finger may be only 31°C (perhaps because the room is cooler); now we record +4°C on the graph. The actual temperature of the injured finger is the same, but the difference (ΔT) is higher. This may indicate an increase in inflammation.

2. Temperatures may be misleading if they are taken immediately after removing dressings or immediately after exercise. The most significant differences are seen after both hands have been exposed to the same environment and at rest for 10 minutes or more.

3. Temperature differentials (ΔTs) are only a sign, not a diagnosis. Normal healing processes result in a raised temperature. This is not bad. Physical activity in a healing hand will elevate the temperature (hyperemia). However, if repeated activity causes progressively increasing temperature that is sustained, this often means that the activity is excessive and is causing progressive inflammation that may lead to stiffness later.

Methods of Temperature Measurement

Assessment by Touch.—A practiced hand can detect a contrast of 1.25°C. Most of us can detect 1.5°C and all can recognize 2.0°C. Most of the significant temperature contrasts in the patient's hand will be more than 2°C. It is a good habit for the therapist to hold the patient's hand and stroke it before treatment. This helps to get the patient relaxed and helps the therapist to be aware of new features of tenderness or localized temperature change. If a skin thermometer is kept in the work area, it will be easy for the therapist to check his or her impression of a hot spot by an actual temperature measurement. Note that objective temperature measurement should not be taken immediately after holding the hand, as this alters the surface temperature.

Thermistor.—There are many skin thermometers on the market. The ones we recommend consist of a handle or probe with a tiny thermistor on the end. This is connected to a battery-operated device that has a temperature dial. Better models may have two probes that can be used simultaneously, one on the affected and one on the normal or reference area.

Advantages.—Thermistors are simple, inexpensive, and accurate.

Disadvantages.—Thermistors are rather slow. The skin needs to be touched (firmly) with the probe.

Selection.—A model with a fast response should be chosen. Specifications will quote a time (e.g., 1.2 seconds), but this refers only to the time taken for the needle of the dial to move two thirds of the distance from 0 to the final resting point of the readout. The needle slows down a lot in the last few degrees before coming to rest, so a "1.2 second" rating may take 10 seconds to give a reading. A second reading, when the probe is moved from one finger to another, will be much faster if the move is made quickly before the probe cools down. We use the Preston TRI-R model. It is sturdy and accurate, it rarely needs new batteries, and is trouble-free. It is moderately fast.

Radiometer.—The radiometer uses an infrared-sensitive cell with an instant readout.

Advantages.—Radiometers are fast. A succession of joints can be scanned as fast as the eye can follow the needle. It does not touch the skin, and it is reliable. We use the Mikron.

Disadvantage.—They are expensive, costing between $1,000 and $2,000.

Thermograph.—Thermographs also use infrared emissions, using the same system as radiometers, but they display the temperature as a total scan six times per second on a CRT (cathode ray tube) screen, giving a moving picture as the hands are moved around during clinical examination.

Advantage.—The thermograph gives a total picture, so unexpected hot spots are picked up early.

Disadvantages.—Thermographs are expensive, costing around $50,000 and will usually be justifiable only if shared with other clinical areas such as general orthopaedics (fractures), plastic surgery (flaps), vascular surgery (scan of dysvascular limbs and stumps), prosthetics and orthotics (scan of stumps to detect early signs of stress), or research. Thermograph equipment must not

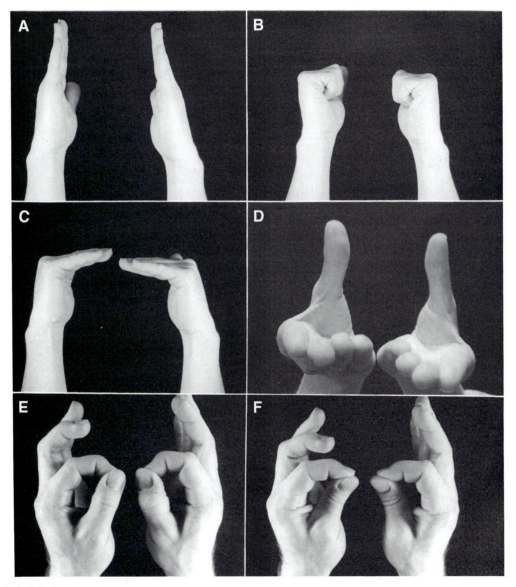

FIG 11–25.
Six positions that we use for photographing preoperative and postoperative hands and for later follow-up records. The patient is asked to imitate these positions.

be allowed to remain in service departments such as radiography; they are *clinical* tools and should be part of the ongoing evaluation of limbs. Unlike in radiography, the hands of the surgeons and therapists must enter the picture, to feel the relationship between hot spots and the subjacent anatomy.

PHOTOGRAPHY

Many preoperative clinical photographs seem to be taken with a view to maximize the apparent disability, whereas postoperative photographs minimize any residual deformity or disability. We strongly advocate having a series of diagrams or photographs of standardized positions of the hand that the patient will be asked to imitate. The same positions and instructions are used before and after treatment (Fig 11–25). The camera distance, focus, and lighting, as well as the background, must be the same every time. In this way a series of photographs becomes a meaningful record of progress (or regress). Of course, intraoperative photographs and immediate postinjury photographs are in a separate category. Modern video cameras are remarkably inexpensive and are a very valuable tool. They are most useful if a patient is photographed performing a standard activity selected to involve the whole hand with emphasis on the injured or paralyzed part. The same activity in the same setting is photographed again after treatment.

REFERENCES

1. Beach, R.B.: Measurement of extremity volume by water displacement, Phys. Ther. 57:286–287, 1977.
2. Bell-Krotoski, J.A.: "Pocket filaments" and specifications for the Semmes-Weinstein monofilaments, J. Hand Ther. 3:26–31, 1990.
3. Bell-Krotoski, J.: Advances in sensibility evaluation, Hand Clini. 7:527–546, 1991.
4. Bell-Krotoski, J.A., and Buford, W.L.: The force/time relationship of clinically used sensory testing instruments, J. Hand Ther. 1:76–85, 1988.
5. Bell-Krotoski, J.A., and Tomanick E.: The repeatability of testing with Semmes-Weinstein monofilaments, J. Hand Surg. 12A:155–161, 1987.
6. Birke, J.A., Cornwall, M.W., and Jackson, M.: Relationship between hallux limitus and ulceration of the great toe, J. Orthop. Sports Phys. Ther. 10:172–176, 1988.
7. Birke, J.A., and Sims, D.S.: Plantar sensory threshold in the ulcerative foot, Lepr. Rev. 57:261–267, 1986.
8. Bowen, V.L., Griener, J.S., and Jones, S.V.: Threshold of sensation: Inter-rater reliability and establishment of normal using the Semmes-Weinstein monofilament, J. Hand Ther., 3:36–37, 1990.
9. Brand, P.W.: Rehabilitation of the hand with motor and sensory impairment, Orthop. Clin. North Am. 4:1135–1139, 1973.
10. Brandsma, J.W.: Secondary defects of the hand with intrinsic paralysis: prevention, assessment and treatment. J. Hand Ther. 3:14–19, 1990.
11. Brandsma, J.W., and Brand, P.W.: Quantification and analysis of joint stiffness. In Proceedings International Conference on Biomechanics and Clinical Kinesiology of Hand and Foot, Madras, India, IIT December, 1985, pp. 65–68.
12. Breger-Lee, D., Bell-Krotoski, J., and Brandsma, J.W.: Torque range of motion in the hand clinic, J. Hand Ther. 3:7–13, 1990.
13. Breger-Lee, D., et al.: Reliability of torque range of motion: a preliminary study. Submitted to J. Hand Ther. 1992.
14. Cantrell, T., and Fisher, T.: The small joints of the hands, Clin. Rheum. Dis. 8:545–557, 1982.
15. Dellon, A.L.: Evaluation of sensibility and re-education of sensation of the hand, Baltimore, 1981, Williams & Wilkins Co.
16. Diamond, J.E., et al.: Reliability of a diabetic foot evaluation. Phys. Ther, 69:797–802, 1989.
17. Flowers, K.R., and Pheasant, D.S.: The use of torque angle curves in the assessment of digital joint stiffness. J. Hand Ther. 1:69–74, 1988.
18. Glanville, HJ.: Objective assessment of treatment of hand injuries: a new method of measurement, Ann. Phys. Med. 7:304–306, 1964.
19. Holewski, J.J., et al.: Aesthesiometry: Quantification of cutaneous pressure sensation in diabetic peripheral neuropathy, J. Rehabil. Res. 25:1–10, 1988.

20. Kolumban, S.L.: The role of static and dynamic splints, physiotherapy techniques and time in straightening contracted interphalangeal joints. Lepr India 41:323–328, 1969.

21. Mathiowetz, V: Grip and pinch strength measurements. In Amundsen, L.R., editor: Muscle strength testing, London, 1990, Churchill Livingstone, pp. 163–177.

22. Mathiowetz, V.: Reliability and Validity of grip and pinch strength. Crit. Rev. Phys. Rehabil. Med. 2:201–213, 1991.

23. McGough, C.E., and Zurwasky, M.L.: Effect of exercise on volumetric and sensory status of the asymptomatic hand. J. Hand Ther., 4:177–180, 1991.

24. Moberg, E.: Evaluation of sensibility in the hand, Surg. Clin. North Am. 40:357–362, 1960.

25. Moberg, E.: The role of cutaneous afferents in position sense, kinaesthesia, and motor function of the hand, Brain 106:1–19, 1983.

26. Omer, G.: Report of the committee for evaluation of the clinical result in peripheral nerve injury, J. Hand Surg. 8:754–759, 1983.

27. Omer, G.E.: Sensory evaluation by the pickup test. In Jewett, D.L., and McCarroll, H.R., Jr., editors: Nerve repair and regeneration: its clinical and experimental basis, St. Louis, 1980, Mosby-Year Book, Inc.

28. Reddy, N.R., and Kolumban, S.L.: The effects of daily and once per week plaster of Paris cylindrical splinting on contracted proximal interphalangeal joints in leprosy patients. Lepr. India 47:151–155, 1975.

29. Schneiderwind, W.: A new method for simultaneously evaluating individual finger function during power grip, thesis, Boston University, 1972.

30. Schultz-Johnson, K.: Volumetrics: a literature review. Santa Monica, Calif., 1988, Upper Extremity Technology, pp.1–36.

31. Semmes, J., and Weinstein, S.: Somato sensory changes after wounds in man, Cambridge, 1960, Harvard University Press.

32. Sosenko, J.M., et al.: Comparison of quantitative sensory-threshold measures for their association with foot ulceration in diabetic patients. Diabetes Care 13:1057-1061, 1990.

33. Szabo, RM., Gelberman, RH., and Dimick, M.P.: Sensibility testing in patients with carpal tunnel syndrome, J. Bone Joint Surg. 66A:60–64, 1984.

34. Waylett-Rendall, J., and Seibly, D.S: A study of the accuracy of a commercially available volumeter. J. Hand Ther. 4:10–13, 1991.

35. Wynn-Parry, C.B., and Salter, M.: Sensory re-education after median nerve lesions, Hand 8:250–257, 1976.

CHAPTER 12

Mechanics of Individual Muscles at Individual Joints

The purpose of this chapter is to direct attention to the mechanical significance of individual muscles and tendons at the joints that they control. We have tried to do this with drawings and diagrams that emphasize the fiber structure of each muscle and with graphs and tables that apply numbers to such things as excursion and moment arms. The tables (Tables 12–1 and 12–2) of resting fiber length, mass fraction, and of tension fraction of all muscles below the elbow are taken from our own studies,[1] published in the *Journal of Hand Surgery*, as is the graph that plots fiber length against tension fraction (Fig 12–1). This graph is one that we have found useful while discussing and planning surgical intervention to compensate for paralysis. For each muscle, in addition to a general mechanical description, we have added bar graphs that compare its performance to the performance of the flexor pollicis longus (FPL). We have chosen the FPL as a standard because its function is well known and because it is one of the very few muscles whose tension capability can be directly calculated from measurements taken noninvasively in any individual. This is done by measurement of flexion torque at the interphalangeal (IP) joint of the thumb. Comparative bar graphs have been prepared for (1) tension fraction, (2) mass fraction, and (3) mean fiber length. Each of these represents an average from several hands.

Also presented are some graphs that show the way in which the moment arms of various tendons change while the joint moves from extension into flexion or from abduction to adduction. Some of these graphs represent the mean of a number of hands. Others are actual direct readouts from our computer. For the latter cases we have chosen a single fresh-frozen hand so we do not claim that these graphs represent an average or mean, but they do represent raw data from a typical good mobile hand. The joint angles are fed into the computer from a potentiometer that records changing joint angles, while other potentiometers record the movement of each tendon at the same time. These are integrated and displayed as a continuous curve of moment arm. At each end of joint motion the record becomes meaningless because of joint or tendon constraints and starting or stopping errors. The smooth curve between these ends is significant and gives insights into such factors as bowstringing of tendons (even inside sheaths). The most constant and level graphs of moment arms are from digital extensor tendons at metacarpophalangeal (MCP) joints. These are so sure and constant that we use them to characterize a hand. The moment arm of the extensor digitorum at the MCP joint of a given hand is dependent on the skeletal diameter of the bone. This may be taken as an index of the size and diameter of all joints of that hand and as a standard for comparison of any hand with any other hand. The digital extensor tendons on the back of the hand are so closely

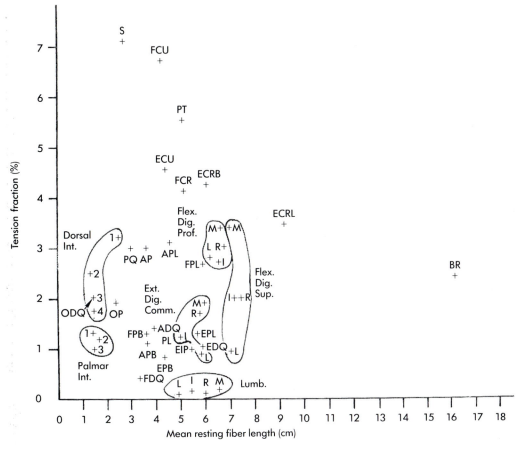

FIG 12–1.
Using values from Table 12–1, this diagram shows the relationship between fiber length and tension-producing capability. Some muscle groups are circled for clarity. *(From Brand, P.W., Beach, R.B., and Thompson, D.E.: J Hand Surg., 6:209-219, 1981. Used by permission.)*

subcutaneous that they are the most readily accessible for measurement of excursion at surgery.

At my age (P.W.B.) the skin of the dorsum of my hand is transparent enough for me to be able to see and measure the excursion of my digital extensors (using anatomic features such as the bifurcation of junctura) and relate them to the angular movement of my finger joints. I can thus be quite precise about my own tendon and joint configuration. Younger surgeons should not be discouraged about their inability to do this. They will improve with time.

The graphs of the moment arms of the wrist-moving tendons were obtained by manual measurements of joint motion rather than by potentiometers. They may look a little different from the graphs of the finger

joints but are equally reliable. They have the advantage that each graph is the mean of seven to nine hands. Note that in all the graphs in which we have plotted moment arms of tendons against the changing angles of joints, we are recording functional moment arms. At some joints there are tendons that are unable to move freely because of some constraint, such as participation in a complex dorsal expansion. In such cases the geometric moment arm may give a false idea of the ability of the tendon to move the joint. Our functional moment arm may be geometrically inaccurate but functionally realistic. In cases in which a constrained component of a complex tendon gives a lower moment arm than one would expect, we have repeated the measurement using a fine wire running through a capillary polyethyl-

256 *Clinical Mechanics of the Hand*

TABLE 12–1.
Reference List of Normal-Expected Values for Fiber Lengths, Mass Fractions, and Tension Fractions in Adults*

Muscle	Resting Fiber Length (cm) Mean†	Mass Fraction (%) Mean†	Mass Fraction (%) SD†	Tension Fraction (%) Mean†
ADQ	4.0	1.1	0.23	1.4
AP	3.6	2.1	0.40	3.0
APB	3.7	0.8	0.18	1.1
APL	4.6	2.8	0.34	3.1
BR	16.1	7.7	2.0	2.4
Shortest fibers	10.9			
Longest fibers	21.3			
First DI				
First metacarpal origin	3.1	0.8	0.25	1.3
Second metacarpal origin	1.6	0.6	0.11	1.9
Total first DI	2.5	1.4	0.29	3.2
Second DI	1.4	0.7	0.17	2.5

Muscle	Resting Fiber Length (cm) Mean†	Mass Fraction (%) Mean†	Mass Fraction (%) SD†	Tension Fraction (%) Mean†
FCU	4.2	5.6	0.66	6.7
FDP				
Index finger	6.6	3.5	0.76	2.7
Middle finger	6.6	4.4	0.94	3.4
Ring finger	6.8	4.1	1.1	3.0
Little finger	6.2	3.4	0.93	2.8
FDS				
Index finger	7.2	2.9	0.64	2.0
Middle finger	7.0	4.7	1.1	3.4
Ring finger	7.3	3.0	0.84	2.0
Little finger	7.0	1.3	0.81	0.9
FDQ	3.4	0.3	0.10	0.4
FPB	3.6	0.9	0.22	1.3
FPL	5.9	3.2	0.42	2.7

Muscle				
Third DI	1.5	0.6	0.19	2.0
Fourth DI	1.5	0.5	0.13	1.7
ECRB	6.1	5.1	1.3	4.2
ECRL	9.3	6.5	0.77	3.5
Shortest fibers	6.3			
Longest fibers	12.3			
ECU	4.5	4.0	0.52	4.5
EDC				
Index finger	5.5	1.1	0.20	1.0
Middle finger	6.0	2.2	0.51	1.9
Ring finger	5.8	2.0	0.35	1.7
Little finger	5.9	1.0	0.41	0.9
EDQ	5.9	1.2	0.35	1.0
EIP	5.5	1.1	0.36	1.0
EPB	4.3	0.7	0.32	0.8
EPL	5.7	1.5	0.48	1.3
FCR	5.2	4.2	0.87	4.1

Muscle				
Lumbricals				
Index finger	5.5	0.2	0.08	0.2
Middle finger	6.6	0.2	0.06	0.2
Ring finger	6.0	0.1	0.06	0.1
Little finger	4.9	0.1	0.05	0.1
ODQ	1.5‡	0.6	0.20	2.0
OP	2.4‡	0.9	0.26	1.9
PI				
First	1.5	0.4	0.12	1.3
Second	1.7	0.4	0.11	1.2
Third	1.5	0.3	0.08	1.0
PL	5.0	1.2	0.34	1.2
PQ	3.0‡	1.8	0.32	3.0
PT	5.1	5.6	1.24	5.5
Superficial fibers	6.5			
Deep fibers	3.7			
Supinator	2.7‡	3.8	0.95	7.1

*From Brand, P.W., Beach, R.B., and Thompson, D.E.: J. Hand Surg., 6:209–219, 1981. Used by permission.
†Data from 15 hands determined the mean and standard deviation of the mass fraction for each muscle. Mass and fiber length measurments from the last five of these hands were used to calculate tension fractions.
‡The fibers of these four muscles cross the joint axes with wide variation of fiber length. The figures quoted here for the mean fiber length of these four muscles are more visual estimates than mathematical averages. The mass fraction is accurate, but the tension fraction for these four muscles is only as true as the fiber length. Mean fiber lengths are included for the shortest and longest fibers of BR, ECRL, and PT because of the large range of fiber lengths. Values are included for the two segments of the first DI as well as total values. The data were not normalized for skeletal size differences.

TABLE 12–2.

Normal Values Abstracted From Table 12–1 and Listed in Order of Magnitude for Mean Fiber Lengths, Mass Fractions, and Tension Fractions for Adults*

Mean Resting Fiber Length (cm)		Mass Fraction (%)		Tension Fraction (%)	
Brachioradialis (BR)	16.1	BR	7.7	Supinator†	7.1
ECRL	9.3	ECRL	6.5	FCU	6.7
FDS (ring finger)	7.3	FCU	5.6	PT†	5.5
FDS (index finger)	7.2	PT	5.6	ECU	4.5
FDS (little finger)	7.0	ECRB	5.1	ECRB	4.2
FDS (middle finger)	7.0	FDS (middle finger)	4.7	FCR	4.1
FDP (ring finger)	6.8	FDP (middle finger)	4.4	ECRL	3.5
FDP (index finger)	6.6	FCR	4.2	FDP (middle finger)	3.4
FDP (middle finger)	6.6	FDP (ring finger)	4.1	FDS (middle finger)	3.4
Lumbrical (middle finger)	6.6	ECU	4.0	First DI	3.2
FDP (little finger)	6.2	Supinator	3.8	APL	3.1
ECRB	6.1	FDP (index finger)	3.5	AP	3.0
EDC (middle finger)	6.0	FDP (little finger)	3.4	FDP (ring finger)	3.0
Lumbrical (ring finger)	6.0	FPL	3.2	PQ†	3.0
EDC (little finger)	5.9	FDS (ring finger)	3.0	FDP (little finger)	2.8
EDQ	5.9	FDS (index finger)	2.9	FDP (index finger)	2.7
FPL	5.9	APL	2.8	FPL	2.7
EDC (ring finger)	5.8	EDC (middle finger)	2.2	Second DI	2.5
EPL	5.7	AP	2.1	BR	2.4
EDC (index finger)	5.5	EDC (ring finger)	2.0	Third DI	2.0
EIP	5.5	PQ	1.8	FDS (index finger)	2.0
Lumbrical (index finger)	5.5	EPL	1.5	FDS (ring finger)	2.0
FCR	5.2	First DI	1.4	ODQ†	2.0
Pronator teres (PT)	5.1	FDS (little finger)	1.3	EDC (middle finger)	1.9
Palmaris longus (PL)	5.0	EDQ	1.2	OP†	1.9
Lumbrical (little finger)	4.9	PL	1.2	Fourth DI	1.7
APL	4.6	ADQ	1.1	EDC (ring finger)	1.7
ECU	4.5	EDC (index finger)	1.1	ADQ	1.4
EPB	4.3	EIP	1.1	EPL	1.3
FCU	4.2	EDC (little finger)	1.0	FPB	1.3
ADQ	4.0	OP	0.9	First PI	1.3
APB	3.7	FPB	0.9	Second PI	1.2
Adductor pollicis (AP)	3.6	APB	0.9	PL	1.2
FPB	3.6	Second DI	0.7	APB	1.1
FDQ	3.4	EPB	0.7	EDC (index finger)	1.0
Pronator quadratus (PQ)	3.0	Third DI	0.6	EDQ	1.0
Supinator	2.7	ODQ	0.6	EIP	1.0
First dorsal interosseus D/I	2.5	Fourth DI	0.5	Third PI	1.0
Opponens pollicis (OP)	2.4	First PI	0.4	EDC (little finger)	0.9
Second palmar interosseus (PI)	1.7	Second PI	0.4	FDS (little finger)	0.9
Third DI	1.5	FDQ	0.3	EPB	0.8
Fourth DI	1.5	Third PI	0.3	FDQ	0.4
ODQ	1.5	Lumbrical (index finger)	0.2	Lumbrical (index finger)	0.2
First PI	1.5	Lumbrical (middle finger)	0.2	Lumbrical (middle finger)	0.2
Third PI	1.5	Lumbrical (ring finger)	0.1	Lumbrical (ring finger)	0.1
Second DI	1.4	Lumbrical (little finger)	0.1	Lumbrical (little finger)	0.1

*From Brand, P.W., Beach, R.B., and Thompson, D.E.: J. Hand Surg., 6:209–219, 1981. Used by permission.
†See ‡ footnote to Table 12–1.

ene tube along the same pathway as the tendon. This unconstrained wire gives a geometrically more accurate moment arm. Because all moment arm graphs that are made by plotting excursion against radians are subject to error from constrained joint movement, we have only used those sections of the graphs that represent the free-moving range of motion on each side of the neutral resting position (see Fig 12–81). In many cases this may be only 70% or 80% of the total maximum range.

STATIC SLINGS

Before beginning on muscles, we need to identify the mechanical significance of one or two static slings that are important in rehabilitation. Very few ligaments or slings are designed to accept regular loads. They serve mostly as emergency structures to prevent subluxation of joints when muscles are caught off guard. However, in cases of paralysis, ligaments sometimes serve to restrain and limit movements of joints that lack muscular control. In such cases the ligament is likely either to give a sensory feedback of discomfort when stretched (resulting in a sense of weakness) or to accept the new stress and become gradually and progressively stretched over months and years of use. Typical of the latter result is the palmar plate of the MCP joints, which, when deprived of the support of the intrinsic muscles, will prevent hyperextension of the joints. However, almost always the attachment of the plate will become stretched over the next few months and allow MCP hyperextension and the resulting clawhand.

A similar situation occurs in ulnar palsy affecting the thumb in cases in which the patient pinches with his index finger pressing against the ulnar side of his thumb. In the absence of an adductor muscle, the ulnar collateral ligament of the MCP joint of the thumb accepts the stress of the pinch for a few months or even years but later stretches to allow abduction and instability (see crank action, Figs 5–34 and 10–19).

Thumb Web

In contrast to the above two passive restraints, which tend to fail over a period of time, is the *dermal-fascial cradle for the thumb*, which includes the thumb web and is of very great value on a continuing basis after paralysis of thumb muscles. This cradle, or sling, is more likely to cause trouble by becoming tighter and shorter than by becoming stretched and attenuated. Since we

have never seen a functional description of this sling that identifies its role following paralysis, we shall now dignify it by a large title—*dermal-fascial cradle of the thumb*. The cradle forms a back-up support for the whole of the first metacarpal and contains within its compass all of the intrinsic muscles of the thumb. It is composed of two main parts. One is attached to the third metacarpal and is hinged at the thenar crease and is about the same functional length in most extended positions of the thumb. It is composed of palmar skin reinforced by palmar fascia and the fascia over the adductor. The other approaches the thumb from around the back and radial side of the second metacarpal. It is highly flexible and changes its functional length considerably according to the position of the thumb (Fig 12–2, A and B). The two limbs are continuous with each other around the back of the thumb, which swings in its arc of circumduction using the palmar side of the cradle as a radius while the dorsal side of the sling opens and closes like a bellows (Fig 12–2,C).

This sling is not of great mechanical significance in the normal hand, since the thumb is under the control of normal muscles and the fascia rarely comes under stretch. The absolutely fixed position of the palmar skin at the thenar crease is caused by fascia and is important because it holds the skin down to allow deep cupping of the palm when the thenar and hypothenar eminences move to oppose each other. In cases of paralysis, the cradle may be a crippling factor or it may be an important help in function.

CRIPPLING FACTOR

In low median palsy the thumb comes to lie constantly in external rotation and supination. In this position the dorsal limb of the cradle is folded, slack, and redundant. In the absence of regular, frequent, full range-of-motion exercises and movements, the dorsal skin and fascia become progressively shortened as the slack is taken up. Subsequently, any attempt to restore opposition and pronation to the thumb is resisted by the now shortened dorsal skin and fascia.

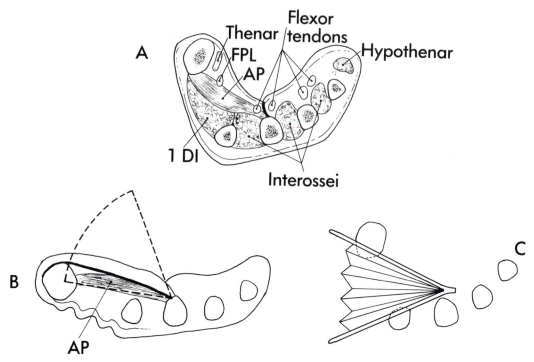

FIG 12–2.
A, diagram of a cross section through the metacarpals of the hand with the thumb pronated and opposed. Both limbs of the dermal-fascial cradle are on the stretch. The fascia over the AP is seen to be attached to the third metacarpal, as is the palmar septum that holds the thenar crease down to serve as a hinge for thumb circumduction. **B,** as in **A,** except that the thumb is now extended in supination. Note that the palmar limb of the dermal-fascial cradle of the thumb is still on the stretch, but the dorsal limb is folded and loose. **C,** the dorsal skin and fascia of the thumb cradle may be seen as the loose foldable side of a bellows that hinges on the third metacarpal. This is a very limited analogy, since in full adduction both limbs of the cradle are loose and the bellows becomes more like a concertina.

This shortening must be corrected, by skin graft if necessary, before tendon surgery. Otherwise the new tendon transfer will face the impossible task of elongating skin and fascia that have large passive moment arms for adduction, external rotation, and supination.

VALUABLE FACTOR

The thumb cradle really becomes useful in ulnar palsy and still more so in really severe multiple paralysis, as in tetraplegia.

Ulnar Palsy

With only the FPL to flex and adduct the thumb at all three joints, it is impossible to control active flexion at all joints in order to pinch against the index or middle fingertips or even for key-pinch. There is adequate moment for flexion at the IP joint, inade-

quate moment at the MCP joint, and absolutely hopelessly inadequate moment for flexion and adduction at the carpometacarpal (CMC) joint.

In most such cases a surgeon will attempt to add some flexion-adduction moment at the MCP joint or else will fuse either the MCP or IP joint so that the FPL can control the other. When the MCP and IP joints are under control, most surgeons do not think much about reinforcing the flexion of the CMC joint. In his early days of surgery P.W.B. naively assumed that somehow CMC flexion was taken care of by the FPL. He did hundreds of tendon transfer operations for low ulnar-median palsy, most of which were successful and for which he gave credit to the transfer of the superficialis tendon for opposition, plus, of course, the remaining FPL. Now he knows that he should have given major credit to the *thumb cra-*

dle. A thumb with ulnar palsy always rests in the fully stretched cradle of the thumb web during strong pinch. The FPL does little to flex or adduct the CMC joint during strong or even firm pinch. The amateur woodsman climbing a tall tree may use his arms and legs for climbing and find it difficult to keep one arm free to manipulate a saw. A professional lumberjack would never do this without a static sling or cradle around his buttocks, which stabilizes his whole body at an angle to the tree so that he can concentrate all his strength and attention on his arms and shoulders, which are free to do the work (Fig 12–3,A).

In Figure 12–3,A, the hip joints of the lumberjack are seen to be like the MCP joint of the thumb. The angle between the legs and the tree can be controlled by a static sling, which takes a backward thrust

when the upper part of the body pushes against the tree. In exactly the same way, the distal two joints of the thumb can push against the index finger if the metacarpal is held by a static sling, which takes the backward thrust that results from flexion of the phalanges against resistance (Fig 12–3,B). The success or failure of a so-called opponens tendon transfer for low ulnar-median palsy is dependent on the integrity and proportions of the thumb cradle. If the cradle is short, the MCP joint may have to be hyperextended to open for pinch. If the cradle is too loose (as is found sometimes in the very mobile hands seen in Thailand and India), the thumb metacarpal might lie too far out for the tips of the thumb and fingers to meet, or the index finger would flex and its tip would reach the thumb only halfway down.

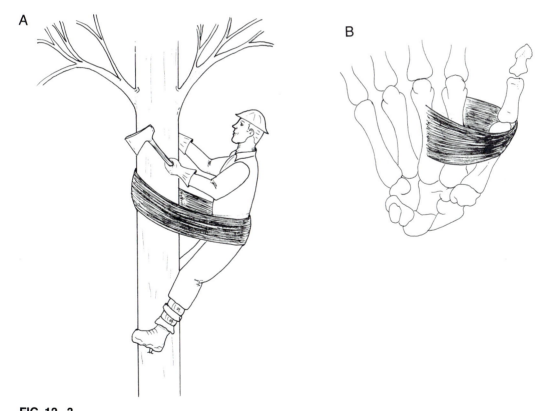

FIG 12–3.
Dermal-fascial cradle for the thumb. **A,** a lumberjack is free to use his arms and trunk because the critical angle between his legs and the tree is supported by a sling around his hips (MCP joint). **B,** when a thumb has lost its adductor and short flexor group of intrinsic muscles, it does not have enough muscle with enough moment arm to hold it stable at its CMC joint. Such patients depend entirely on the dermal-fascial cradle to hold the metacarpal, preventing hyperextension and hyperabduction of the CMC joint. The FPL and transferred tendons are then free to mobilize and control the MCP and IP joints.

If a lumberjack had a constant length on one side of his cradle and an adjustable length on the other side, he could turn left or right by shortening or lengthening his adjustable strap. In intrinsic palsy a tight dorsal web results in supination that no tendon transfer can correct. A loose dorsal web (perhaps after webplasty) permits pronation, because the palmar side of the web does not usually change very much.

Tetraplegia

In really severe paralysis one active tendon may have to control more than one joint. However, the attempt to do so may mean that no joint is under control. With only an FPL tenodesis to produce a pinch (or key-pinch), it is tempting to hope that it can flex the CMC, MCP, and IP joints of the thumb. That is a false hope. No FPL by itself could possibly flex the CMC joint (moment arm, 1.5 cm) against external stress with a 10-cm moment arm at the end of the thumb. Alternatively, some surgeons will fuse the CMC joint.[2] However, if one realizes that the CMC joint can be securely held by the cradle, it is safe to forget the CMC joint and plan only for control of the MCP or IP joints. The function of the cradle must be further understood if it is to be used intelligently in severe paralysis.

THE CRADLE IN PULP PINCH (TIP-TO-TIP PINCH)

When the normal thumb is opposed for pinch, it is usually held in an abducted and pronated position. Both the palmar and the dorsal skin and fascia are nearly tight. When pinching in this position, most people move their metacarpal very little (maybe 5 degrees). Complete absence of metacarpal movement during pinch is not very disabling. In severe paralysis it is acceptable to have the metacarpal resting in the web cradle. The main movement of pulp pinch is finger movement, plus some MCP and IP movement of the thumb. Thus pulp-to-pulp pinch may be powered entirely by flexion of the finger against an immobile thumb and even with immobile IP joints of the fingers, so long as the MCP joints of the fingers are fully mobile and under control.

If the intended prehension is to be by a *key-pinch*, however, *the thumb has to do all the moving*, except that the fingers must be in a flexed position to receive the lateral thrust of the thumb. In the semipronated position of the thumb necessary for key-pinch, the thumb cradle is *not* tight except in the wide-open preparatory position, and most normal people move the CMC joint more than they move the MCP or IP joints to effect the pinch. In fact, key-pinch is based mainly on the CMC joint. Therefore, in cases of severe paralysis the accomplishment of key-pinch cannot make use of the thumb cradle and has to have a motor with a large moment arm at the CMC joint of the thumb (see Chapter 10 and Figs 10–29 to 10–32).

DEDICATED WRIST-MOVING MUSCLES

The muscles dedicated to moving the wrist are the extensor carpi radialis longus (ECRL), extensor carpi radialis brevis (ECRB), extensor carpi ulnaris (ECU), flexor carpi radialis (FCR), flexor carpi ulnaris (FCU), and palmaris longus (PL). We call these muscles "dedicated" wrist movers to distinguish them from the muscles whose tendons cross the wrist on their way to the digits. The latter are also wrist movers and wrist stabilizers, but they cannot be relied on for their wrist function because synergism usually demands a conflict at the wrist between primary finger movers and primary wrist movers.

In the cross section of the wrist at the level of the axes of flexion-extension and of radial-ulnar deviation, these muscles group themselves as the following: (1) wrist extensors and radial deviators (ECRL and ECRB), (2) wrist extensor and ulnar deviator (ECU); (3) wrist flexor and radial deviator (FCR); (4) wrist flexor and neutral deviator (PL) and (5) wrist flexor and ulnar deviator FCU.

WRIST EXTENSORS AND RADIAL DEVIATORS

Extensor Carpi Radialis Longus

Functional and Mechanical Anatomy.—
The ECRL is named as an extensor of the wrist, but it has larger moment arms for flexion of the elbow and for radial deviation of the wrist (Fig 12–4). The origin runs along the lateral supracondylar ridge of the humerus from a point 4 or 5 cm above the epicondyle down to and including the upper aspect of the epicondyle. The bulkiest part of the muscle seems to center on an origin about 2 cm proximal to the axis of the elbow (Fig 12–5,A). The upper fibers of the muscle are long, because of the excursion required by the elbow (Fig 12–5,B), while the lowest fibers, which have minimal effect on the elbow, are almost as short as the fibers of the ECRB (Fig 12–6). Since the elbow has other stronger flexors, the effect of the ECRL is significant mainly on the wrist. It is most effective as a wrist extensor when its

radial deviation effect is balanced by an ulnar deviator such as the ECU (Figs 12–7 and 12–8) and when the elbow is extended to put the upper fibers on the stretch.

It is a useful muscle to transfer in median or ulnar palsy, because it has a long tendon of insertion, making change of direction rather easy, and because wrist extension may be handled well by the ECRB and radial deviation may be handled by the abductor pollicis longus (APL). It is not difficult to reeducate, although in about 20% of our specimens we found that the ECRB and ECRL formed a common muscle belly in their proximal part. There is sometimes a separate slender muscle between the two, the extensor intermedius, which has sometimes been used as a transfer[4, 5] and is especially useful when there are few remaining muscles, as in tetraplegia.

Quite commonly, a strand of tendon that has an origin as the ECRL may be inserted with the ECRB or vice versa. If the tendon of either the ECRL or ECRB is divided

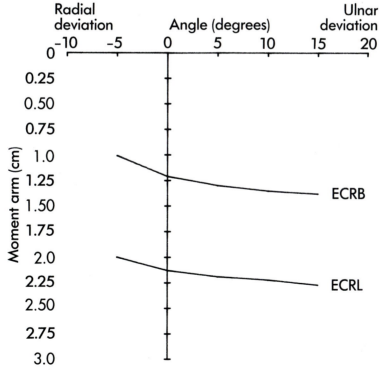

FIG 12–4.
Moment arms for radial deviation of the ECRL and ECRB at the wrist (mean of 11 hands) showing the moment arm with changing degrees of deviation. Flexion and extension are neutral.

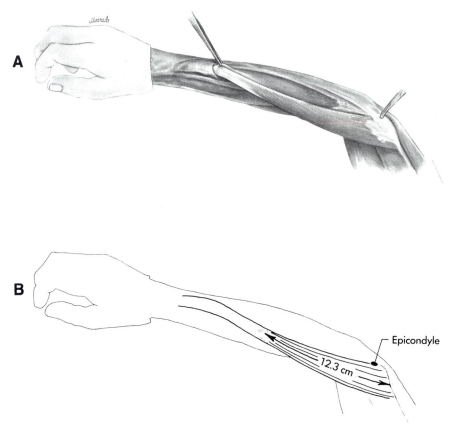

FIG 12–5.
A, ECRL displayed in situ with a hemostat point marking the position of the epicondyle. **B,** muscle fibers that arise from the epicondyle affect only the wrist joint and are as short as ECRB fibers. Those that arise 5 cm up the humerus are elbow flexors and may be twice as long. They are superficial.

distally and then difficulty is experienced in pulling it out proximally, the reason may be that such a crossover tendon slip needs to be identified and divided or treated separately.

Surgical Significance.—In radial palsy, to provide wrist extension, some surgeons have used the tendon of the ECRL as insertion for the pronator teres (PT), either alone or with the tendon of the ECRB. This is mechanically unsound. It makes it very difficult to extend the wrist without radial deviation. In these cases the FCU is the only remaining ulnar deviator, and the FCU has a large moment for wrist flexion. Thus the only way to get balanced wrist extension is to contract the transferred PT and FCU together. Thus an already weak wrist extension is further weakened by being opposed by a flexor. A better procedure is to use the ECRB alone

as the tendon of insertion for the PT or to use the ECRB with the ECRL, but to reroute the ECRL tendon to the ulnar side of the dorsum of the wrist to give balanced extension.[3]

Any tendon that crosses the wrist must be held by a retinaculum to prevent bowstringing. At one time, we rerouted a wrist extensor, extended by free grafts subcutaneously across the dorsum of the wrist to serve the fingers for intrinsic replacement. The resulting bowstringing of the tendon produced hyperextension of the wrist which required a second operation for its correction.

The ECRL and FCU are a pair of reciprocal antagonists that work together by alternation in very many common activities such as hammering nails or fishing with a rod and line, both casting and hauling in the fish. This alternation of radial wrist extension and ulnar wrist flexion is as basic as simple flex-

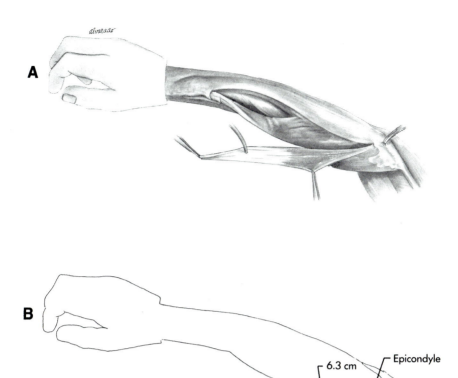

FIG 12–6.
The dorsal edge of the ECRL has been picked up and pulled over to show the underside of the muscle where the tendon receives the short fibers arising from the epicondyle.

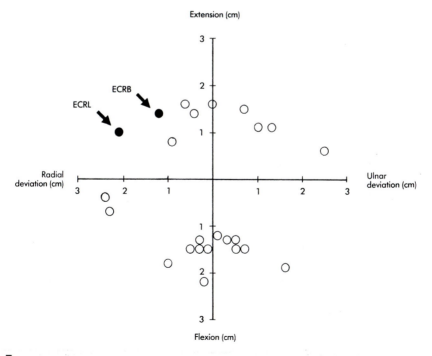

FIG 12–7.
Diagram of moment arms of all muscles that cross the wrist (wrist neutral, forearm midprone) with the ECRL and ECRB identified.

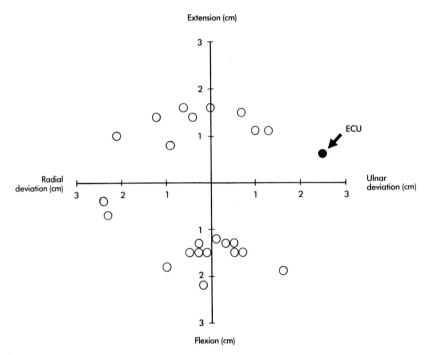

FIG 12–8.
Moment arm diagram at wrist, midprone, with the ECU identified.

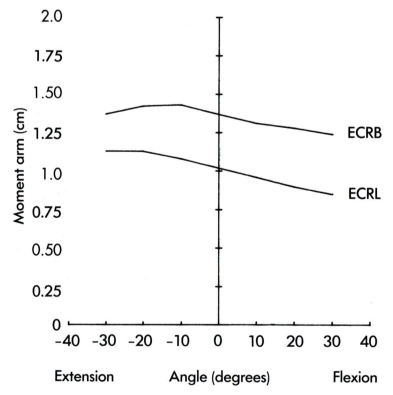

FIG 12–9.
Wrist extension moment arm, forearm midprone (mean of 11 hands). Both the ECRL and the ECRB have somewhat larger moment arms in wrist extension than in wrist flexion.

ion-extension without deviation (also see Figs 12–4, 12–9, and 12–10).

Extensor Carpi Radialis Brevis

Functional and Mechanical Anatomy.— The ECRB, smaller in mass than the ECRL, is a more effective wrist extensor because (1) it has a larger tension fraction than the ECRL (see Fig 12–10), (2) it has a larger moment arm for wrist extension (see Fig 12–9), (3) it has a smaller moment arm for radial deviation (see Fig 12–4), and (4) the whole muscle is available for wrist movement because it has no effect on the elbow and is not affected by elbow position.

The tendon of origin is long and flat, arising from the lateral epicondyle. It lies on the deep surface of the muscle (Fig 12–11,A). The tendon of insertion lies on

the surface of the muscle. If it is cut and then lifted away from the arm, it will swing out and the muscle will be displayed as a palisade of fibers running parallel to one another and all of about the same length (Fig 12–11,B).

The required excursion of the ECRB in the average adult is only about half the potential excursion, as measured by fiber length. Thus the muscle can exercise almost full and equal tension in all positions of the wrist. Also, it has the capability of being effective over a larger excursion if required after transfer. This muscle should rarely be used for transfer, however, because it is needed as the key extensor of the wrist. It may be used for distal insertions on the fingers, provided that (1) the tendon continues to cross the wrist on the dorsal side, (2) the tendon remains in the sheath, so it cannot

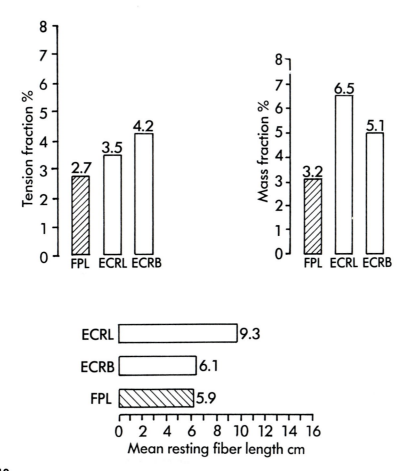

FIG 12–10.
Bar graphs of tension fraction, mass fraction, and mean fiber length of the ECRL and ECRB, with FPL for comparison.

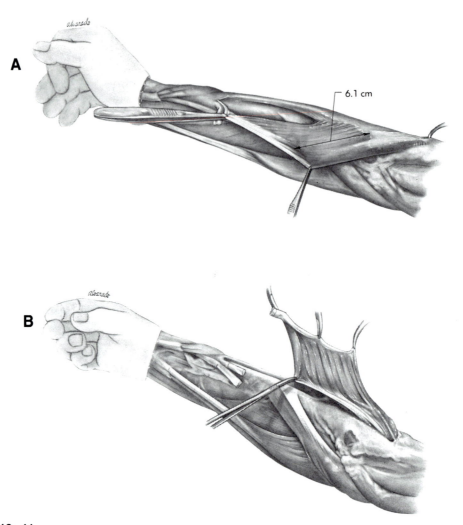

FIG 12–11.
A, ECRB fibers displayed after the ECRL is removed but before the ECRB tendon is cut. **B,** the tendon of insertion has been cut and lifted away from the arm while the tendon of origin is held down to display the row of equal-length muscle fibers running obliquely between the tendons.

bowstring at the wrist, and (3) the distal action is synergistic with wrist extension. It has been used with a tendon graft through an intermetacarpal space to replace the adductor pollicis (AP).[7] It has also been used with a four-tailed graft, to correct intrinsic paralysis of the fingers[6] (see also Fig 10–16).

WRIST EXTENSOR AND ULNAR DEVIATOR

Extensor Carpi Ulnaris

Functional and Mechanical Anatomy.— The ECU is a long muscle with rather short fibers. The tendon of origin is deep and has a glistening surface. The tendon of insertion is superficial distally but dives into the muscle proximally where many superficial muscle fibers originate from the investing fascia. The muscle is enclosed in a very strong fascial compartment (Fig 12–12, A and B). This has occasionally resulted in ischemic pain and even necrosis of the muscle following heavy exertion (compartment syndrome). A unique feature of the distal tendon is that it has a constant relationship with the head of the ulna bone. All other tendons that cross the wrist move with the radius as it moves around the ulna into pronation and

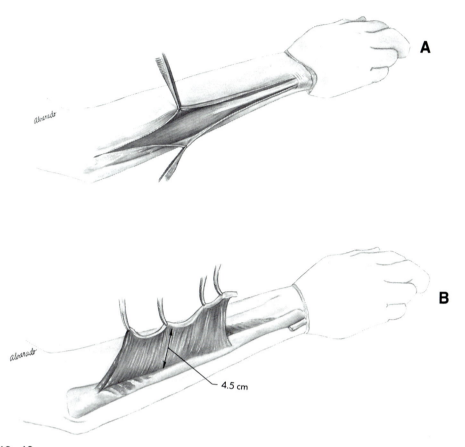

FIG 12–12.
A, the tough sheath that holds the ECU in its isolated compartment is here shown incised. The same sheath gives origin to muscle fibers further proximally. **B,** ECU tendon cut, muscle fibers displayed.

supination. Since the radius carries the carpal bones, with the axes of flexion-extension and abduction-adduction, it follows that the ECU changes its relationship to the axes of wrist movement and is an effective wrist extensor only in supination and may not be an extensor at all when the forearm is pronated (Fig 12–13). It is a strong ulnar deviator in pronation and supination. The flat tendon of the ECU serves also to stabilize the head of the ulna in relation to the radius and to the wrist.

Surgical Implications.—The ECU is well placed for transfer around the ulnar border of the wrist for abduction-pronation of the thumb or for pronation of the forearm or to serve as a motor for digital flexors. Caution should be exercised in disturbing the ECU in rheumatoid arthritis, where it may even need to be reinforced if the radial-ulnar joint

is tending to be subluxated. In radial palsy the ECU should not be used in situ as one tendon of insertion for a new motor for wrist extension (e.g., pronator), because it has a small and variable moment arm for wrist extension, which would hinder the free use of the larger moment arm of the ECRB (see Figs 12–18, 12–14, 12–15, and 12–16).

WRIST FLEXOR AND RADIAL DEVIATOR

Flexor Carpi Radialis

Anatomy and Mechanics.—The FCR is a well-defined fusiform muscle that arises from the medial epicondyle and from the common flexor origin, especially from a thick fibrous band that it shares with the PT. Some of the fibers arise above the axis of the elbow, and the whole muscle moves perhaps

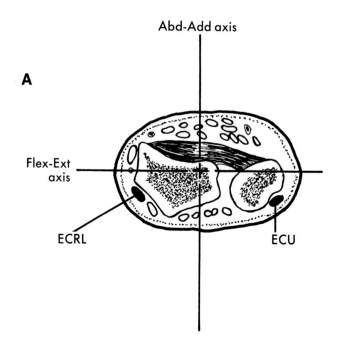

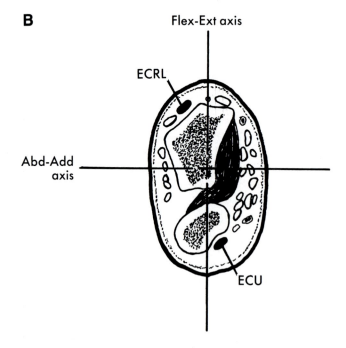

FIG 12–13.
A, in supine to midprone positions the ECRL and ECU are both wrist extensors. **B,** as the radius rotates around the ulna into the full prone position, the axis of flexion-extension rotates with the radius. The ECU may now lose its moment arm for extension.

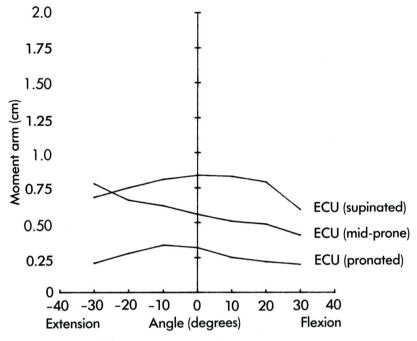

FIG 12–14.
Wrist extension moment arm (mean of ten hands). There is a marked reduction of the extensor moment arm of the ECU at the wrist as the wrist moves into pronation.

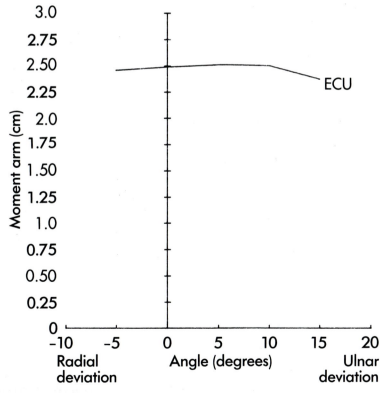

FIG 12–15.
ECU, forearm midprone (mean of ten hands). Moment arms for ulnar deviation of the wrist do not vary much with different positions of deviation.

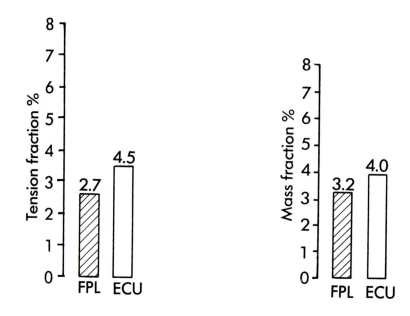

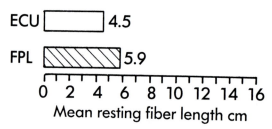

FIG 12–16.
Bar graphs of tension fraction, mass fraction, and mean fiber length of the ECU, using values of the FPL for comparison.

5 mm with full elbow movement, but this is not really significant. It is a prime wrist mover (Fig 12–17,A). The tendon of insertion is superficial to the muscle distally, but it plunges into the muscle proximally with muscle fibers coming into it from all sides. At the wrist the tendon is rounded and firm and has a sheath as it burrows into the wrist. Its moment arm for wrist flexion is about 1.5 cm (Fig 12–18), while for radial deviation it is 1.0 cm (Fig 12–19). Its tension fraction is 4.1% (Fig 12–20). It can exert almost as much tension as the ECRB. It is 50% stronger than the FPL but is only 60% as strong as the FCU.

In radial palsy we have often used the FCR as a finger extensor because it has longer fibers and better potential excursion than the FCU and can be more easily spared from the radial side of the wrist than the FCU can from the ulnar side. However, it is probably better to use a superficialis tendon for finger extension and move the FCR to the insertion of the APL. Note that when the FCR is used to power the APL, the whole tendon should be moved, not split and used as a yoke between the FCR and APL. If the latter is done, the APL action will be lost in radial deviation of the wrist and weakened in wrist extension. When the FCR pulls only on the APL insertion, it retains half of its flexor moment at the wrist but doubles its moment for radial deviation. This is acceptable only if the FCU remains to balance it as an ulnar deviator.

The split tendon of the FCR is sometimes

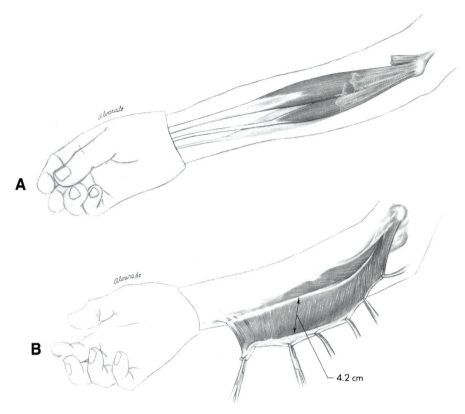

FIG 12-17.
A, FCR *(above)* and PL in situ. **B,** FCU tendon divided and lifted away from its origin to display muscle fibers.

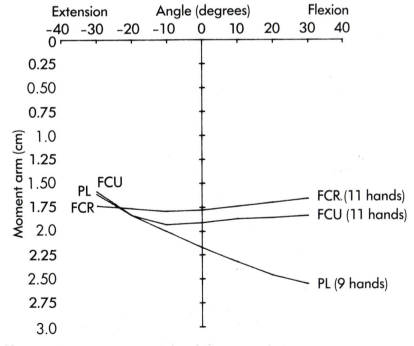

FIG 12-18.
Flexion moment arms of the FCU, FCR, and palmaris longus (PL) during flexion of the wrist (forearm midprone, mean values). Note that the PL increases its moment arm by bowstringing during flexion.

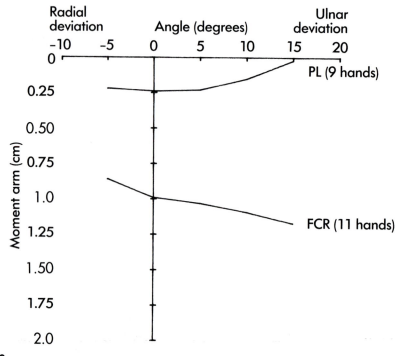

FIG 12–19.
Moment arms for radial deviation of the FCR and palmaris longus (PL) (mean values). Note that the PL is almost a nondeviator of the wrist.

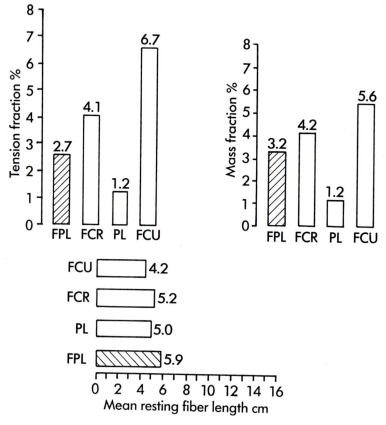

FIG 12–20.
Bar graphs for tension fraction, mass fraction, and mean fiber length for the FCR, PL, and FCU using the FPL for comparison. Note the high tension and low fiber length of the FCU.

used as a static sling or ligament to stabilize the trapezium following arthroplasty. Only a half-thickness of the tendon should be used, leaving the rest of the tendon in continuity (Fig 12–21).

WRIST FLEXOR AND NEUTRAL DEVIATOR

Palmaris Longus

Anatomy and Mechanics.—The PL is inconsistent and variable; it may be absent in about 10% of hands. It arises from the common flexor origin and from septa between it and its neighbors. It is inserted into the palmar aponeurosis distally (see Fig 12–17,A). The palmaris is a pure flexor of the wrist, crossing the wrist near the axis for radial and ulnar deviation (see Fig 12–21). It assists in the cupping of the palm, through its extension into the longitudinal fibers of the palmar fascia. The palmaris fans out at its insertion and may give an extension that runs over the thenar eminence and serves to assist thumb abduction. This feature inspired Camitz[11] and Littler[15] to use the palmaris

and its palmar fascial extension as an abductor for the thumb in low median palsy. The palmaris is also well placed to be transferred to act as a thumb extensor in radial palsy. In fully mobile clawhands the palmaris is often quite adequate for powering a two-tailed,[3, 8] or even a four-tailed, intrinsic replacement being passed through the carpal tunnel. This is now the routine practice of Fritschi[12, 13] when the hand has good mobility.

Unfortunately, the palmaris is often thought of merely as a source of tendon graft material. It is this, but we must remind surgeons that there are other equally good or better sources for graft material, and the palmaris tendon of a damaged hand should be used as a free graft only if the damage is very localized and unlikely to spread or to require revision. We have seen cases in which we have needed to use the muscle-tendon unit for transfer but another surgeon had previously used the tendon as a free graft in another operation. What a waste! In leprosy the most advanced cases of paralysis have only three muscles remaining below the elbow: the pronator, FCR, and palmaris. If the palmaris tendon has been removed

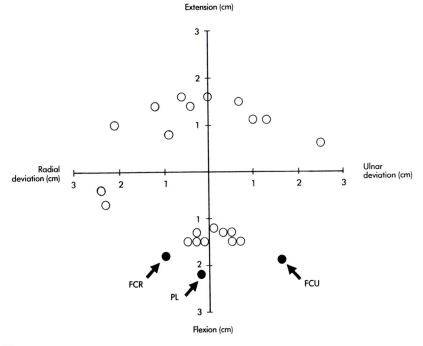

FIG 12–21.
Moment arm diagram of tendons around axes of flexion-extension and abduction-adduction of the wrist (wrist neutral, forearm midprone). The FCR, FCU, and PL are identified.

earlier, there is no way that muscle can be used.

The palmaris tendon is good grafting material, because it is long (12–15 cm) and flat, so revascularization is easy. Its tendon fibers are loosely held together and largely parallel, so that lateral traction on the tendon edges by needle-toothed forceps results in a very marked broadening and shortening of the tendon so that it looks like a membrane and can be used to surround the cut end of a motor tendon to which the palmaris is being attached (Fig 12–22).[10] It may also be easily stripped out either through two small transverse incisions or even through one[9] by using a narrow tendon stripper that is chisel-sharp (not knife-sharp) at its end. The stripper cuts into the muscle belly around the tendon and cores out the tendon from within the muscle.

Note that exactly the same method can be used for the plantaris tendon, which is longer and has all the same qualities. The use of the plantaris avoids the need to further weaken an already damaged hand[14, 16] (see also Figs 12–18, 12–19, and 12–20).

WRIST FLEXOR AND ULNAR DEVIATOR

Flexor Carpi Ulnaris

Anatomy and Mechanics.—Of all the muscles that cross the wrist the FCU can exert the highest tension and has the smallest potential excursion. It has the greatest cross-sectional area of its muscle fibers and of its tendon and it has the shortest muscle fibers. The FCU is the muscle that powers the karate chop and that resists the upward jump of a pistol barrel when it is fired. It powers the axe and the hammer stroke. In high ulnar palsy, even when the clawhand has been corrected, patients complain of a sense of indefinable weakness in heavy manual work. This is often a result of the loss of the FCU and the ulnar half of the profundus. Both of these muscles may be paralyzed without manifest deformity and therefore no corrective surgery is commonly suggested.

The FCU is a long fleshy muscle that could be described as fusiform and gives the impression of a strap muscle like the brachioradialis. The opposite is true. The tendon of insertion and the tendons and fasciae of origin run parallel to one another and overlap through most of the muscle while short fleshy muscle fibers run obliquely from one to the other all the way down. The tendon of insertion of the FCU runs into the muscle for about 20 cm. Its proximal tip reaches to within 5 cm of the medial epicondyle of the humerus. Most of the muscle fibers arise from a strong aponeurotic sheet of tendon on the deep side of the muscle and from a fan-shaped tendon investment on the ulnar superficial surface. The muscle fibers average 4.2 cm in length and have a tension fraction of 6.7% of all muscles below the elbow (compare with the brachioradialis at 2.4%, though the latter muscle has a greater bulk) (see Fig 12–17,B).

A feature of the FCU is its insertion to a somewhat mobile bone, the pisiform, which holds the tendon away from the axes of the wrist and serves to give the FCU enhanced moment arms. The pisiform has an independent movement of almost 1 cm and gives a little interplay between the FCU and the abductor digiti quinti (ADQ) and other muscles of the hypothenar eminence. If the FCU is paralyzed, it weakens the ADQ a little, by moving its origin distally.

Surgical Significance.—The FCU is often transferred in radial palsy to power the finger extensors. We object to this. Not only does the muscle lack 30% of the needed excursion, but it can ill be spared from its key position in the wrist. The FCR can be better spared for finger extension or, better still, the flexor superficialis to the middle finger. Some of those who use the FCU for finger extension in radial palsy[17] prefer to remove the distal muscle fibers, supplied by the lowest motor nerve branch, thus avoiding the need to have a bulky muscle winding around the forearm.

The FCU may be an overwhelming deforming force in some spastic hands. It is sometimes necessary to inactivate it as a wrist flexor, perhaps by using it as a yoke to stabilize the wrist,[18] before the hand can begin to be reeducated (see Figs 12–18, 12–20, 12–21, and 12–23).

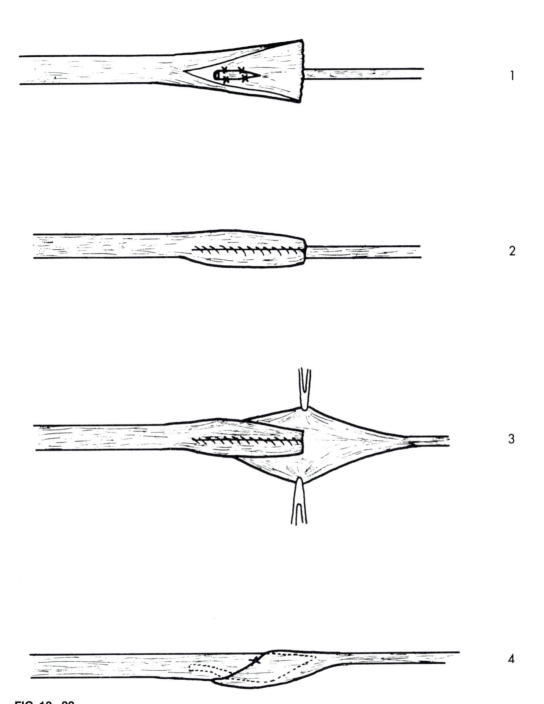

FIG 12–22.
Attachment of a palmaris tendon graft to a motor that has a larger diameter. *1,* the motor tendon is split open, and the graft enters the open area through a small slit. It is sutured into position. *2,* the split tendon is closed with a fine running stitch. *3,* the palmaris tendon is held by two needle-pointed forceps and firmly stretched into a membrane. *4,* the spread membrane is used to cover the cut end of the motor, which is cut obliquely at the end to avoid a sudden narrowing. (*Note:* The same technique may be used on a plantaris and on a strip of fascia lata. In the case of the plantaris, it may be folded into two and sutured to the motor at the fold leaving two exiting grafts, one of which is used to wrap the end of the motor.)

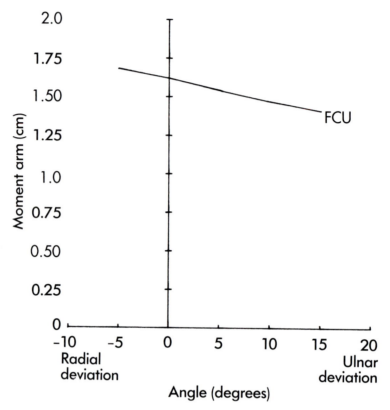

FIG 12–23.
Wrist ulnar deviation, moment arm of FCU, forearm in midprone position (mean of 11 hands).

THUMB-MOVING MUSCLES

MOVEMENTS OF THE THUMB

The thumb consists of four bones linked by three joints (Fig 12–24, A and B). There is a small amount of mobility from the joints between the trapezium and other carpal bones, but the vast majority of thumb motion occurs in the CMC, MCP, and IP joints. The CMC joint is at the base of the thumb between the first metacarpal and the trapezium, and has two axes of rotation, one in the trapezium and the other in the metacarpal. The MCP joint also is a two-axis joint. The flexion-extension axis is in the metacarpal passing under the epicondyles and the abduction-adduction axis passes between the sesamoids just proximal to the beak of the proximal phalanx. The axis of rotation of the IP joint is just palmar and distal to the epicondyles on the proximal phalanx. None of these axes is perpendicular to the bones or to one another.[23a, 23b, 24a, 25a]

The classic description of the CMC joint is that it is a saddle. It took the insight of Kuczynski[24, 25] to call the articular surface of the trapezium "the saddle of a scoliotic horse." This gives a vivid picture, but is of limited value in visualizing the joint in the

FIG 12–24.
Facing page. **A,** the thumb has three joints. The CMC joint has two axes of rotation, a flexion-extension axis in the trapezium and an abduction-adduction axis in the metacarpal. The MCP joint has two axes, a flexion-extension axis in the metacarpal and an abduction-adduction axis which passes through the volar plane just distal to the sesamoids. The flexion-extension axis of the IP joint is in the condylar head of the proximal phalanx. None of the axes are perpendicular to the bones or to one another. **B,** these cross sections of the joints are at the level of the flexion-extension axis and show the relationships of the muscles and tendons to the axes of rotation of each joint. The abduction-adduction axis is actually distal to the flexion-extension axis of the MCP and CMC joints and is shown at the same level here for simplicity. The joints are drawn to scale.

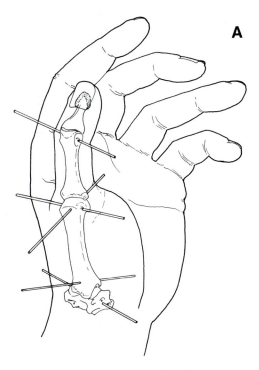

A

Joints of the Thumb

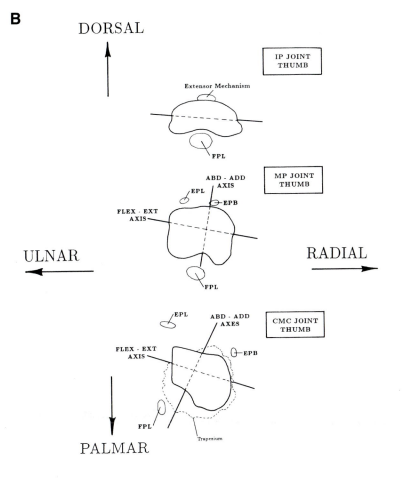

B

intact hand because the metacarpal bone, which rides in the saddle, is facing sideways, not forward, on the horse. In flexion of the CMC joint (Fig 12–25,A), the metacarpal leans over to one side (palmar) of the saddle and, in extension, to the other (dorsal). In abduction, the rider leans back in the saddle (radial), and leans forward for adduction (ulnar). The main lesson we learn from this model is that the thumb cannot rotate at its base any more than a rider can rotate on a horse. In an ordinary saddle, the axis of forward and backward motion is perpendicular to the horse and perpendicular to the axis for sideways motion. The axes of rotation of

the CMC joint are tilted relative to the bones and to each other, so functional rotation occurs with both motions (Fig 12–25,B). When the rider on the right CMC joint leans back, he (the masculine pronoun is used here for convenience) turns to the right; when he leans forward, he turns to the left. He has in effect turned his body through an arc because his back has moved around a cone. A similar situation applies to the MCP joint, where the flexion-extension and abduction-adduction axes are tilted so that the proximal phalanx turns right when the MCP joint is abducted and left when the joint is adducted.[25a]

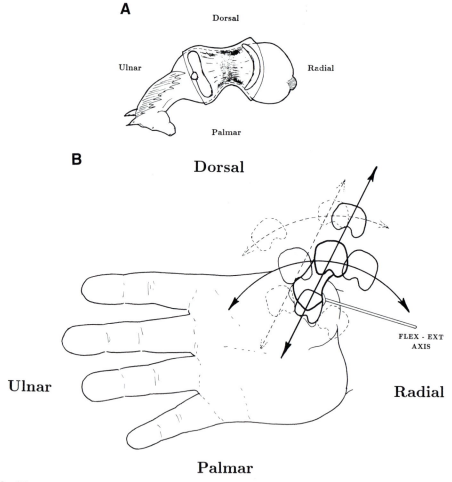

FIG 12–25.
A, the trapezial surface of the CMC joint is a saddle for a scoliotic horse because the axes of rotation for the joint are not perpendicular to each other or to the bones. The metacarpal faces sideways in the saddle. **B,** the movements of the metacarpal about each axis of rotation for the CMC joint are not in anatomic planes because neither axis is in an anatomic plane. Flexion and extension occur in a plane that runs from the hypothenar eminence to the dorsum of the hand. Abduction-adduction movements are in arcs about the tilted abduction-adduction axis. There is a greater range of abduction in extension than in flexion of the CMC joint.

Motions of the thumb like circumduction and opposition are combinations of movements of both the CMC and MCP joints.[20] The muscles that cross the MCP joint also cross the CMC joint. Those muscles that abduct one joint abduct the other and the muscles that adduct the CMC joint adduct the MCP joint. This arrangement of the muscles allows the two joints to act together as a unit in the normal thumb. Paralysis of the individual thumb muscles affects both joints, and the effect on each joint must be considered individually (Fig 12–26). (See Chapter 4).

SURGICAL SIGNIFICANCE OF INTRINSIC MUSCLES OF THE THUMB

Since the fibers of all these muscles cross more than one joint axis, every muscle fiber has a unique action based on its own individual position within the muscle. Also, since many of the muscles merge with one another at their margins, so that it is not always clear where one starts and the other leaves off, it is sometimes convenient to

look on these muscles as one continuous sheet of unnamed muscle fibers that fan out from the thumb and give it mobility and stability wherever it may currently be necessary.

Looked at this way, it is quite striking that most fibers seem to be aimed toward the center of the base of the proximal phalanx, and most seem to approach that point from a wide fan base covering the full right-angle sector of a circle starting at the center of the wrist crease and paralleling the thenar crease until it intersects the second metacarpal. As the thumb moves from one function to another, a ripple or wave of activity sweeps across this fan of muscle fibers distributing the balance of power and support in a flexible response to the changing patterns of external stress. When these muscles are paralyzed or destroyed, it becomes necessary to evaluate the thumb to determine how its balance has been disturbed and then be ready to act to restore the balance as far as possible.

At one end of the quadrant fan the action is flexion-abduction, and at the other end it is flexion-adduction. One end is supplied by

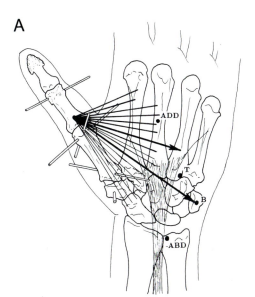

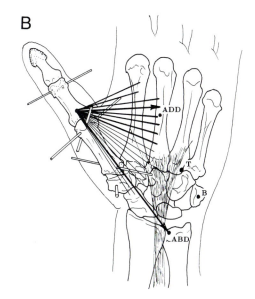

FIG 12–26.
A, this fan of muscles represents the APB, FPB, and AP. The closely spaced lines represent fibers that are usually ulnar nerve–supplied, and the more widely spaced lines are supplied by the median nerve. The *long arrow* is the direction Bunnell[20] suggested for a single replacement tendon for low ulnar-median palsy. The *short arrow* is the direction for a single tendon transfer for the same, as suggested by Thompson.[28] **B,** the same diagram with two separate tendons used together for rebalancing after low ulnar median palsy, one to replace the adductor-flexor group and the other the abductor group of fibers.

the median nerve and the other by the ulnar nerve (see Fig 12–26). The division between the two is variable. In median palsy, there is rarely enough remaining of the abductor function to avoid the need to transfer a tendon to abduct and flex.[21] In ulnar palsy, there may be enough median-supplied flexor brevis to position the thumb for pinch, but when power is needed, the lack of the adductor usually results in hyperextension of the MCP joint and in hyperflexion of the IP joint. A replacement for the adductor is needed.

It is in ulnar and median palsy that there has been the most disagreement on procedure. Much of it has been caused by a failure to understand the mechanics for balance of the thumb. The flexion of the tip of the thumb (Froment's sign) has been thought of as a failure of the extensor. Therefore, transfers of the abductor-opponens function have been attached to the extensor tendon in the hope that this extra pull would result in the extension of the IP joint during pinch. Not so. The major extensor of the IP joint is normal. The weakness is on the flexor-adductor side of the MCP joint, and the unwanted flexion of the IP joint is corrected as soon as flexor power is added to the MCP joint.

Many surgeons seem to have overlooked the dominance of the adductor-flexor group of muscles and the need for a transfer to replace them when they are paralyzed. The arrows in Figure 12–26,B show how the fan of intrinsic muscles, when paralyzed, may be replaced by one or two different muscles. When two are used, the stronger needs to be near the adductor end of the fan (see also under Static Slings below).

When John Agee[23] was at Carville he set up a number of fresh-frozen cadaver hands so that the thumbs would move as a single unit, fixed at the MCP and IP joints and free only at the CMC joints. The thumb was then guided around an arc of a cone whose axis was along the third metacarpal bone and whose apex was at the CMC joint. The dorsum of the thumb (the nail) was kept tangential to the cone, and the thumb pulp was kept facing the axis. Movement of the thumb in this configuration seems to occur naturally, with the CMC joint congruent

and free, and it seemed to simulate the normal sequence of circumduction of the thumb. At every 5-degree interval of this movement around the cone, the thumb was moved inward toward the axis and then out again along a radius of the cone (Fig 12–27). Thus a pinch movement was simulated from each position of wide circumduction. (*Note:* In this experimental study of the activity of the muscles that control the movements of the CMC joint of the thumb, there has been no attempt to define the effect of each muscle in the movement around any one axis of the CMC joint. Most movements of the thumb involves some rotation around each of the axes, both flexion-extension and abduction-adduction.)

As with any two-axis joint, one may produce a circumduction of one bone on another without actual rotation. Muscles crossing such a joint may have a moment arm that affects both axes, as is shown by the arrows in the diagrams. The crossing of second or third joints may modify the action at the CMC joint and vice versa. Therefore these diagrams should not be assumed to be true for the thumb as a whole, but only for the CMC joint.

On these hands fine wires were attached to the insertion of each muscle that affects the CMC joint. These wires ran back along the line of action of each muscle toward the anatomic origin of each and then on to a pulley at which a potentiometer measured its exact excursion. It was thus possible to determine the moment arm (or effective leverage) of each muscle at every position of the thumb. Since most of these muscles are unconstrained by pulleys at the CMC joint, it was expected that the effect of each muscle would be different at different joint positions.

With so much data it became a real challenge to display it in a meaningful way. Austin Ou[26] of the Louisiana State University Department of Mechanical Engineering, guided by Professor David E. Thompson, used this study as a Ph.D. thesis and the arrow diagrams used in this chapter are from his thesis. Each diagram represents the varying moment arms and directions of effective action at the CMC joint of a single

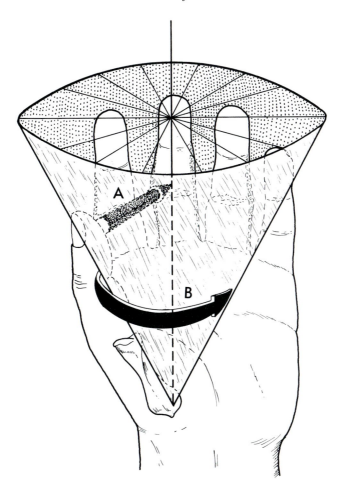

FIG 12–27.
The cone model of movement at the CMC joint. The *heavy black arrow (B)* represents circumduction and is measured as the *e* angle (vertical axes) in the arrow diagrams of Agee[23] and Ou[26] in this chapter. The *shaded tubular arrow (A)* represents the motion of the thumb that determines the size of the cone (B angle). This is the horizontal axis in the arrow diagrams (see Fig 12–28). *(From Ou, C.A.: The biomechanics of the carpometacarpal joint of the thumb. Dissertation, Louisiana State University, Baton Rouge, December 1979. Used by permission.)*

muscle. The results are expressed for each muscle as a series of arrows covering the whole range of positions of the CMC joint of the thumb. Each arrow is at the intersection of two coordinates that define the position of the thumb in relation to (1) its position around the arc from full supination (0 degrees) to full pronation or opposition (90 degrees) and in relation to (2) its position on the radius of the cone, from out at the periphery inward toward the center. The first was called the T angle, and the second was called the B angle. At each of these defined positions the arrow represents the moment arm of that muscle in relation to the CMC joint. The length of the arrow represents the

length of the moment arm, and the direction of the arrow is the direction of the moment. The direction may be toward further movement around the cone (into more pronation or supination) or it may be toward or away from the axis. Most arrows are oblique and can be seen as the resultant of vectors for tangential movement and radial movement.

For example, as shown in Figure 12–28, the biggest moment arm of the FPL muscle occurs when the thumb is in full supination and it tends to move the thumb toward a narrowing cone angle (i.e., flexion and adduction). When the thumb is in full pronation, the FPL is least effective as a flexor or adductor of the CMC joint; it has a moment

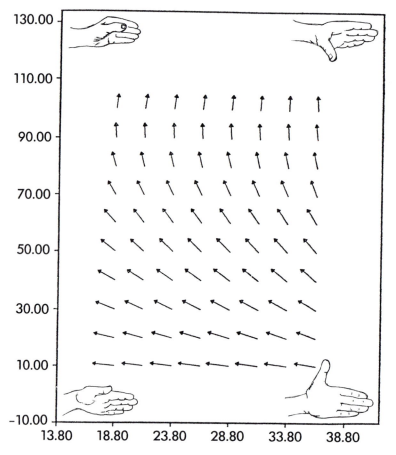

FIG 12–28.
Vectorial moment arms of the FPL at the CMC joint. In supination of the thumb the FPL at the CMC joint is mainly a flexor. In pronation it becomes mainly a pronator. (*Note:* Each arrow diagram represents one muscle. Each arrow represents the effect of that muscle at one defined position of the CMC joint. The *position* of the arrow defines the angular posture of the CMC joint. The vertical scale defines its posture around the cone (circumduction) from supination (low on the scale) to pronation (high). The horizontal scale defines the angle from the axis of the cone: a wide-open cone angle to the right, and a closed angle (flexed and adducted) to the left. The *length* of each arrow represents 50% of the length of the moment arm of the muscle at the CMC joint. The *direction* of each arrow marks the direction of the moment of the muscle (see Fig 12–27). (*From Ou, C.A.: The biomechanics of the carpometacarpal joint of the thumb. Dissertation, Louisiana State University, Baton Rouge, December 1979. Used by permission.*)

for further circumduction into more pronation. Contrast this with the arrow diagram for the AP (Fig 12–29). The adductor has much larger moment arms and a relatively consistent direction toward adduction-flexion.

ABDUCTOR POLLICIS BREVIS

Anatomy and Mechanics

The abductor pollicis brevis (APB) is rather small and weak in the scale of size and tension of all muscles. However, it holds a significant position in the scale of im-

portance because it is uniquely able to position the thumb in the opposed position for pinch and for grasp against the fingers. The muscle appears more bulky than it is because it is thrust forward by the opponens beneath it (Fig 12–30). The contraction of the opponens pushes the APB farther from the axis of the CMC joint of the thumb and therefore increases the moment arm and the effectiveness of the APB. Agee[27] has designed a silicone prosthesis to restore bulk to the thenar eminence in low median palsy and also to lift an opponens transfer away from the metacarpal and improve its mo-

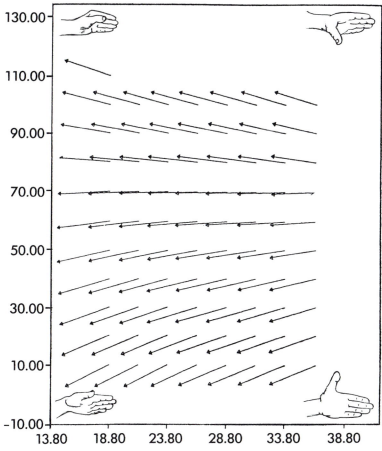

FIG 12–29.
Vectorial total moment arms of the AP at the CMC joint. At the CMC joint of the thumb the AP is entirely a flexor adductor (it narrows the cone angle). It has a large moment arm. The distal fibers have even larger moment arms (see Figs 12–26 and 12–28). *(From Ou, C.A.: The biomechanics of the carpometacarpal joint of the thumb. Dissertation, Louisiana State University, Baton Rouge, December 1979. Used by permission.)*

ment arm. With a tension fraction of 1.1% (Fig 12–31) the APB is weaker than any of the interossei and is about the same as the extensor proprius of the index and the little fingers.

The muscle is fan-shaped. Its tendon crosses the MCP joint and its fibers cross the CMC joint, at which every fiber has, therefore, a slightly different action. The most radial fibers may arise from a portion of the tendon of the APL, and these fibers are abductors of the CMC joint. The most distal fibers are immediately adjacent to the flexor pollicis brevis (FPB) and are really flexor-abductors of the joint.

Figure 12–32 shows how the moment for abduction of the APB increases at the CMC joint as the thumb abducts. This diagram represents only one typical fiber in the middle of the muscle.

The moment arms at the MCP joint are much less variable, because it is as a tendon that the muscle crosses this joint. It is a true abductor with a moment arm of 7.5 mm, but its required excursion at the MCP joint is only about 2.5 mm, because the range of motion of the joint is often only about 20 degrees of abduction-adduction. The muscle is relatively weak because the action of abduction of the thumb is not one that is usually done against resistance. It positions the thumb for action rather than performing the action itself. The muscle that most nearly opposes the APB is not the adductor, the direction of which is about 90 degrees from the APB, but the extensor pollicis longus

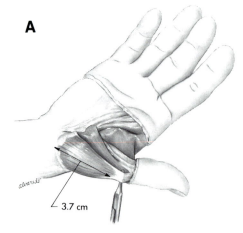

A

3.7 cm

FIG 12–30.
A, APB. *Arrow* marks fiber length. **B** and **C,** the APB is an abductor of both the CMC and MCP joints, having the largest moment arm of any muscle for abduction of both joints. The muscle is flexor of the CMC joint. Its tendon moves as the MCP joint flexes and extends, being a weak flexor with MCP flexion and a weak extensor with MCP hyperextension. It inserts into the extensor mechanism and thereby acts to extend the IP joint.

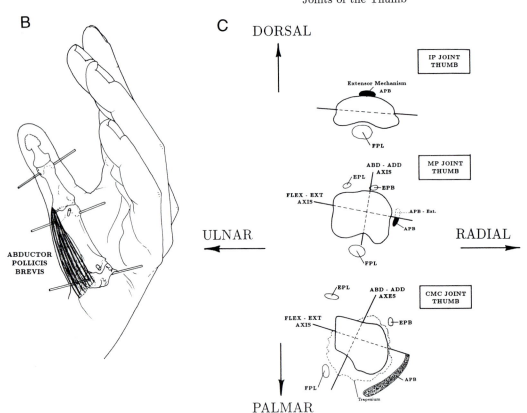

Joints of the Thumb

B

ABDUCTOR POLLICIS BREVIS

C

DORSAL

IP JOINT THUMB

Extensor Mechanism
APB

FPL

MP JOINT THUMB

ABD - ADD AXIS
EPL
EPB
FLEX - EXT AXIS
APB - Ext.
APB
FPL

ULNAR

RADIAL

CMC JOINT THUMB

EPL
ABD - ADD AXES
EPB
FLEX - EXT AXIS
APB
FPL
Trapezium

PALMAR

(EPL), which is about 150 degrees from the APB in terms of the axis of rotation around a cone.

Surgical Replacement

Many different muscles have been used to restore abduction to the thumb in median palsy. Most have had adequate strength and many have been far stronger than necessary.

This is because surgeons have blamed their failures on the weakness of the transferred muscle when they should have blamed themselves for their own failure to ensure sufficient passive range of motion and sufficient skin and fascia on the dorsum of the thumb web. In other cases the failure has been a result of the persistence of the habit pattern of the EPL contraction as a pinch

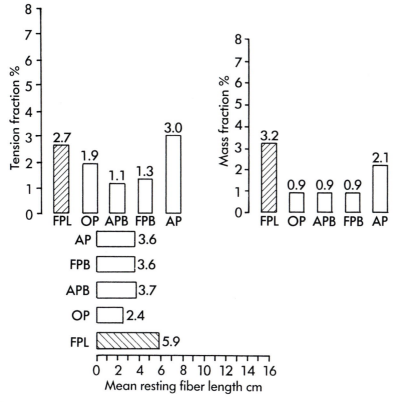

FIG 12–31.
Bar graphs of tension fraction, mass fraction, and fiber length of the OP, APB, FPB, and AP, with the FPL for comparison.

substitute, a habit developed while the patient was waiting for the median nerve to recover. For simple replacement of the APB the extensor indicis proprius (EIP) is of about the same tension capability. However, if one is to provide a substitute for the APB plus opponens, then the flexor digitorum superficialis (FDS) from the ring finger is appropriate.

The tendon of the APB crosses the MCP joint and serves to abduct that joint also. The tension of this tendon is low because the APB is only a fraction of the thenar muscle mass. However, when surgeons transfer a tendon to restore opposition to the thumb, they commonly attach it to the tendon of the APB. This gives the MCP joint more abduction torque than it ever had before paralysis. If there is also ulnar paralysis robbing the MCP joint of adduction torque, there is likely to be an abduction deformity.

OPPONENS POLLICIS

Anatomy and Mechanics

The opponens pollicis (OP) is an interesting muscle that inserts all along the radial surface and ridge of the first metacarpal. Its most publicized function is that of opposition, for which it is named. It is able to rotate the thumb into pronation because of its eccentric insertion on the metacarpal bone. However, careful inspection of the muscle (Fig 12–33) shows considerable variation in fiber length. The proximal fibers may be about 1.5 cm long and its distal fibers 3.6 cm long.

If the function of opponens were for simple rotation of the metacarpal, the excursion of every muscle fiber would be about the same, and the muscle fibers would be the same length. The variable fiber length is justified by movement around an axis that is oblique to the rotational axis. Figure 12–34

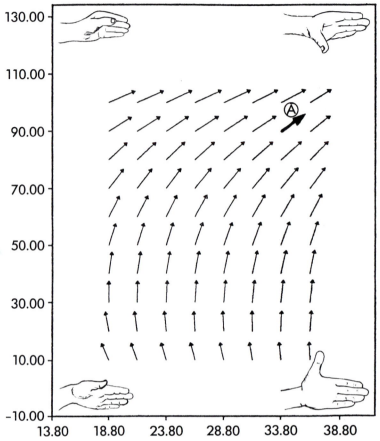

FIG 12–32.
Vectorial total moment arms of the APB at the CMC joint. APB is mainly a pronator. Note the largest moment arm is when the thumb is already pronated. In this position *(A)* the APB is moving the CMC joint toward more pronation and wider abduction (see Figs 12–27 and 12–28.) *(From Ou, C.A.: The biomechanics of the carpometacarpal joint of the thumb. Dissertation, Louisiana State University, Baton Rouge, December 1979. Used by permission.) (Note: For a diagram of CMC joint axes see Chapter 4.)*

shows that the opponens has a strong vector for movement around a cone whose apex is at the CMC joint and whose axis lies close to the third metacarpal. It also has a vector for movement toward the axis of the cone. Thus the OP is a pronator or opposer of the thumb, as its name implies, rather than a simple rotator of the metacarpal. However, its rotational function is important and is the basis for its ability to stabilize the CMC joint in a torsional sense.

The muscle has a greater tension capability than the APB, but it is difficult to be sure of the accuracy of our estimate for strength when the fiber length is so variable. Since the APB and the OP are both median nerve–supplied and are both lost in low me-

dian palsy, most surgeons think of them together and attempt to restore their action, after paralysis, with a single transfer. This composite muscle should therefore (1) abduct the metacarpal, (2) abduct the proximal phalanx at the MCP joint, (3) rotate the metacarpal around the cone axis (pronate the thumb), and (4) flex the metacarpal. However, if there is no ulnar palsy, the thumb has adequate flexors, so only three functions need be considered for restoration. Of all tendon transfers for median palsy, those that direct the tendon from the pisiform bone are acting like the APB, while the Thompson transfer (of Royle and Thompson)[28] is the nearest to being a replacement for the OP (see Fig 12–31).

FLEXOR POLLICIS BREVIS

Anatomy and Mechanics

The FPB is a straplike muscle with rather long parallel fibers that is a flexor both of the MCP joint and of the CMC joint. There is not always a clear division between the FPB and the opponens, except that by definition any fibers that cross the MCP joint must be FPB and those that end on the metacarpal bone must be opponens (Fig 12–33,A).

The function of the FPB is continuous, in a sense, with the function of the opponens on its proximal side and with the adductor on its distal side. It is continuous with the opponens because both flex the metacarpal on the carpus; the FPB also flexes the MCP joint. It is continuous with the adductor because both are MCP flexors. The FPB is a flexor-pronator and the adductor is a flexor-supinator. Both contribute to the dorsal expansion for extension of the IP joint, and both are strong flexors of the CMC joint (Fig 12–35).

The flexor is the most variable of all muscles with regard to nerve supply. It is sometimes ulnar, sometimes median, and often a bit of both. This accounts for the variable effect on the thumb of either median or ulnar palsy. Sometimes in complete ulnar palsy there is enough FPB function to partly compensate for loss of the adductor and provide an acceptable, though weak, pinch. Sometimes in complete median palsy a little pro-

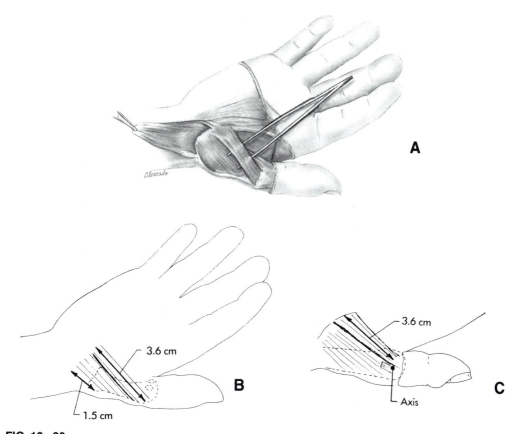

FIG 12–33.
A, FPB (over forceps) and OP (APB has been retracted to the left). **B,** the drawing is from a thumb in external rotation. **C,** the diagram is true lateral. **D** and **E,** the OP crosses only the CMC joint. It is a flexor and abductor, and the combination of these movements produces opposition of the metacarpal. **F** and **G,** the FPB acts primarily as a flexor of the CMC joint. Its fibers lie on both sides of the abduction-adduction axis at the CMC joint. It is also an MCP abductor since it inserts into the radial sesamoid and has a large moment arm for MCP flexion. The sesamoid gives the FPB a larger moment arm for MCP flexion than the FPL tendon.

(Continued.)

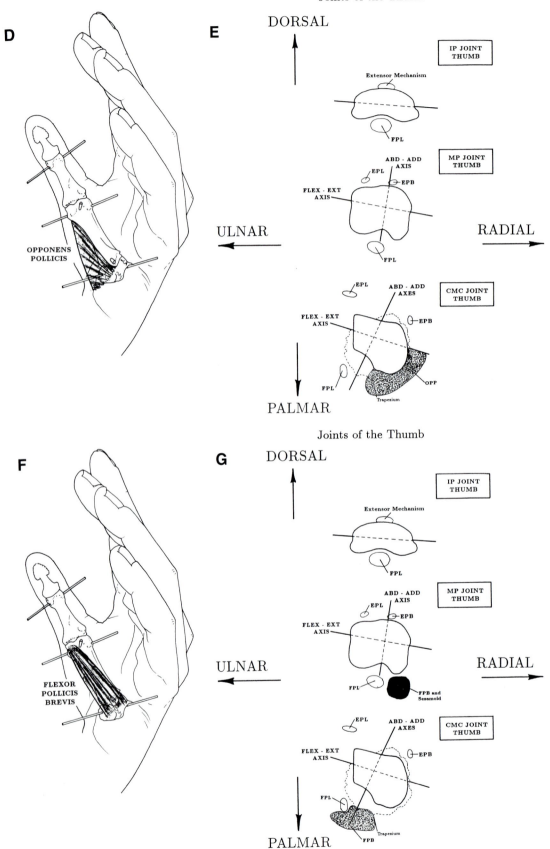

Joints of the Thumb

D

DORSAL

E

IP JOINT THUMB

Extensor Mechanism

FPL

ABD - ADD AXIS

EPL

EPB

MP JOINT THUMB

FLEX - EXT AXIS

ULNAR

RADIAL

FPL

OPPONENS POLLICIS

EPL

ABD - ADD AXES

CMC JOINT THUMB

FLEX - EXT AXIS

EPB

FPL

OPP

Trapezium

PALMAR

Joints of the Thumb

F

DORSAL

G

IP JOINT THUMB

Extensor Mechanism

FPL

ABD - ADD AXIS

EPL

EPB

MP JOINT THUMB

FLEX - EXT AXIS

ULNAR

RADIAL

FPL

FPB and Sesamold

FLEXOR POLLICIS BREVIS

EPL

ABD - ADD AXES

CMC JOINT THUMB

FLEX - EXT AXIS

EPB

FPL

FPB

Trapezium

PALMAR

FIG 12–33 (cont.).

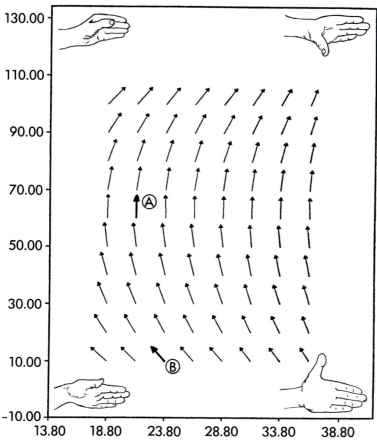

FIG 12–34.
Vectorial total moment arms of the OP at the CMC joint. The OP has rather a simple action. It encourages cir-cumduction in the direction of pronation *(A)*. In full supination it has a vector for flexion *(B)* (see Figs 12–27 and 12–28.) *(From Ou, C.A.: The biomechanics of the carpometacarpal joint of the thumb. Dissertation, Louisiana State University, Baton Rouge, Department of Mechanical Engineering, December 1979. Used by permission.)*

nating effect may be observed from the ra-dial insertion of an ulnar-supplied FPB. Since, however, the major effects of FPB are in sequence with those of the larger AP, we will consider these muscles together (see Fig 12–31).

ADDUCTOR POLLICIS WITH FLEXOR POLLICIS BREVIS

Anatomy and Mechanics

The AP is the pinching muscle. There is a real sense in which all other muscles affect-ing the thumb are just positioners and syn-ergists to allow the AP to pinch. In making this oversimplification one should include the FPB with the AP. The APL and APB and OP all hold the thumb in the proper po-sition of rotation and opposition so that the pinchers can then pinch. The FPL (at 2.7%), our standard muscle for comparison of all muscles, is weaker than the AP (3.0%) and much weaker than the AP and FPB together (4.3%). Beyond their strength (or tension ca-pability) the real effectiveness of these two muscles is in their very large moment arm at the CMC joint, larger than any other muscle has at any joint in the hand (Fig 12–36).

In the action of grasp, when the hand holds an object in the circle of its digits, each segment of the thumb and fingers is in-dividually opposed by the object in the hand. In that case the AP plays a minor role. In pinch, where the opposing force is all at the end of the digits, stability demands much larger moments at proximal joints than at distal ones (see Figs 5–2 and 5–3).

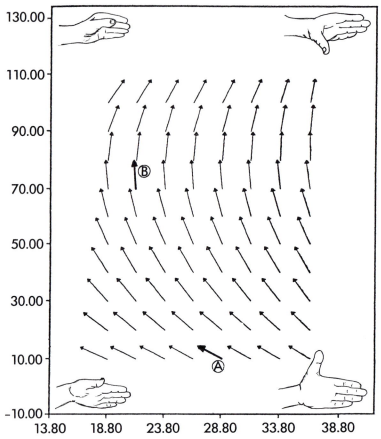

FIG 12–35.
Vectorial total moment arms of the FPB at the CMC joint. The FPB is a flexor of the MCP joint in all positions of the thumb. However, the origin of the FPB is more superficial than that of the AP and it has an action at the CMC joint that is somewhat akin to its immediate neighbor, the opponens pollicis. It is mainly a flexor in supination *(A)*, and it augments pronation in pronation *(B)* (see Figs 12–27 and 12–28). *(From Ou, C.A.: The biomechanics of the carpometacarpal joint of the thumb. Dissertation, Louisiana State University, Baton Rouge, Department of Mechanical Engineering, December 1979. Used by permission.)*

Here the thumb is lost without its AP and could more easily spare its long flexor.

Only a small part of this muscle inserts into the lateral band, and this part of the muscle helps to extend the IP joint. After loss of the EPL, this action, through the lateral band, together with an active APB, may provide a useful extension capability to the IP joint of the thumb. The rest of the muscle is inserted into the base of the proximal phalanx and the medial sesamoid. The AP is a supinator of the thumb from the pronated (opposed) position. Since the thumb is rarely completely pronated in pinch, the AP is not a serious supinator, because the medial-anterior border of the skeleton of the thumb is the advancing edge for normal pinch. However, it would hinder further pro-

nation, and may resist full pronation after a pure median palsy.

The AP is a fan-shaped muscle (Fig 12–37). All of its fibers cross the axis of the CMC joint, at which they have widely differing moment arms. All the muscle fibers converge on the insertions, which are bunched together, distal to and close to the MCP joint. The moment arms at the MCP joint are therefore not very variable. Thus by selecting which muscle fibers to activate, a person can use the AP to have relatively more effect on the CMC joint (distal fibers) or more on the MCP joint (proximal fibers). In grasp the proximal fibers might be more active and in pinch the distal (or all) fibers are needed.

When the AP and FPB are both para-

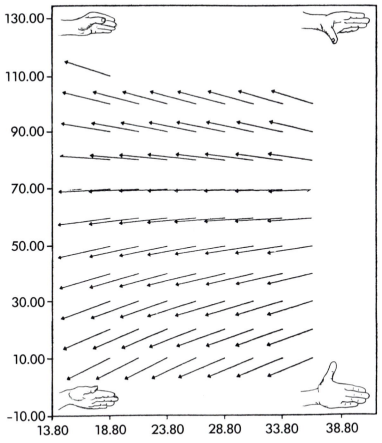

FIG 12–36.
Vectorial total moment arms of the AP at the CMC joint. Here is a muscle that is wholly devoted to the narrowing of the cone angle. It has very little effect on the circumduction angle. In making these calculations of moment arms, Agee used a single wire in about the middle of the AP. The distal fibers would have shown much longer arrows, and the proximal fibers much shorter ones, but the vector directions would have been about the same (see Figs 12–27 and 12–28). *(From Ou, C.A.: The biomechanics of the carpometacarpal joint of the thumb. Dissertation, Louisiana State University, Baton Rouge, December 1979. Used by permission.)*

lyzed, the thumb would be quite useless for pinch if it were not for the restraint of the skin and fascia of the thumb web. Resting in the hammock of this passive support, the first metacarpal can simulate having strength enough to oppose the force of the fingers in pinch. It is only strong in the way a drunken man is strong who leans against a wall for support. You cannot push him over unless the wall is also pushed over. The ulnar-paralyzed thumb shows its weakness mainly at the unsupported MCP joint, where it needs a flexor-adductor transfer. In a case of severe shortage of muscles the thumb may still function as a post for pinching against if the MCP joint is fused just short of full extension. The whole thumb may then rest back

in the passive support of the web and fascia. This substitution of a skin-fascial cradle for active adductor muscle will not work if the web is too wide (see Fig 12–3).

Surgical Significance

Not enough attention has been given to the replacement of the action of the AP in ulnar palsy. Its loss grossly weakens the pinch. Surgeons who do not measure the actual strength of pinch or observe the usefulness of the hand at work after surgery may be content after restoring the appearance and range of motion of thumb action. It is strength and stability, however, that working men and women need.

For the restoration of adduction in ulnar

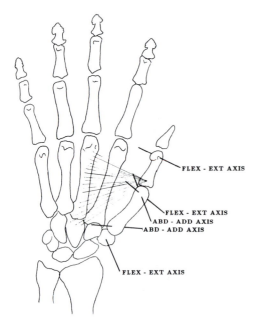

FLEX - EXT AXIS

FLEX - EXT AXIS
ABD - ADD AXIS
ABD - ADD AXIS

FLEX - EXT AXIS

FIG 12–37.
The fan of adductor–short flexor muscle fibers. Note the rather constant moment arm at the MCP joint while at the CMC joint distal fibers may have more than double the moment arm of the proximal fibers.

palsy, muscles such as the EIP have been used for transfer but are much too weak and wholly inappropriate. Boyes[29] has used the brachioradialis (BR) extended with a graft and taken through the third intermetacarpal space. This is a good approach. We have used the flexor superficialis from the middle finger and brought it up through the palmar fascia, which serves as a pulley. We have also taken the same muscle back through the interosseous membrane in the forearm and then across the dorsum of the wrist and through the third intermetacarpal space (as suggested by Boyes). More recently, Smith[30] has used the ECRB extended by a graft and taken through the intermetacarpal

space. Whatever motor is used, it must be strong and must have at least 2 cm of excursion beyond what is needed to accommodate to wrist movement. It should approach the thumb from the midpalm, toward the radial side. (See also Surgical Significance of Intrinsic Muscles of the Thumb and Figs 12–31 and 12–38.)

FIRST DORSAL INTEROSSEUS

This unique muscle is usually thought of as an index finger muscle, but it is also a very important thumb muscle (Figs 12–39,A and B). The majority of its fibers have their origin either on the first metacarpal shaft or the ligament between the bases of the metacarpals, and so act on the first metacarpal. When the muscle exerts tension, it tightens the intermetacarpal ligament and pulls the metacarpal bases closer together. The first dorsal interosseus (DI) is very different from the other thumb muscles because it does not cross the CMC joint from proximal to distal and does not compress, but distracts the joint when it is used; an unusual role for a muscle. The muscle's line of pull is nearly parallel to the flexion-extension axis of the CMC joint, and the majority of its insertion is close to the abduction-adduction axis, so it adds little to moving the thumb or giving it power. This is why the muscle is frequently overlooked by those who study thumb function. However, the first DI serves a *very* important role in stabilizing the CMC joint. During lateral pinch and power grip, the adductor-flexor muscle groups exert a powerful force pushing the metacarpal base dorsally and radially, tending to subluxate the CMC joint. The first DI pulls strongly ulnarly and dis-

FIG 12–38.
Facing page. **A,** transverse and oblique fibers of the AP. **B,** anterior view of the AP with flexor tendons and lumbrical lifted away. **C** and **D,** the AP acts on all three thumb joints. It is a flexor and adductor of the CMC and MCP joints. The moment arm for adduction varies for both joints because of the muscle's wide origin and insertion, and is quite large at the CMC joint. CMC adduction is a combination of AP, FPL, and some of the fibers of the FBP. The AP is the only adductor of the MCP joint, but this absence is often overlooked in patients with ulnar nerve palsy because the weakness in CMC adduction from the loss of the AP and the FPB prevent the use of the thumb in positions that would abduct the MCP joint. The moment arm for most AP fibers for MCP flexion is larger than that of the FPL because of the insertion onto the ulnar sesamoid. The insertion of some of its fibers into the extensor mechanism allow it to act as an IP extensor.

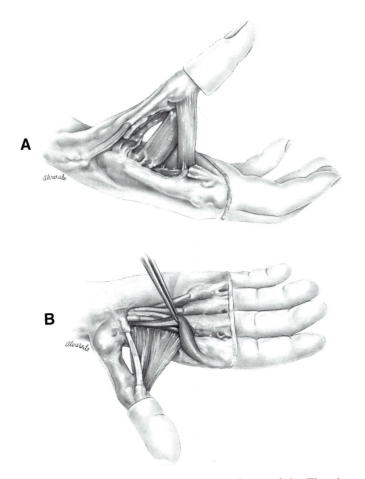

A

B

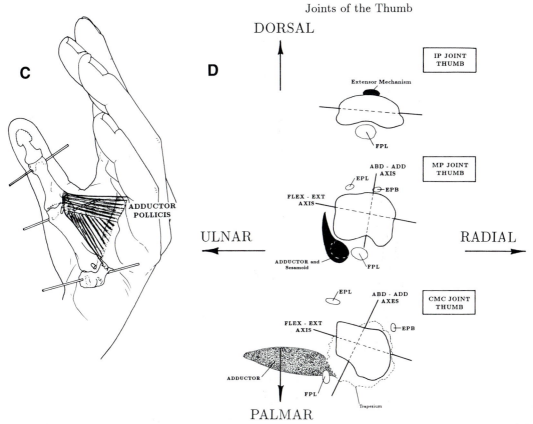

C

ADDUCTOR
POLLICIS

D

Joints of the Thumb

DORSAL

IP JOINT
THUMB

Extensor Mechanism

FPL

MP JOINT
THUMB

ABD - ADD
AXIS

EPL

EPB

FLEX - EXT
AXIS

ADDUCTOR and
Sesamoid

FPL

ULNAR

RADIAL

EPL

ABD - ADD
AXES

CMC JOINT
THUMB

FLEX - EXT
AXIS

EPB

ADDUCTOR

FPL

Trapezium

PALMAR

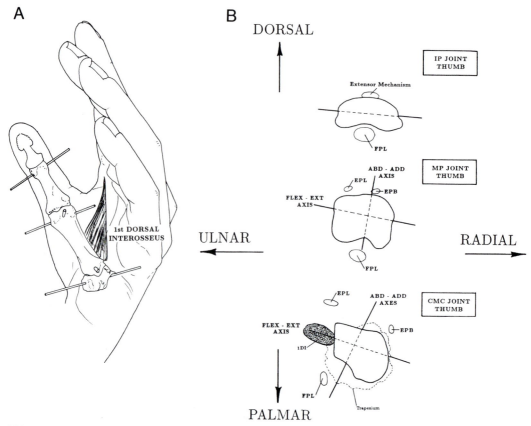

A

B

DORSAL

IP JOINT THUMB

Extensor Mechanism

FPL

MP JOINT THUMB

ABD - ADD AXIS

EPL

EPB

FLEX - EXT AXIS

FPL

ULNAR

RADIAL

1st DORSAL INTEROSSEUS

CMC JOINT THUMB

EPL

ABD - ADD AXES

FLEX - EXT AXIS

EPB

1DI

FPL

Trapezium

PALMAR

FIG 12-39.
A and **B,** the first DI acts on the CMC joint of the thumb. Its direction of pull is nearly 90 to 120 degrees from that of the AP, and it acts to distract the CMC joint and to pull the metacarpal distally and ulnarly. It has fibers on both sides of the abduction-adduction axis and lies nearly parallel to the flexion-extension axis, so it does little to move the joint. It is a very important stabilizer of the CMC joint, acting against the dorsal radial subluxating forces of the AP and the FPB.

tally. In our laboratory, using fresh cadaver hands in which the muscles of the thumb were loaded in the position of lateral pinch, removing tension from the first DI resulted in radial subluxation of the CMC joint, which relocated following restoration of the tension to the first DI.

While the first DI is vital to function in the normal thumb, it is never replaced in surgery for the paralysis. For one thing, the location of the pulleys would make the operation technically challenging. For another, it is rarely needed, because even the best surgery for paralysis restores only a fraction of the normal thumb power, and most transfers do not adequately replace the very strong adductor–short flexor group. The intrinsic bony stability of the CMC joint and the dermal-fascial sling seem adequate to provide CMC stability in these cases.

ABDUCTOR POLLICIS LONGUS

Anatomy and Mechanics

The APL is a difficult muscle to evaluate but it has some important features for the hand surgeon. The muscle runs as a spiral from the dorsal and around the radial aspects of the radius to be inserted on the lateral aspect of the first metacarpal (Fig 12–40). Its muscle fibers are variable in length. Such variability occurs when fibers cross a joint axis and vary in their moment arms to it. However, the APL is wholly tendinous at the points where it crosses the wrist and thumb joints. It does not even significantly affect pronation and supination proximally, although it helps transmit supination across the wrist. Therefore its fiber length variability must be a reflection of the greater circumference of the spiral

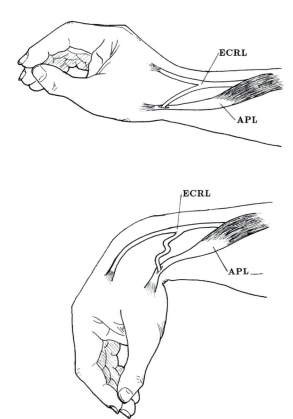

FIG 12–40.
We do not recommend splitting either the ECRL or FCR to make a yoke to power a paralyzed APL. As seen here, flexion of the wrist will inactivate the contribution of ECRL to APL. If the FCR is used, its contribution to the APL would be lost when the wrist was extended.

traversed by the more superficial fibers.

The tendon of the APL is often multistranded and often has more than one insertion. Surgeons have sometimes used one strand of tendon with its muscle fasciculi as a transfer, leaving the remainder to continue its original function[32] (Fig 12–41). However, not all separate strands of tendon have a complete cleavage back to muscle. The only way to be reasonably sure while operating is to pull a strand distally and feel a muscle that yields proximally without affecting the other parts of the distal tendon.

The main tendon attaches to the lateral aspect of the bone of the first metacarpal, but other strands may attach to local carpal bones or may run across the thenar eminence serving as the origin for some part of the APB.

This muscle is named for thumb abduction, but this does not reflect its true action. We prefer the older anatomic name "extensor ossis metacarpi pollicis." The tendon pulls the thumb laterally or radially. This is abduction in body terms but not in terms of the thumb. It pulls on the back or extensor surface of the thumb and thus *extends* the thumb rather than abducts it (Fig 12–42). If the thenar muscles are paralyzed, the APL is usually quite unable to abduct the thumb, i.e., to bring it forward from the plane of the palm, at right angles to the plane of flexion-extension of the thumb. However, we have known cases in which patients have learned that by flexing the wrist they can make the APL tendon bowstring forward enough to cause some true thumb abduction.

The APL is also an abductor and flexor of the wrist (Fig 12–43). It is the only significant wrist flexor that is supplied by the radial nerve. The APL is very important to the thumb as it maintains the arch of the pinch by preventing the collapse of the metacarpal under the influence of the adductors. It is a strong muscle, stronger than the FPL. It has to be strong in order to hold the metacarpal out against the adducting force of the strong adductor, which has a much greater moment arm at the CMC joint. By opposing the adductor and short flexor at the CMC joint, the APL allows them to flex the MCP joint effectively (Fig 12–44).

Surgical Significance

The loss of the APL results in a collapse of the arch of the thumb. The first metacarpal falls in toward the second metacarpal, and the patient has to hyperextend at the MCP joint to be positioned for pinch. This collapse may occur from paralysis of the APL or it may result from proximal shift of the base of the metacarpal as a result of the loss of its stable axes in arthritis. A proximal shift of the insertion of APL may be compensated for by shortening of the muscle fibers over a period of time, but the loss of the stable axes of the CMC joint renders the muscle useless. It just pulls the metacarpal bodily in a proximal direction. If this happens it is no use just advancing the now slack insertion of the APL. The only way to

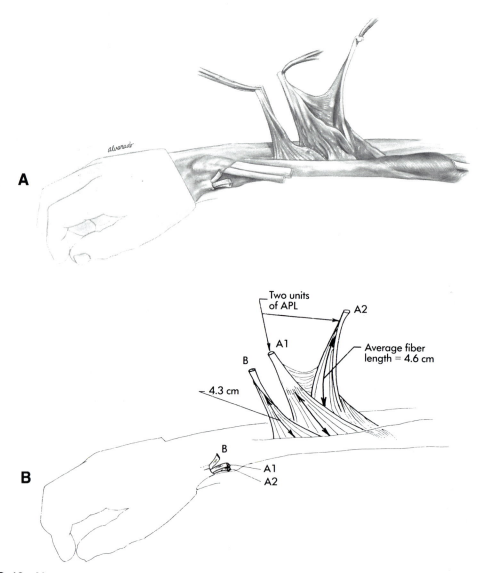

FIG 12–41.
A, the APL and EPB displayed in dissection. **B,** in this cadaver hand the APL had two well-defined divisions *(A₁ and A₂)* that would allow a surgeon to use one as a transfer for the first DI and leave the other to extend the first metacarpal. *B* is the EPB.

restore the effectiveness of the APL in such cases is to restore the axes to the CMC joint so that a pull on its outer border results in angulation rather than migration (or translation).

In cases of paralysis of the APL, as in radial palsy, it is of no use transferring a weak muscle to take its place. It has to be strong enough to be effective with its small moment arm for abduction against the short flexor-adductor mass, which has a large moment arm. It also has to have fibers that are long enough to allow for the excursion it

needs for radial-ulnar deviation of the wrist for which its moment arm is five times greater than that for abduction of the thumb.

The FCR is about 30% stronger than the APL. In radial palsy it may be transferred to the APL insertion where it will still retain much of its moment for wrist flexion. This is an excellent transfer, but only if an ulnar deviator for the wrist (FCU) remains in situ. If the FCU has been used for finger extension, the transferred FCR will become an overwhelming and unbalanced radial deviator

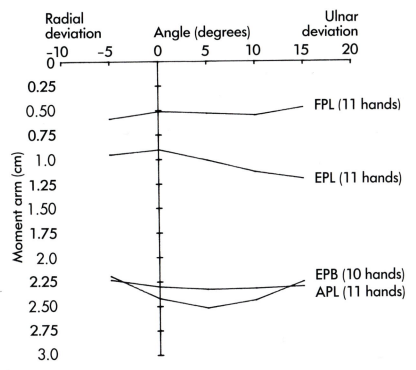

FIG 12–42.
Moment arms for radial deviation at the wrist of the FPL, EPL, EPB, and APL (forearm midprone, mean values).

with its new enhanced moment arm for deviation. We do not recommend splitting the tendons of either the ECRL or FCR to make a yoke with the APL (see Fig 12–39). When this is done, the position of the wrist will determine whether one half of the yoke transmits all the force (e.g., in the case of a split ECRL the thumb will be deprived of all APL support when the wrist is flexed).

The APL is rarely used as a motor for transfer. Edgerton and Brand[31] described a way to reroute the APL to make it approach the base of the first metacarpal from the anterior side, so that it would be a true thumb abductor in median palsy. We do not use this transfer now because the tendon is rather short for the new route and therefore requires high tension to reach its insertion, resulting sometimes in imbalance, with the thumb constantly abducted.

Neviaser et al.[32] have used a single strand of the APL with a graft to restore abduction to the index finger when the first DI is paralyzed. This may be good, but only if the strand of tendon can be shown to represent a truly separate muscle segment, and only if it is realized that this will strengthen a key-

pinch, but will be of doubtful benefit in tip pinch.

EXTENSOR POLLICIS BREVIS

The extensor pollicis brevis (EPB) is closely associated with the APL (Fig 12–45) and has similar actions at the wrist (Fig 12–46) and CMC joint (Fig 12–47). It is smaller and weaker than the APL, and it is only worthy of separate mention because of its primary action as an extensor of the MCP joint of the thumb. Whereas it is very important that the thumb has an independent ability to flex the MCP joint, it is less important that MCP extension is independently controlled. Therefore it is probably unnecessary to replace the EPB in the event of paralysis; a good EPL will extend both joints. Conversely, the EPB may be spared for use elsewhere in any situation where a rather weak muscle is adequate (0.8% tension fraction) (Fig 12–48). It is too weak to be an effective first DI in terms of pinch, but could serve as a positioning abductor for the index finger in ulnar palsy (e.g., for piano players or typists).

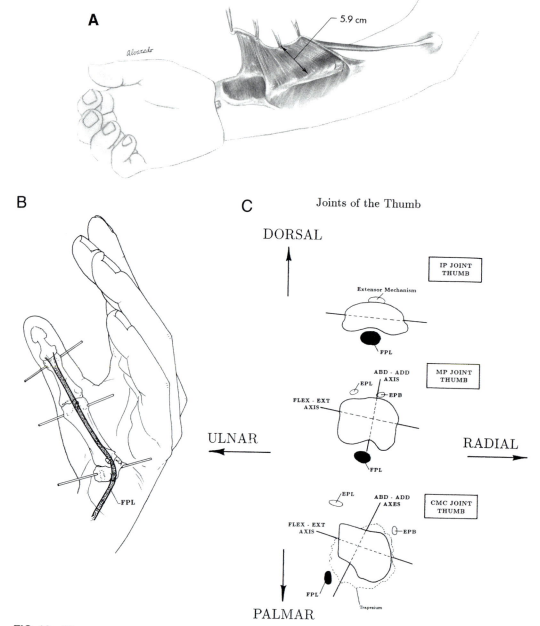

FIG 12–43.
A, the FPL lifted away from its origin, showing the A-frame arrangement of fibers, and Ganzer's muscle. **B** and **C,** the FPL is a flexor of all three joints. It is the only IP joint flexor. It has a relatively small moment arm for MCP and particulary for CMC flexion when compared to the intrinsic muscles. It crosses the MCP abduction-adduction axis and is a pure MCP flexor, but is a weak CMC adductor.

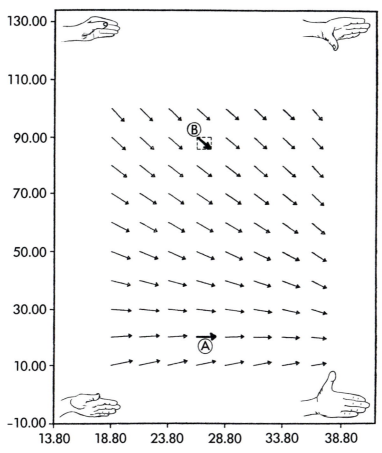

FIG 12–44.
Vectorial total moment arms of the APL at the CMC joint. The APL widens the cone angle of the CMC joint in all positions of the thumb *(A)*. It has a vector for supination that increases as the thumb is pronated *(B)* (see Figs 12–27 and 12–28). *(From Ou, C.A.: The biomechanics of the carpometacarpal joint of the thumb. Dissertation, Louisiana State University, Baton Rouge, December 1979. Used by permission.)*

The tendon of the EPB has been used as a graft for some other muscle to provide abduction of the thumb. We object to the sacrifice of a good thumb muscle when the thumb is already weakened, just to use the tendon as graft material. It would be legitimate if the MCP joint were to be fused (see Figs 12–42 and 12–49).

EXTENSOR POLLICIS LONGUS

Functional and Mechanical Considerations

The muscle fibers of the EPL arise from the middle third of the ulna and from the interosseous membrane. It is a linear origin and the fibers run obliquely to end along the underside of the tendon of insertion. The lowest fleshy part of the muscle lies under the retinaculum. The EPL has its own com-

partment in the retinaculum on the dorsum of the radius and emerges on the back of the wrist (Fig 12–50,A). There is a point of stress at the angle where the tendon of the EPL turns laterally around Lister's tubercle on the radius. In the presence of rheumatoid arthritis this may be a point of rupture of the tendon. From there the tendon runs freely to join the thumb from the dorsilateral side. It crosses the MCP joint on the lateral aspect of the dorsum and goes on to the middle of the back of the terminal phalanx.

The action of the EPL is complex and important. At the IP joint of the thumb it is a simple extensor. At the MCP joint it is an extensor. At the CMC joint, it is an adductor and extensor. The moments for supination and adduction increase in supination

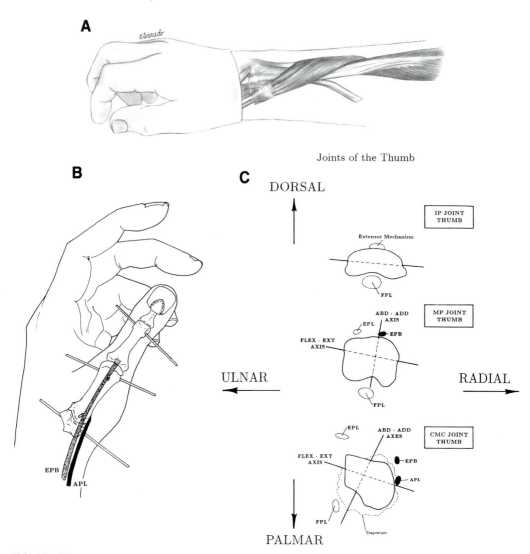

FIG 12–45.

A, the APL and EPB in situ. The EPB is the more dorsal of the two spiral muscles. **B** and **C,** the EPB and the APL are extensors and abductors of the CMC joint. The APL has a good moment arm for CMC abduction and is just dorsal to the flexion-extension axis. The EPB is an extensor of the MCP joint, but crosses the MCP joint abduction-adduction axis, making it a pure MCP extensor.

while the moment for extension of the CMC joint decreases. These changes occur because the tendon is free to bowstring. In adduction the EPL is also an external rotator (or supinator) of the thumb metacarpal. Thus it moves the metacarpal around the cone of CMC circumduction (Fig 12–51). The EPL is an extensor and a radial deviator of the wrist (see Fig 12–46).

In the normal hand the EPL is mainly an extensor of the phalanges and is also used to pull the whole thumb back when a flat open hand is needed (for clapping, slapping,

pushing, and holding up traffic). However, when the thenar muscles are paralyzed and the thumb cannot be used for opposing the fingers, the EPL takes on a whole new function. This is the *lateral squeeze* between the thumb and second metacarpal (Fig 12–52).

In intrinsic palsy the patient quickly learns that if he contracts the long flexor and long extensor of the thumb simultaneously, they both have a vector for adduction at the CMC joint. This is used as a very strong and effective sideways pinch or squeeze, which becomes the most useful type of prehension

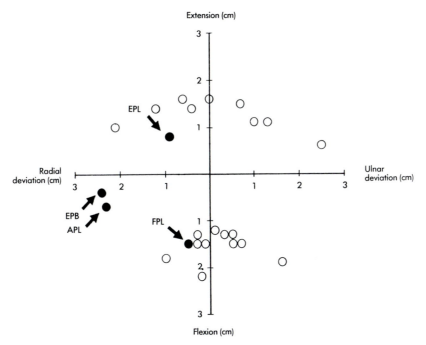

FIG 12–46.
Moment arm diagram of the thenar extrinsic tendons at the axes of the wrist (wrist neutral, forearm midprone).

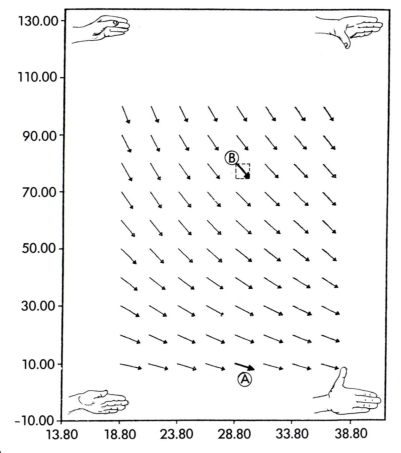

FIG 12–47.
Vector diagram (total moment arm) of the EPB at the CMC joint of the thumb. It is similar to the APL at the CMC joint (A), but it has a greater vector for supination when the CMC is pronated (B) (see Figs 12–27 and 12–28). (From Ou, C.A.: The biomechanics of the carpometacarpal joint of the thumb. Dissertation, Louisiana State University, Baton Rouge, December 1979. Used by permission.)

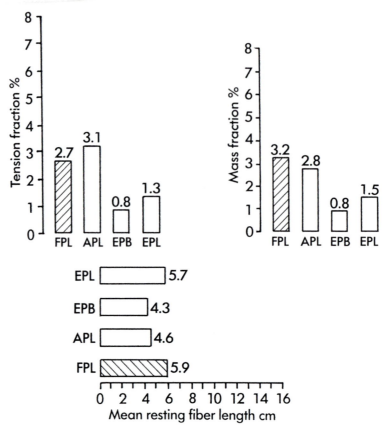

FIG 12–48.
Bar graphs of tension fraction, mass fraction, and mean fiber length of the APL, EPB, and EPL, with the FPL for comparison.

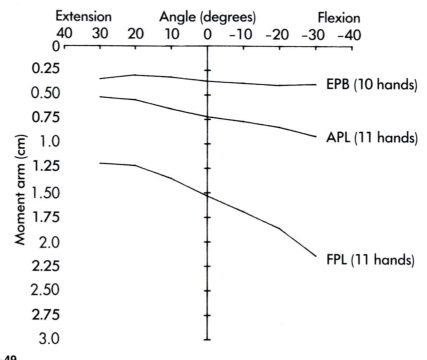

FIG 12–49.
Changing moment arms for wrist flexion as wrist moves into flexion (forearm midprone, mean values). The FPL has a tendency to increase its moment arm in full flexion of the wrist.

that many of these patients have. Unfortunately, this pinch or squeeze is only effective if the EPL and FPL contract together. Thus the position of each joint is fixed by the relative moments of the two tendons at each joint. At the IP joint the moment of the flexor for flexion is greater than the moment of the extensor for extension, so the IP joint is strongly flexed. The same is true at the MCP joint, so that joint is fully flexed. At the CMC joint both tendons are adduc-

tors, the EPL more than the FPL, and strong adduction is achieved. But the EPL has more of a moment for extension than the flexor has for flexion at this joint (in that position), so the metacarpal is adducted in hyperextension (see Figs 12–51 and 12–52).

Thus the EPL becomes the prime mover in this much used "trick" movement. This might be commendable but for two unfortunate results. First, the pulp tissues are not used in this sideways pinch; the unpadded

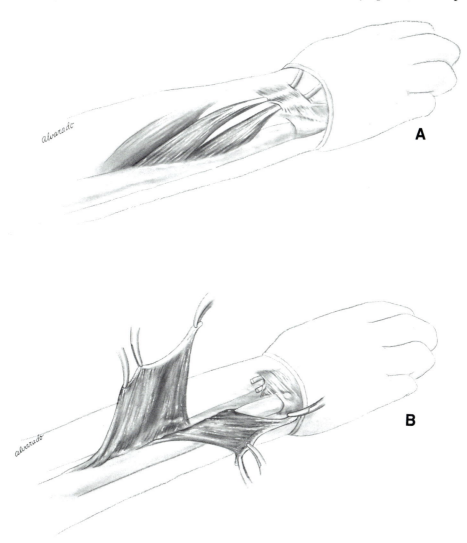

FIG 12–50.
A, EPL *(above)* and EIP in situ. **B,** EPL *(left)* and EIP muscles, with their tendons divided and lifted away from the bone to demonstrate the uniformity of their fibers and unipennate arrangement. The EIP muscle is a little smaller. **C** and **D,** the EPL is an extensor of all three thumb joints. It is ulnar to the center of both the MCP and CMC joints, and has a smaller moment arm for extension at both joints than does the EPB. It is an adductor of both the MCP and CMC joints, and slides further ulnarward at both joints with adduction, increasing its moment arm for adduction. This increase in adduction moment arm at both joints is used by patients with intrinsic palsy in side pinch (see Chapter 4). *(Continued).*

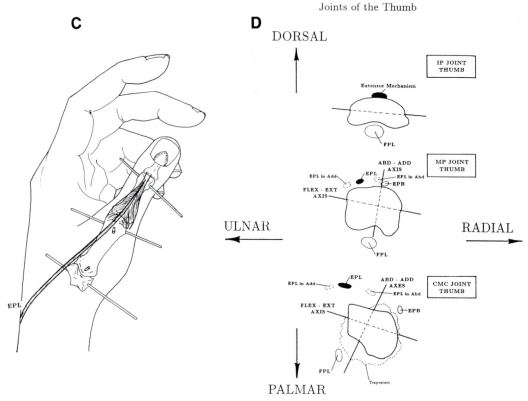

C

D

Joints of the Thumb

DORSAL

IP JOINT
THUMB

Extensor Mechanism

FPL

ABD - ADD
AXIS

MP JOINT
THUMB

EPL in Add   EPL   EPL in Abd
EPB

FLEX - EXT
AXIS

ULNAR   RADIAL

FPL

EPL

EPL in Add   ABD - ADD
AXES

CMC JOINT
THUMB

EPL in Abd

FLEX - EXT
AXIS   EPB

EPL

FPL   Trapezium

PALMAR

FIG. 12–50 (cont.).

dorsal-type skin of the cleft between the first and second rays is used. This often breaks down from constant repetitive stress if there is sensory as well as motor paralysis. Second, once it is established, the *habit pattern* of use of the EPL becomes dominant and carries over to the postoperative phase of opponens replacement operations. This often causes failure of the transfer and a return to the old lateral squeeze type of pinch.

Surgical Considerations

In radial palsy, the action of the EPL should be restored by a tendon transfer of its own, although in severe combined palsies it may share a motor with the extensor of the index finger or even with all fingers. The preferred motor for the EPL in radial palsy is the PL when it is present. The EPL tendon should be divided at the musculotendinous junction and withdrawn distally near the MCP joint of the thumb. It may then be retunneled just in front of the APL and attached end to end to the rerouted palmaris tendon. In the absence of the palmaris a

flexor superficialis may be used to power the long extensors of thumb and index fingers together.[33] In the case of isolated paralysis or rupture of the EPL the EIP may be used as a replacement.

The EPL is not often used as a motor to transfer to other situations because of its unique value where it is. However, in cases of longstanding thenar paralysis, in which the EPL has long been used for a lateral squeeze prehension, a good operation is to divide the EPL proximal to the MCP joint, withdraw it on the back of the forearm, and pass it through the interosseous space and back to the thumb again from the front of the wrist, to be reattached to its own tendon where it had been divided. Thus the EPL remains an extensor of the thumb, but becomes an abductor-pronator instead of an adductor-supinator in relation to the axes at the CMC joint.

In poliomyelitis or other diseases where the EPL may be lost along with the thenar muscles, the tendon transfer for opposition of the thumb may be attached to the EPL

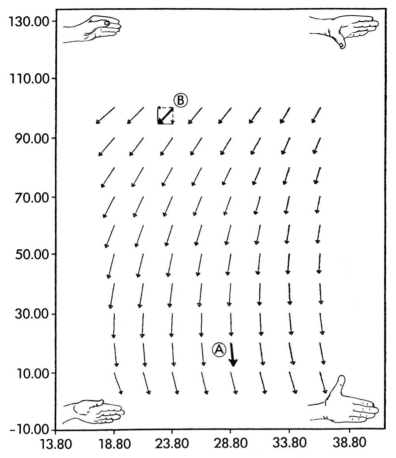

FIG 12–51.
The vectorial total moment arms of the EPL at the CMC joint. This diagram was a surprise to us at first. It shows that the EPL is not a significant extensor of the CMC joint in the supinated position *(A)*. It is an active supinator of the CMC joint in all positions and is also an extensor when the thumb is pronated (opposed) *(B)*. Looking at the back of the hand when the thumb is maximally extended and supinated, it is clear that the bony bump of the base of the thumb is lateral to the ridge of the EPL tendon. The EPL is an extensor to the MCP and IP joints. The APL is an extensor to the CMC (see Figs 12–27 and 12–28). *(From Ou, C.A.: The biomechanics of the carpometacarpal joint of the thumb. Dissertation, Louisiana State University, Baton Rouge, December 1979. Used by permission.)*

tendon on the dorsum of the thumb where it will serve to oppose and extend. In such cases any adductor-flexor replacement should approach the thumb from the radial side to maintain pronation during the flexion of the thumb (see Figs 12–42, 12–48, and 12–53).

FLEXOR POLLICIS LONGUS

Anatomy and Mechanics

The FPL tendon is the most radial of the tendons in the carpal tunnel, where it has its own sheath. In the lower forearm it lies deep to the median nerve. All of its muscle fibers lie to the deep and radial side of the tendon and arise from the bone of the radius and part of the interosseous membrane (Fig 12–54). When the tendon of insertion is lifted forward away from the bone, there seem to be two sheets of muscle fibers forming an A-frame tent that straddles the radius (see Fig 12–43). A slender separate muscle slip, sometimes called Ganzer's muscle, arises from the coronoid process of the ulna and joins the tendon of the FPL at its proximal end.

The FPL is one of the few muscles in the arm that is usually completely independent and that has an action (flexion of the IP joint

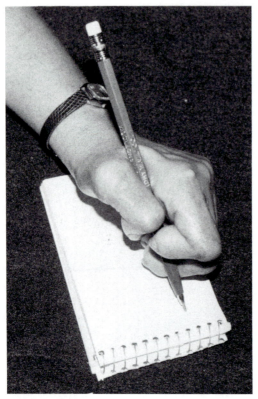

FIG 12–52.
The lateral squeeze pinch, which is used by persons who have low median-ulnar paralysis of the thumb. The thumb squeezes with all joints in flexion except the CMC joint, which is in extension and supination.

of the thumb) that cannot be contributed to by any other muscle. This is one reason we have chosen the FPL as the standard muscle for comparison with all other muscles. A measurement of the torque or moment of IP flexion of the thumb under standard conditions gives a direct measure of the tension capability of the FPL and is a good baseline from which to evaluate the actual tension capability of each of the muscles below the elbow, using Tables 12–1 and 12–2.

The FPL acting alone would seem to be in a good position to produce a key-pinch (i.e., a pinch between the pulp of the thumb and the side of the partially flexed index finger). This is not so. As soon as any force is used, the IP joint will flex and the MCP joint will tend to hyperextend, unless there is an FPB or AP to stabilize the MCP joint. The FPL is really effective during the act of pinch *only* as a flexor of the IP joint. At the MCP joint it is less important than the ad-

ductor and short flexor, and it provides a small fraction (less than one fourth) of the effective muscle power at the CMC joint. An analysis of the moment arms of the FPL at the CMC joint is shown in Figure 12–55.

If the FPL is to be the only muscle to power pinch, and if pulp-to-pulp pinch is desired, it is best to fuse the MCP joint and let the fingers do the moving to pinch against a thumb that rests in the cradle of the web. If a key-pinch is the goal, then the IP joint should be fused. The MCP then becomes the active joint or the FPL must be wholly rerouted to give it a moment at the CMC joint (see Fig 10–32,C).

It is a mistake to regard the FPL as the main strength of the thumb in pinch. The adductor and short flexor together have one third more tension capability than the FPL and they have a far greater moment arm at the CMC joint.

In tetraplegia the FPL tendon has sometimes been anchored to the radius in the forearm,[34] so that a key-pinch will result from active wrist extension. This will give only a weak pinch unless (1) arthrodesis (or tenodesis) of the IP joint is performed and (2) the moment arm at the CMC joint is markedly increased by rerouting the tendon (see Figs 12–43, 12–46, 12–49, and 12–55).

FINGER-MOVING MUSCLES

The only general comments we have on this group of muscles, before discussing each in turn, relates to their *interconnectedness.* Except for the fact that the index finger has a special independence, enabling it to point and to pinch while the other fingers are fully flexed, there is a considerable degree of constraint on all fingers (1) to move together in concert, and (2) to flex and extend the joints of each finger in a predetermined sequence, with some independence of the MCP joints but very little independence of the IP joints.

The limitation of independence of movement of the various fingers is well understood and is related to the (1) interdigitation of muscle fibers of the flexor profundus in

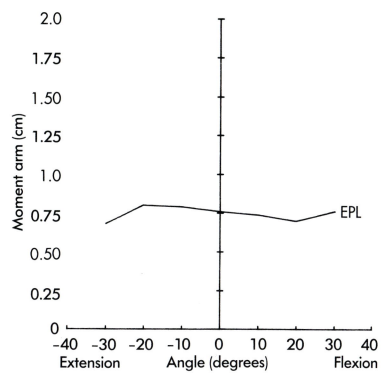

FIG 12–53.
The EPL moment arm for wrist extension, as the wrist extends (forearm midprone, mean of 11 hands).

the forearm and tendinous connections between profundi in the proximal palm and carpal tunnel area, (2) interconnections of sublimis muscles in the forearm in the digastric arrangements of the index and little finger muscles, and (3) juncturae between the extensor digitorum communis (EDC) tendons on the back of the hand and sharing of muscle bellies of the EDC in the forearm.

The constraints within each finger that determine the sequence of flexing and ex-

tending the MCP and IP joints are also well known and have been described effectively by Bunnell,[41] Landsmeer,[55] Kaplan,[63] and others. However, each has tended to identify structures (such as the oblique retinacular ligament) and attribute clear-cut functions to each. As we have instrumented numbers of fingers and have tried to record the contribution of each muscle to each part of each movement, we have been impressed with the *variety* of structural detail and yet

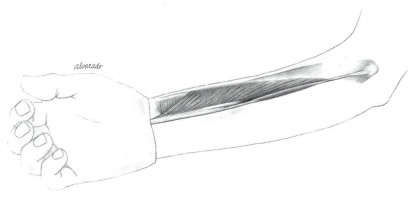

FIG 12–54.
The FPL in situ.

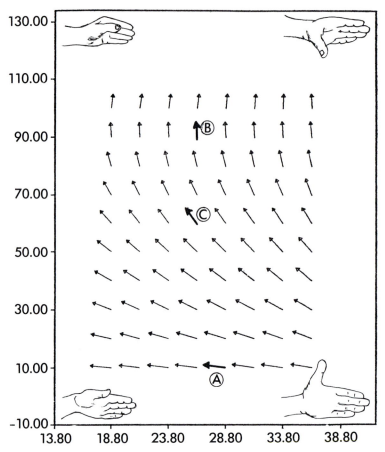

FIG 12–55.
Vectorial total moment arms of the FPL at the CMC joint. Note that in full supination of the thumb the FPL works to narrow the cone angle *(A)*. This is flexion. In full pronation the FPL stabilizes pronation and is not a CMC flexor *(B)*. In the middle areas, in the positions used in key-pinch, the FPL is a flexor-pronator *(C)* (see Figs 12–27 and 12–28). *(From Ou, C.A.: The biomechanics of the carpometacarpal joint of the thumb. Dissertation, Louisiana State University, Baton Rouge, Department of Mechanical Engineering, December 1979. Used by permission.)*

with the *constancy* of the overall result. There is sometimes an isolated band of fibers in one finger while in other fingers it seems that the same function is taken up and shared in a wider sheet or area of the dorsal expansion. A tendon from an interosseus muscle will join the dorsal expansion and seems to extend into a lateral band. In reality, however, it will fan out and different fibers will become tense in different positions of the various joints. If one group of fibers is wired up to record its excursion in relation to a given joint, it will be found that its free movement is dependent on the motion of adjacent fibers and is constrained by the position of distal and proximal joints. It is as though each finger were enclosed in a tube of an interwoven aponeurosis, each fi-

ber of which has a distinct though limited ability to move independently of the others. Muscles feed their tendons into this tube of fibers at various points while actions and effects are fed outward by the same or other fibers as they become attached to segments of the skeleton or as they cross joints in various relationships to joint axes.

Agee and Guidera[35] called this fibrous investment of the finger a "flexible torque tube" because of the way it seems to act as a whole to stabilize and mobilize a finger. They emphasized the importance of maintaining this tube filled out or dilated to the right extent, since the obliquity of the fibers changes considerably if the tube narrows or widens. For example, if a resection arthroplasty results in shortening, lengthening, or a change of diameter of the torque tube, it will no longer be able to transmit faithfully the tension forces that are fed into it. This may be the reason many arthroplasties produce a finger that is unstable in torsion and why its flexor and extensor mechanisms become relatively incompetent. Other examples include the inability of the central slip portion of the extensor expansion to transmit adequate tension to extend the proximal interphalangeal (PIP) joint in a patient with a fracture of the proximal phalanx that has shortened only a few millimeters on its dorsal side. This is because of the interrelationship of the transverse fibers at the MCP joint level to the volar plate, to the transverse metacarpal ligament, and to other structures. This prevents compensation by adjustment of muscle fiber lengths as would occur in most other areas of the body.

We are alternately frustrated at the difficulty of identifying and isolating a structure as a clear-cut cause of a definable movement, and fascinated and entranced to see that it all works. It works with such a glistening lubricated smoothness and efficiency that we have to apologize for intruding with a strand of wire to bypass the real system in order to get a number to read out as an unconstrained excursion or mathematical moment arm. Our numbers have little significance in the milieu of a normal living torque tube, unless we qualify them by noting the position of adjacent joints; but we need the numbers for the time when we have to try to restore motion after the normal fibers have been cut or torn or have lost their ability to move freely on one another, because trauma and infection have bound them all together in a motionless matrix. It is important for the surgeon not to have too firm a preconception of the mechanics of any bit of the extensor expansion or torque tube, which may be exposed at surgery. He or she should use a fine needle-pointed pair of thumb forceps to grasp and hold one part after another while imposing passive motion on distal joints and while stimulating proximal muscles. The excursion of the fibers the surgeon has hold of in relation to the joint motion gives the surest index of the present function of those fibers. The finger knows

more than the surgeon does. Surgeons must take time to ask it mechanical questions and match the answers with what they know of the normal function.

After surgery is complete, and bands of fibers have been shifted or grafted and sutured, the finger again must be gently moved to see the actual mechanical effect of the changes and to see whether the sutures have bound fibers together that should have freedom to move a little on one another.

Our studies of the mechanics of the extensor expansion and lateral bands and of the flexible torque tube are still in the formative stage, and we are not including them in this book. When we have needed numbers for moment arms of parts of the complex, we have usually threaded a fine steel wire through the tendon and measured its excursion in response to joint movement. This gives us a number that is true of an unconstrained tendon, but not always true of the actual effect in the intact hand.

FLEXOR DIGITORUM PROFUNDUS

Functional and Mechanical Anatomy

The flexor digitorum profundus (FDP) makes up 12% of the tension capability of all the muscles below the elbow. It is 50% stronger than the FDS, although in the middle finger the FDP and FDS can exert about equal tension. The FDP to the little finger may be three times as strong as the FDS.

The four segments of the FDP lie side by side in the deep part of the forearm with their tendons of insertion superficial to the muscles. The muscles are formed of sheets of parallel unipennate fibers that look like palisades when exposed with the tendons lifted away from the forearm (Fig 12–56). Together with the FPL the FDP forms a sheet of muscle that is the muscular floor of the anterior forearm. The individual muscles become less independent of one another as one goes from the radial to the ulnar side. The FPL is usually completely separate while the index finger profundus is almost separate, having light areolar connective tissue between it and the middle finger profundus.

The middle, ring, and little fingers have

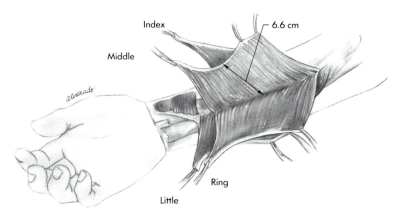

FIG 12–56.
Profundus muscles exposed after section of tendons of insertion. All muscle fibers are about the same length.

little independence from one another. The middle finger profundus has its own distinct muscle fibers, but there is close connective tissue binding it to the ring finger profundus. The ring and little finger profundus muscles are intricately linked, with some fibers common to both. Even the tendons may be difficult to assign to a single finger, because they form a sheet of multistranded tendon material in the distal forearm.

The interconnectedness of the profundus muscles is supplemented by tendinous cross-connections at the level of the carpal tunnel and by the fact that in the palm the lumbricals to the ring and little fingers usually arise from the adjacent sides to two profundus tendons.

The muscle fibers vary between 6 and 7 cm in length and are on average 0.5 cm shorter than the FDS. Although the FDP has one more joint to cross than the FDS, both the short moment arm and the limited range of motion of that distal interphalangeal (DIP) joint make for a small required excursion. The profundus has a slightly larger moment arm and larger excursion than the superficialis at the PIP joint. These extra requirements are compensated for by the fact that at the MCP joint and at the wrist joint the profundus has a smaller moment arm than the superficialis (Figs 12–57 through 12–60), so it has a smaller required excursion. In total, the required excursion of the profundus at all joints of a medium-sized middle finger and wrist is 6.0 cm.

The median and ulnar nerves share the

innervation of the profundus; the ulnar nerve may supply more or less than the muscles to the ring and little fingers. In palsy of either nerve, the three ulnar side muscles—middle, ring, and little—tend to move together because of intermuscular and intertendinous connections. The index finger profundus rarely has enough lateral linkage to benefit from, or contribute to, the other profundus muscles in cases of palsy. In the palm each profundus tendon gives rise to the lumbrical muscle (or parts of lumbrical muscles), which is inserted into the radial lateral band of its own finger. This is a device for diverting some profundus power away from PIP flexion and DIP flexion to the extensor side of these joints, thus allowing MCP flexion to precede and dominate the early part of the sequence of finger closure.

The profundus tendon traverses the palmar aspect of the length of each finger through a series of pulleys (see Figs 5–2 and 5–3). These pulleys serve to keep the moment arms constant within about 1 or 2 mm from extension to flexion. Since the profundus is deep to the superficialis at the MCP level (Fig 12–61) and superficial at the PIP level, it is a more effective PIP flexor than the superficialis and a less effective MCP flexor, given equal tension. At the PIP joint the two strands of superficialis lie one on either side of the profundus and in acute flexion may lift off the bone and achieve as good a moment arm as the profundus. In a cadaver an isolated pull on the profundus may

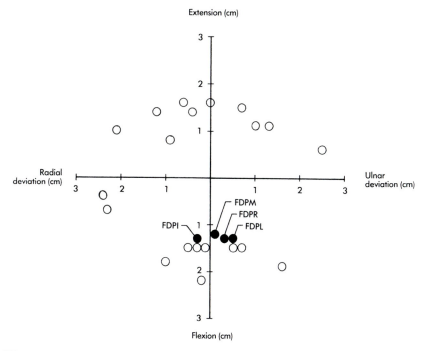

FIG 12–57.
Moment arm diagram of the FDP at the axes of the wrist joint (wrist neutral, forearm midprone). The index is a radial deviator and the others are usually ulnar deviators.

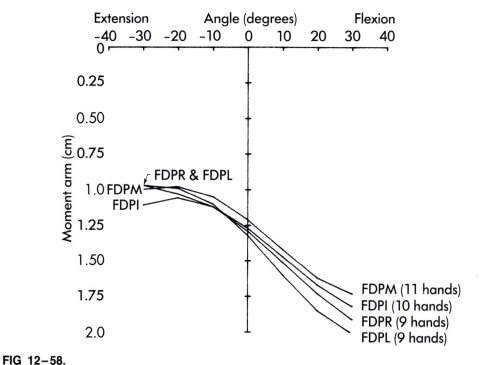

FIG 12–58.
Wrist flexion moment arm (forearm midprone, mean values). The moment arms of the FDP tendons at the wrist increase sharply in full wrist flexion, because the carpal tunnel is loose enough to allow bowstringing. (FDS tendons not loaded.)

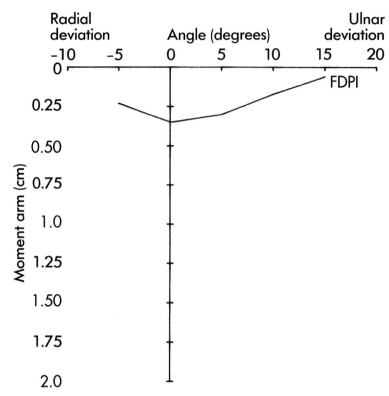

FIG 12–59.
Wrist radial deviation moment arm, FDP of the index finger, forearm midprone (mean of ten hands).

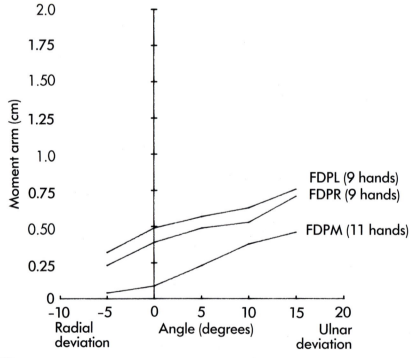

FIG 12–60.
Wrist ulnar deviation moment arm, forearm midprone (mean values).

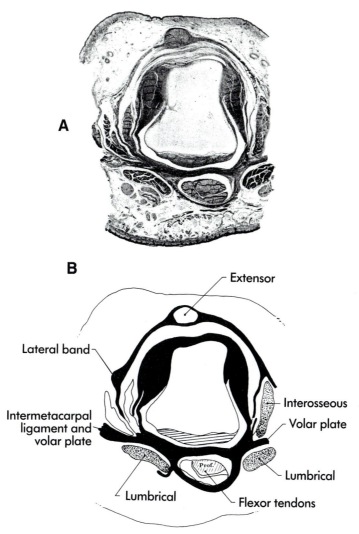

FIG 12–61.
A, cross section at axis of the MCP joint of the middle finger. **B,** diagram identifies profundus tendon. **(A** From Landsmeer, J.M.F.: Atlas of anatomy of the hand, Edinburgh, 1976, Churchill Livingstone. **B** Used by permission.)

result in the tendon slipping to the superficial position of the sheath during MCP flexion, pushing the slack superficialis out of the way and increasing its moment arm. This probably does not happen in life.

The electromyographic (EMG) studies of Long et al.[37] have shown that the profundus is the muscle that performs most of the unloaded flexion movement of the fingers and that the superficialis comes in when more strength is needed. The superficialis also powers most individual finger motion. In most mammals with prehensile hands and in the marsupial opossum, the profundus is a powerful muscle and its tendons join to-

gether into a single tendon in the carpal tunnel and separate again distally in the palm. The superficialis tendons are very much smaller and independent throughout. In the cat tribe the profundus has extra moment at the DIP joint and is thus a claw flexor and a fighting muscle while the superficialis flexes the toes without flexing the claws and is the muscle for play and toilet and in walking. In digging animals like the armadillo and pangolin, the profundus dominates the forearm.

The moment arms of the profundus tendons and the range of motion in radians multiplied give the required excursion of each joint. We have tabulated these values, reduced to round figures, for an average index finger (Table 12–3). A very flexible hand of average size may require up to 7 cm of excursion for full motion of all joints, which is about the same as the fiber length or potential excursion of the FDP.

The profundus with its tendon in its intact sheath is probably a pure flexor of all the finger joints. It may sometimes have a small vector for ulnar deviation of the index MCP joint or of radial deviation of the little finger MCP joint, but these are probably a result in part of the rotated position of the fingers as part of the metacarpal arch. However, in injury or disease of the sheath, as in rheumatoid arthritis, the flexors may bowstring at the MCP joint and develop a significant vector for deviation that comes unavoidably into play whenever the finger is strongly flexed.

Surgeons who advance the pulley of the flexor tendon sheath at the MCP joint, with the idea of balancing a hand that has paralysis of the intrinsic muscles, should realize that they may do harm. They allow bowstringing not only at the axis of flexion-extension[36] but also at the axis of abduction-adduction.

Surgical Implications

The tendons of the FDP are rarely used for transfer. This is because of their importance to finger flexion, and also because it is difficult to isolate them from one another. However, in cases of profundus paralysis various other muscles have been used to provide power to the profundus tendons in the forearm. In high ulnar palsy, the tendons of the ring and little finger profundus may be sutured side to side to the middle finger profundus in the forearm. In high me-

dian palsy the middle finger tendon or the index and middle finger tendons may be similarly sutured to the ring and little fingers to use their power for full hand flexion. In high median and ulnar palsy the ECRL is often used to power all the profundus tendons. In very severe paralysis of almost all forearm muscles, a single muscle may activate wrist extension while tenodesis of the profundus tendons into the radius is performed so that active wrist extension results in finger flexion. By this means about 2.5 cm of excursion may be made available to the finger joints.

It is worth noting that the sum of the required excursions of the profundus at the finger joints is about 3.25 cm, while the required excursion at the wrist is 2.5 cm. This is important in severe paralysis, when there is tenodesis of the profundus tendons in the forearm so that a remaining wrist extensor may flex the fingers by extending the wrist. In such cases 120 degrees of wrist movement can produce only 75% of finger motion. If the DIP joints were fused or tenodesis done, then the profundus tenodesis could result in fairly full closure of the fingers, but only if the wrist could be moved through 120 degrees. If the profundus tendon is taken up and attached to the tendon of the active wrist extensor, then just 75 degrees of wrist movement would provide a full range of finger flexion, because to the required excursion of the profundus at the wrist would be added the required excursion of the wrist extensor for 75 degrees.

Note that in severe upper motor neuron paralysis in which the muscles are spastic, as in tetraplegia, the tonus of the profundus muscle may be preferred to the tenodesis, and it may be controlled by a wrist extensor in the same way. The disadvantage of using

TABLE 12–3.

Moment Arms, Ranges of Motion, and Required Excursions of Profundus Tendons

Joint	Moment Arm	Range of Motion	Required Excursion
Wrist	1.25 cm	120 degrees	2.5 cm
MCP joint	1.00 cm	90 degrees	1.5 cm
PIP joint	0.75 cm	100 degrees	1.25 cm
DIP joint	0.50 cm	60 degrees	0.5 cm
Total			5.75 cm

only spastic tonus is that the grasp is less secure, but the advantage is that the full independent flexibility of the wrist is preserved for use when the limb is used for body support and transfer (Fig 12–62).

FLEXOR DIGITORUM SUPERFICIALIS

Anatomy and Mechanics

Most hand surgeons think of the FDS as four more or less independent muscles. This is far from true and this view may lead to problems when these tendons are transferred. Neither Gray[46] nor Cunningham[43] refer to the common digastric arrangement of the FDS that has been described by the German anatomists and Arnold Henry.[47] It is described in Gray et al.[45] but has been ignored in the writings of most hand surgeons. Chase[42] is an exception.

Central to the FDS muscle mass in the forearm is a large flat common tendon that connects a single proximal muscle belly to two or three separate distal muscles, thus forming a complex digastric muscle (Fig 12–63,A). The muscle to the middle finger

is wholly independent. The ring finger muscle is sometimes independent, but usually derives some of its muscle fibers from the common tendon. The index and little fingers are closely linked to each other by arising entirely from the tendon of the common proximal muscle and having two separate distal muscles (Fig 12–64). Thus, when the index and little fingers need to move independently of each other, they can only use their distal fibers; these are short, allowing only enough excursion for the finger joints. If either of these two tendons were transferred to flex another digit, there would be no problem, but if one were transferred to the dorsal side of the forearm or hand, only the distal fibers would transfer and the intermediate tendon would straddle the separation. The proximal muscle belly would be useless, because it would need to be simultaneously a flexor and an extensor.

There is a wide variation in the tension capabilities (cross-sectional areas) of the superficialis muscles. The middle finger is the strongest. Its tendon lies superficial and receives its muscle fibers from the radial side

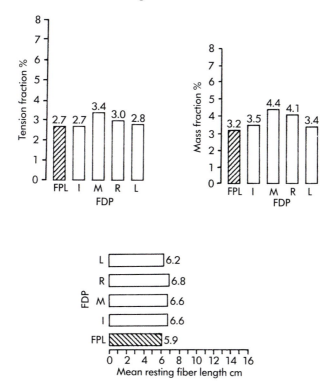

FIG 12–62.
Bar graphs of tension fraction, mass fraction, and fiber length of all FDP muscles, with the FPL for comparison.

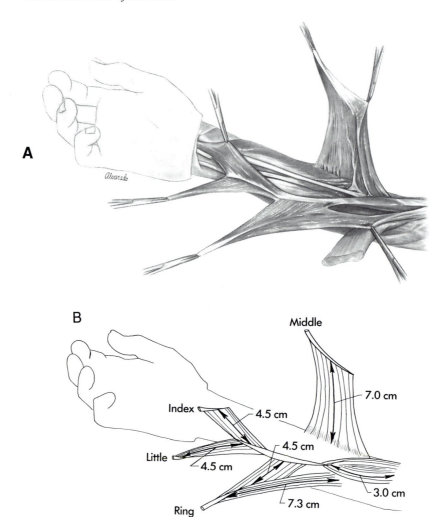

FIG 12–63.
A, drawing of the FDS after three of the four tendons have been cut and spread apart. **B,** the same dissection simplified by a diagram showing fiber lengths.

FIG 12–64.
Highly diagramatic schema of the FDS.

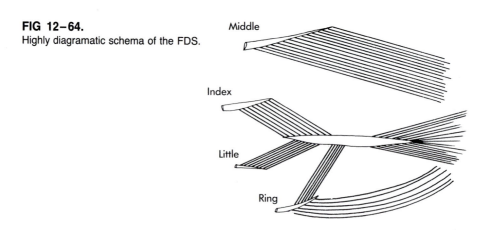

where they form a great unipennate sheet of fibers sweeping across from a linear origin on the radius and from arcuate fibers that bridge over the interosseous structures proximally. These muscle fibers are the longest of any muscle that is truly below the elbow (the BR and ECRL have longer fibers, but they cross the elbow joint).

The ring finger tendon lies to the ulnar side of the middle finger tendon; its tendon of insertion is longer and its muscle fibers are attached further proximally than the middle finger muscle. It is bipennate. The larger muscle mass arises on the ulnar side from the common flexor origin. The smaller mass from the radial side may also arise from the proximal common origin or may arise in part from the common central tendon of the digastric system.

The index and little finger tendons at the wrist lie deep to the other two and are side by side. The index finger tendon is larger and flat; the little finger tendon is slender and may be quite vestigial. If these two tendons are cut distally and held apart, it is quickly realized that each has short muscle fibers with a unipennate arrangement and that all their fibers arise from the large, flat, common central tendon. It takes a lot of dissection up in the depths of the proximal forearm to identify all of the stout proximal common muscle that has its origin deep near the elbow to the ulnar side and that inserts into the common digastric tendon. The fibers of this common muscle are about the same length as the short fibers of the distal separate muscles, and the sum of the lengths of the proximal common muscle fibers and the fibers of the distal separate muscles is about the same as the length of the long fibers of the independent middle and ring finger superficialis muscles (Fig 12–63,B). The cross-sectional area of the proximal muscle is about the same as the sum of the cross sections of the distal muscles that arise from its tendon. Table 12–4 shows the wide variation in tension of the superficialis muscles and the relative uniformity of the profundus.

Effect at the Wrist.—All of the flexor superficialis tendons have moment arms for

TABLE 12–4.

Relative Tension of Finger Flexors

Muscle	Finger			
	Index	Middle	Ring	Little
FDS	2.0%	3.4%	2.0%	0.9%
FDP	2.7%	3.4%	3.0%	2.8%

wrist flexion of nearly 1.5 cm (Fig 12–65), and they are grouped around the axis of deviation, the ring and index fingers being ulnar and radial deviators, respectively (Fig 12–66). All these tendons have a degree of freedom within the carpal tunnel, and if any one tendon is in tension while the others are slack, it will tend to move across the tunnel to take the shortest available line from origin to insertion. Thus if the wrist were held in ulnar deviation while only the index finger was strongly flexed, the tendon would move toward the ulnar side of the tunnel and probably lose its moment for radial deviation.

Effect at the MCP Joints.—The FDS tendons have the largest moment arm for flexion of the MCP joint, as they lie superficial to the profundus (Fig 12–67). The long flexors probably exercise no significant moment for deviation of the fingers as long as the pulleys and ligaments are intact. However, the strain on the restraining system is so great during flexion against resistance that if there is a weakness in the attachments of the ligaments, as in rheumatoid arthritis, a freedom of lateral and palmar movement may develop leading to deviation or palmar bowstringing and subluxation or all three.

At the PIP joints the split sublimis tendons lie on either side of the profundus and they hug the skeletal plane on their way to their insertion around the proximal part of the middle phalanx. Thus the FDS has a smaller moment arm than the profundus at this joint. However, in flexion of the PIP joint, the two halves of the sublimis are able to bowstring a little, rising up on either side of the profundus and adding 1 or 2 mm to their moment arm.

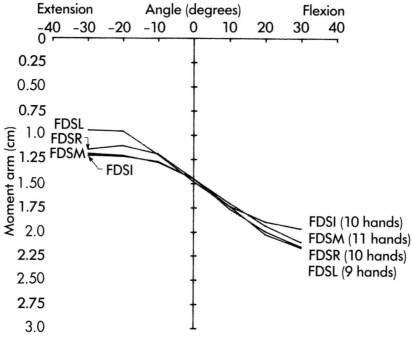

FIG 12–65.
The moment arms of the FDS tendons at the wrist joint increase sharply in wrist flexion as the carpal tunnel gives room for movement and bowstringing of tendons. Wrist flexion moment arm, forearm midprone (mean values). (FDS tendons not loaded.)

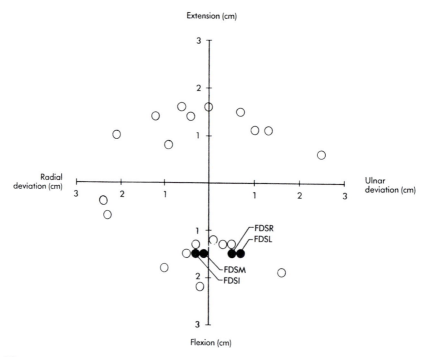

FIG 12–66.
Moment arm diagram of the flexor digitorum superficialis at axes of wrist movement (wrist neutral, forearm midpoint).

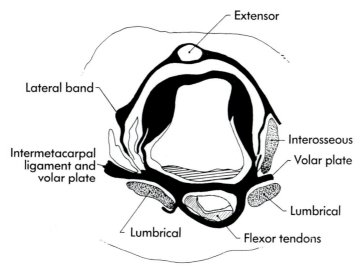

FIG 12–67.
Cross section of the MCP joint showing flexor tendons (FDS *shaded*). *(Redrawn from Landsmeer, J.M.F.: Atlas of anatomy of the hand, Edinburgh, 1976, Churchill Livingstone Ltd.)*

Surgical Significance

The flexor superficialis tendons are frequently used for transfer to other locations in cases of paralysis. They are popular for transfer because all fingers have a profundus tendon that crosses the same joints, so that the superficialis, according to some surgeons, "will not be missed." Also, the tendons are long and allow considerable change of direction while leaving the muscles in situ, and the superficialis muscles have a lot of independence of action and can easily be retrained. Thus superficialis tendons have been used for correction of clawed fingers,[40, 41, 49] for extension of fingers in radial palsy,[39] for opposition of the thumb,[49, 50] for adduction of the thumb,[44] and for extension of the wrist.[38]

Superficialis Minus.—It must be stated, however, that the superficialis muscles are definitely missed when they are removed. The loss is significant and transfer should not be undertaken unless the benefit is greater than the loss. A finger that has lost its superficialis tendon may suffer in various ways.

1. In a very mobile finger the posture of the resting finger will change. There is a tendency for hyperextension of the PIP joint and hyperflexion at the DIP joint (swan-neck deformity.)

2. In any finger subjected to force at the tip, as in pinch, the DIP will tend to hyperflex and the PIP to remain straight or hyperextended (superficialis minus finger) (see Chapter 5). Thus the superficialis should rarely be removed from the index finger.

3. A flexion contracture may develop at the PIP joint. This is caused by tenodesis across the joint by the stump of the cut superficialis tendon. This is a common deformity that is not seen early and is missed by surgeons who do not follow up their cases on a long-term basis. It tends to be progressive as the band of scar contracts.

Surgeons should be aware of the very marked differences in strength and available excursion of the superficialis in the four fingers. The middle finger is 75% stronger than the ring or index fingers while the little finger sublimus has less than half the strength of either of them and sometimes much less. The middle finger superficialis is the only one that is always independent and wholly transferable. The ring finger is largely independent, and the index and little finger tendons are closely linked and have independence of only their distal fibers (see Fig 12–64). If the ring finger superficialis is to be transferred to the dorsum of the hand, it would be wise to divide the band of muscle fibers that often links it to the superficialis digastric tendon in midforearm.

The FDS tendons have an investing synovial sheath over the length that moves through the carpal tunnel. Snow and Fink[48] have pointed out that this sheath will remain with the tendon if it changes direction in this area. We have confirmed the fact that this makes it safe to bring a superficialis tendon out through a hole cut in the carpal tunnel roof—an excellent pulley for a transfer for thumb opposition (Figs 12–68 and 12–69).

EXTENSORS OF THE DIGITS

The tension capability of all of the forearm muscles that extend the fingers is about 38% of the tension of all the forearm muscles for finger flexion. The average fiber length is about 85% that of the flexors. Thus the work capacity of the extensors is less than one third of that of the flexors. There

are two more or less independent finger extensors that we will consider first.

Extensor Indicis Proprius

The EIP is called the "pointing muscle," because it allows independent index finger extension with the rest of the fist clenched. It is fairly constant, rather weak (1% tension fraction), and one of the most used muscles for transfer because the index finger can still be extended after the EIP is removed and the muscle can be rather easily reeducated. It also has a long tendon. It is inserted on the ulnar side of the dorsal tendon of the index finger and may sometimes be capable of producing some adduction.

Patients who are to have this tendon transferred should be warned that not only will they lose some independent index finger extension but also they may even have a slight lag of the index finger when all fingers are extended. This will not usually occur if any defect in the hood is well repaired and the extensor communis tendon is centralized. Some surgeons excise the junctura that links the EDC of the index finger to the

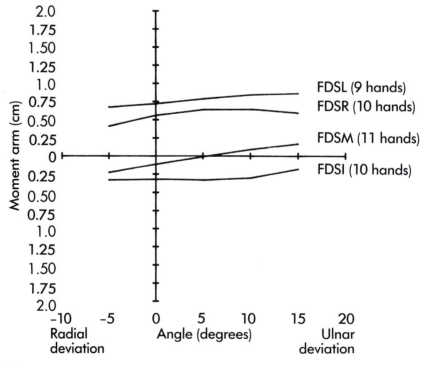

FIG 12–68.
The moment arms for wrist deviation of the tendons of the FDS (forearm midprone, mean values), showing changing values as the wrist moves from radial to ulnar deviation.

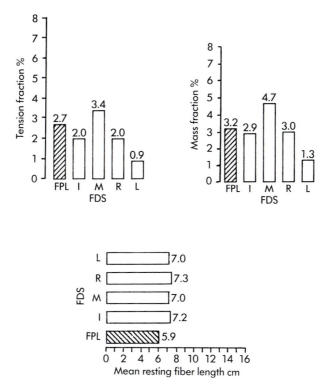

FIG 12–69.
Bar graphs of tension fraction, mass fraction, and fiber length of the FDS, with the FPL for comparison.

middle finger in order to give more independence to index finger extension. We do not know if this actually confers benefit or not. The tendon of the EIP may have some small but tough connections on the back of the hand with the extensor digitorum. These must always be identified and divided if there is any resistance to proximal withdrawal of the tendon, because the musculotendinous junction of this slender muscle is fragile and will be torn if any force is used during withdrawal. The whole muscle is deep and distal in the forearm. Attempts to free it proximally may endanger the nerve supply. The muscle fibers join the tendon so far distally that the proximal edge of the extensor retinaculum may have to be incised in order to see the tendon and get hold of it.

Because this muscle is so often used for transfer, we must comment that while it is excellent for solving positioning problems, like clawed fingers or thumb opposition, it is not strong enough to compensate for lack of real power, as in finger flexion or thumb adduction, or even to replace the first DI.

Extensor Digiti Quinti

The extensor digiti quinti (EDQ) is of the same order of size and strength as the EIP and is another useful muscle for transfer. The muscle is slender and lies to the ulnar side of the extensor digitorum muscles with which it may be connected. Its tendon has its own compartment under the extensor retinaculum and divides into two tendons there or proximally. The two tendons of the EDQ may have some degree of independent movement related to their different moment arms at the MCP joint where they may affect radial and ulnar deviation of the little finger. The little finger is commonly regarded as having two extensors, as the index finger has, one of its own and one from the extensor digitorum. However, in our experience, the contribution from the extensor digitorum is usually inadequate to extend the finger by itself. In most cases the digitorum tendon is oblique and looks more like a junctura than a tendon. In such cases if the EDQ is removed there will be quite a gross lag in extension of the little finger. A slip

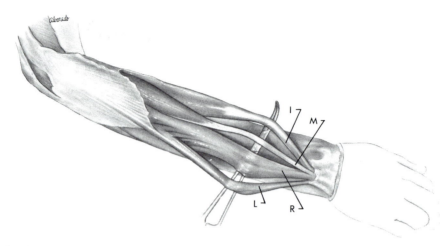

FIG 12–70.
General view of the digital extensor muscles and tendons. *I* = index; *M* = middle; *L* = little; *R* = ring.

from the ring finger extensor needs to be dissected back to the level of the metacarpal base and redirected to the stump of the divided EDQ tendon for extension of the little finger.

Extensor Digitorum Communis

Functional and Mechanical Considerations.—The strength of the extensor muscles of all the digits amounts to about 10% of all the muscles below the elbow. The two long flexors of the middle finger alone are together as strong as all of the extensors to all four fingers. The individual extensor muscles interact with one another at two levels, having incomplete separation between the muscle bellies in the forearm (Fig 12–70) and also some interdigitation and tendinous connections on the back of the

hand. For this reason the EDC is often looked on as one muscle with four tendons and has rarely been split up for purposes of transfer. However, it deserves a more careful examination.

The EDC to the index finger is usually distinct and is separable from the fibers of the middle finger muscle. The flat index finger tendon runs back, mainly on the surface of the muscle, and receives fibers along a considerable distance, making this a rather long thin muscle (Fig 12–71). The EDC to the middle finger, by contrast, has a well-defined rounded tendon that receives no muscle fibers in the distal half of the forearm. In the proximal forearm the muscle belly is fusiform and well defined, having commonality with the ring finger muscle only very proximally (Fig 12–72). The ring finger EDC has more distal fibers and has a

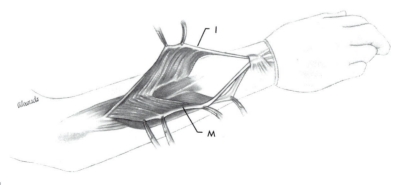

FIG 12–71.
Muscle fibers of individual digital extensors after light connective tissues have been divided. *I* = index; *M* = middle.

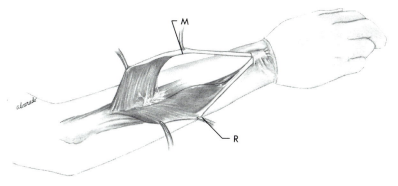

FIG 12–72.
After index finger extensors are removed, the middle finger extensor *(M)* is seen as a well-defined proximal group of fibers with a long clean tendon. The ring finger extensor *(R)* is muscular throughout the forearm.

flat stranded tendon with many connections with the little and middle fingers (see Fig 12–72). The EDC to the little finger is quite variable, both in its muscle belly and in its insertion. It is often impossible to separate it from the ring finger muscle belly proximally, and it often inserts as an oblique strand of tendon that reaches the little finger as a branch from the multistranded ring finger tendon (Fig 12–73). The main extensor of the little fingers is always the EDQ, which is usually a distinct muscle with a double tendon.

Juncturae Tendinum.—These oblique tendinous connections between adjacent extensor tendons on the back of the hand have usually been ignored by surgeons. Some have thought of them as unfortunate restrictions on individual finger movement, and in the past pianists have had them excised in

an attempt to improve their performance. Others have thought of them as significant in maintaining the proper spacing of the tendons in relation to one another and to their position in relation to each joint axis.

Force Redistribution by Extensor Tendons and Their Associated Junctura System.—Agee and Guidera[51] described a system of force redistribution that helps to dynamically stabilize the MCP joints of the fingers. They described two opposing force systems, one of which forces the metacarpal heads apart in radial and ulnar directions with the other force system being capable of collapsing the metacarpal heads toward each other by tension forces transmitted into the junctura system by the EDC muscles. The biomechanical effect of this interrelationship is that the combination of metacarpal spread and proximal traction by the finger extensor

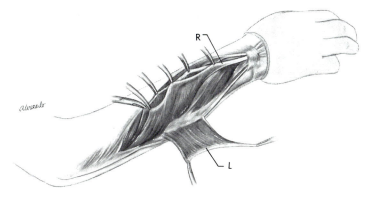

FIG 12–73.
The extensor digitorum to the little finger, displayed after dividing the tendon of insertion. Note uniformity of fiber length. *R* = ring; *L* = little.

musculature working against fingers stabilized in a flexed position results in tension forces being created in the junctura system; these forces are secondarily transmitted into the radial transverse lamina of the index and long fingers and the ulnar transverse lamina of the little finger, thereby dynamically stabilizing these digits against deviating forces from their respective borders of the hand (Fig 12–74).

Surgical Significance.—Rarely used in tendon transfers, the EDC often has to receive replacement motors in case of radial palsy. In such cases one motor has usually been used for all four fingers, though sometimes the index finger has been accorded a separate motor or one to share with the thumb. Whenever the EIP or the EDQ tendons are removed for transfer, it is essential that the surgeon determine in every case whether the remaining tendon from the EDC is adequate to extend the deprived finger. In the case of the little finger, most commonly the oblique tendon from the

EDC will be inadequate, and a strand of EDC from the ring finger must be split off from the ring finger and redirected to the little finger where it must be attached under matched tension. In such cases the patients must be warned, particularly if they belong to British aristocracy, that they will no longer be able to keep their little finger individually extended while they hold their teacup with the remaining digits.

Extensor Digitorum Communis to Middle Fingers.—We must take this opportunity to bring this potentially useful muscle to the attention of surgeons. It has a longer free tendon than any muscle on the back of the forearm. This allows it to be easily redirected around the radius or ulna, or through the interosseous space. As a motor for thumb abduction and opposition or as an intrinsic muscle replacement it has much to commend it. It requires open dissection on the back of the hand to detach the juncturae tendinum and to separate a strand of extensor from the ring finger to take its place.

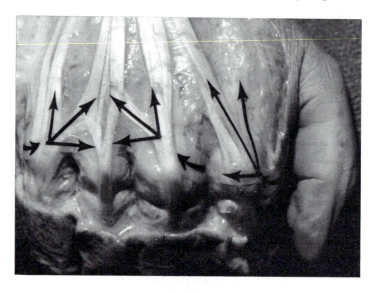

FIG 12–74.
Vector analysis of tension forces defines the force redistribution. The tension force in the index communis tendon produces two forces: a large one that extends the MCP joint and a smaller ulnarly directed one that is transmitted into the radial transverse lamina of the extensor hood mechanism. Similarly, analysis of tension forces in the oblique junctura interconnecting the long, ring, and little finger communis mechanisms produces proximally directed "finger extensor" forces and laterally directed transverse laminar "finger stabilizer" forces. These laterally directed forces are transmitted into the appropriate transverse laminae of the long and little finger MCP joints to help stabilize these fingers from deviating forces from their respective borders of the hand. *(From Agee, J., and Guidera M.: The Functional Significance of the Juncturae Tendinae in Dynamic Stabilization of the Metacarpophalangeal Joints of the Fingers. Presented at the Annual Meeting of the American Society for Surgery of the Hand, Atlanta, 1980. Used by permission.)*

However, these are rather simple maneuvers and it is surprising that this muscle has not been more widely used. Its tension is almost the same as the sum of the tensions of the EIP and EDQ.

The tendons of digital extensors all finally converge on the dorsum of the MCP joints. At this level the tendon becomes part of a complex sleeve around the finger and works in conjunction with intrinsic and other muscles, as well as with restraining bands, to create balanced movement of the finger.

The interplay of the components of this sleeve or tube of tendon, sagittal bands, and palmar plate is beyond the scope of this book. It is important for the surgeon to realize that there is a real tube that surrounds the MCP joint; Agee and Guidera[51] call it a "flexible torque tube" to emphasize some of its mechanical implications. This tube serves to keep the various component tendons in their proper relationship to their axes (e.g., to prevent the long extensor tendon from slipping off the knuckle into the interdigital gutter), and it also limits the freedom of longitudinal movement of the tendons that

blend with it. For example, the extensor tendons cannot pull through to affect the PIP joint when the torque tube is in an unfavorable position, having reached the limit of its proximal shift. This proximal limit of the dorsal aspect of the tube is in part determined by the *distal* shift of the palmar part of the tube, which is attached to the anterior lip of the proximal phalanx. Figures 12–75 through 12–78 illustrate various aspects of the digital extensors.

LUMBRICALS

Anatomy and Mechanics

The lumbrical muscles are unusual in that the origin moves further than the insertion (Fig 12–79). The muscles are made up of longitudinally running fibers. The tendons of origin and insertion may each occupy up to 1.5 cm of muscle leaving the middle 3 to 4 cm as a cylindrical muscle with parallel fibers. Thus any cross section near the middle of the muscle will be a cross section of all its fibers.

The excursion of the origin will be the

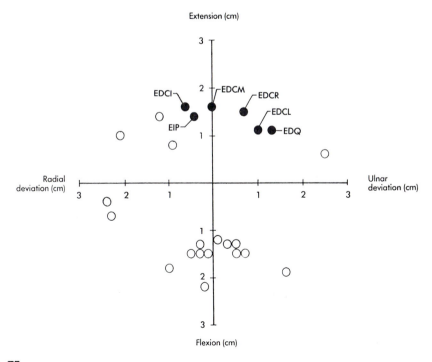

FIG 12–75.
Moment arms at the axes of flexion-extension and radial-ulnar deviation of the wrist (wrist neutral, forearm midprone). All tendons that cross the wrist are shown, and the long extensors of the fingers are identified.

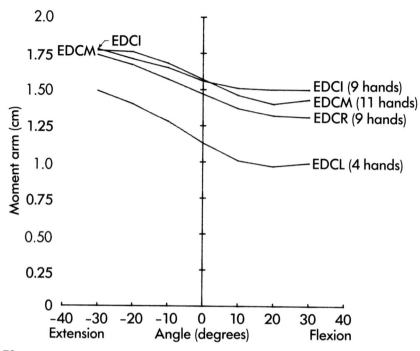

FIG 12–76.
Changing moment arms for wrist extension of tendons of extensor digitorum, recorded while the wrist moves into extension (forearm midprone, mean values).

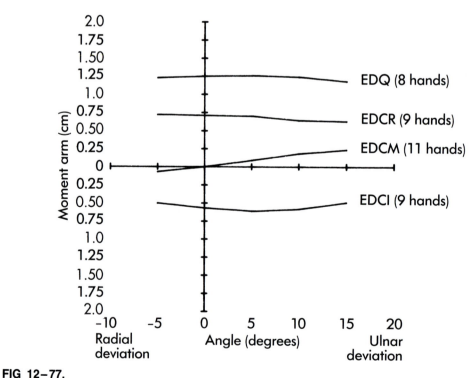

FIG 12–77.
Moment arms for wrist deviation of four extensor tendons showing changing values as the wrist moves from radial to ulnar deviation (mean values).

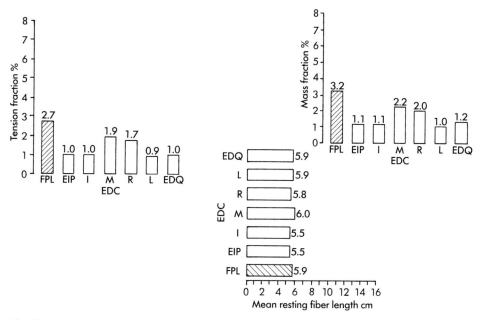

FIG 12–78.
Bar graphs of tension fraction, mass fraction, and mean fiber length of the digital extensor muscles, with the FPL for comparison.

sum of the excursions of the flexor profundus for the MCP, PIP, and DIP joints, perhaps 3.5 cm, while the excursion of its insertion will be the sum of the excursions of the lateral band for MCP flexion, PIP extension, and DIP extension plus, possibly, the excursion of the lateral band for radial deviation. This might total about 1.75 cm. However, since the profundus and the lumbrical move together at the MCP joint and have comparable excursions, the required excursion of the lumbrical is the sum of profundus excursion at the PIP and DIP joints and the excursion of the lateral bands at the same joints. This might add up to about 2.75 cm.

In careful dissections of the lumbrical

tendon as it crosses the palmar side of the intermetacarpal ligament, a fibrous "adhesion" is usually observed, which connects the tendon to the structures of the ligament and volar plate area. This little fibrous attachment allows only a restricted movement of the tendon on its way to the lateral band. It is a reminder of the fact that the whole complex of volar plate, A1 pulley flexor sheath, and intermetacarpal ligament moves proximally with the base of the phalanx during MCP flexion (Fig 12–80). Thus almost the only movement that is needed by the lumbrical tendon over the intermetacarpal ligament is that related to the extension of the IP joints.

FIG 12–79.
The four lumbrical muscles.

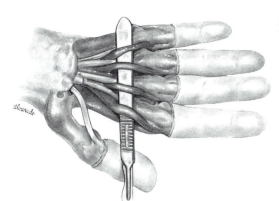

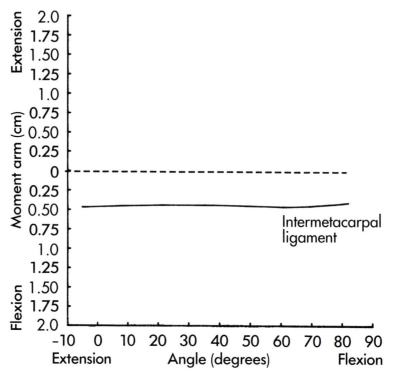

FIG 12–80.
Moment arm of the intermetacarpal ligament ring finger MCP joint with distal and proximal joints in extension. This graph was made by measuring the proximal movement of the transverse metacarpal ligament complex during flexion of the MCP joint of a ring finger. This demonstrates that a tendon transfer for intrinsic paralysis may be attached to the edge of the plate or to the ligament and be an effective MCP flexor.

Probably the maximum length of the lumbrical would occur with a fully closed fist, and the shortest length with the fully open hand. The function of the lumbrical is not what it seems to be. It seems to act as an MCP flexor because it crosses the MCP joint with a good moment arm on the flexor side. However, it cannot exert any tension without diminishing the tension in the profundus tendon by the same amount, because it pulls its origin on the profundus distally with the same force that it exerts on its insertion. Now, since the profundus is also an MCP flexor, and because it has a moment arm about the same as the lumbrical or even larger (Figs 12–81 and 12–82), the profundus can flex the MCP joint as well by itself as it can with the help of the lumbrical. The profundus at the MCP joint is like a father carrying his lumbrical child on his shoulders. The child may offer to carry the bag that is in the father's hand, but the father's legs will not notice the difference.

It is at the IP level that the lumbrical has a real function. Here it is an extensor, and it

also diminishes the flexion moment of the profundus at the same joint. In the action of controlled unopposed flexion of the finger, the lumbrical may dominate the sequence of closure.[52, 55] It ensures that the MCP joint flexes ahead of the IP joints, allowing the hand to surround a large object. In the case of pinch the lumbrical also helps to concentrate finger flexion force toward the base of the finger where it is needed to oppose the moment from the thumb. The lumbricals are also radial deviators of the MP joints, but in this respect they are inferior to the interossei both in strength and moment arm (Figs 12–82 through 12–86).

The index and usually the middle fingers have a single origin for the lumbricals, from their own profundus tendons, while the little and ring finger lumbricals often originate from two adjacent profundi. The tendon of the index finger lumbrical is lateral to the base of the phalanx; it is therefore assumed to be an abductor of the MCP joint. However, the fleshy fibers of the muscle lie on the palmar side of the metacarpal head,

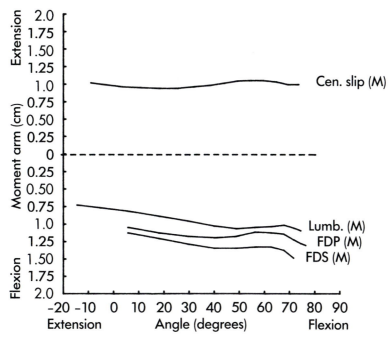

FIG 12–81.
Moment arms for flexion and extension of the long flexors and extensor at the MCP joint of the middle finger with the IP joints in extension. (*Note:* The excursion of the extensor was measured by using a fine wire running through the central slip of the tendon, to avoid the effect of restraints on the real tendon caused by sagittal bands.)

right beside the flexor sheath. Thus at the level of the MCP joint axis the index finger lumbrical has about three times as much moment arm for flexion as for abduction. Because it is not restrained by a sheath, the moment arm for MCP flexion increases with flexion, but the difference may not be very significant because all structures in the palm are held in place by septa and there is little scope for bowstringing within the lumbrical canal.

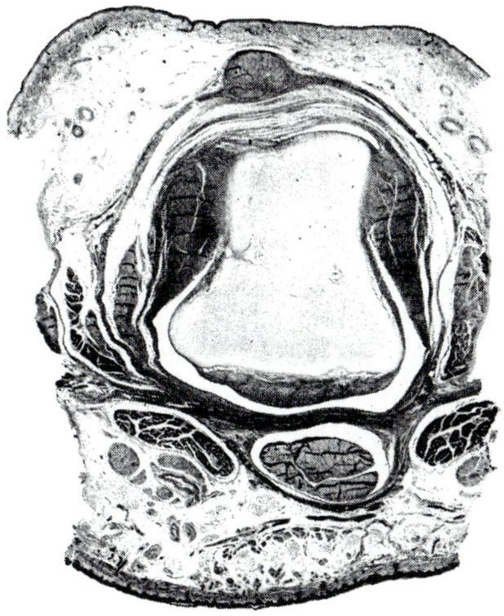

Surgical Significance

Lumbricals are rarely used for transfer, but many operations have been used to re-

FIG 12–82.
Cross section of the MCP joint at the level of the axis of flexion-extension, showing the relative moment arm of flexor tendons and intrinsic muscles. (*From Landsmeer, J.M.F.: Atlas of anatomy of the hand, Edinburgh, 1976, Churchill Livingstone Ltd. Used by permission.*)

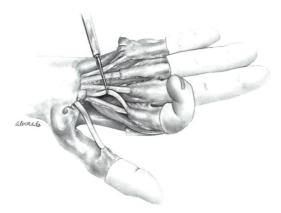

FIG 12–83.
Lumbrical muscle to the index finger, showing the length of the attachment to the tendon at origin and insertion.

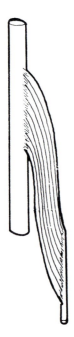

FIG 12–84.
Diagram to show the extent of origin and insertion of lumbrical fibers. Note that the middle third or more of the muscle is parallel-fibered and without tendon overlap. (Contrast with the interosseus, Fig 12–89,B.)

place them when they are paralyzed (clawhand) or to correct their overaction. In both cases the operations should be thought of as corrections for intrinsic minus or intrinsic plus, rather than lumbrical minus or lumbrical plus since it is not easy to isolate the ef-

fect of the lumbrical from the other intrinsic muscles. One common condition that is a pure lumbrical plus is the result of traumatic division of the flexor profundus distal to the lumbrical origin. The origin of the lumbrical is pulled proximally and the result is lumbrical lengthening and constant tension on the radial lateral band. This must be corrected by restoring the profundus tendon to its normal position. The same imbalance may be caused by surgeons who use the fleshy lumbrical to wrap around a flexor tendon suture in the palm.

In correcting an intrinsic minus clawhand, we prefer to pass the tendon transfer along the path of the lumbrical to its insertion in the lateral band, as suggested by Bunnell.[54] The reason for this may be seen in Figure 12–82 which is a tracing from Landsmeer's beautiful book, *Atlas of Anatomy of the Hand.*[56] In this transverse section taken at about the axis of the MCP joint, it will be seen that the lumbrical has more of a flexion moment arm at this joint and less of a radial-deviating moment arm than those of the other intrinsic muscles.

If one is to attempt to restore lateral deviation, it should be to both sides of the finger. Any single replacement for paralyzed intrinsics should tend to keep all fingers side by side and therefore toward the middle fin-

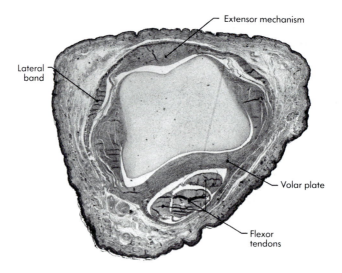

FIG 12–85.
At the level of the PIP joint axis of flexion, the lateral band is an extensor, but not as much as the middle band of extensor expansion. *(From Landsmeer, J.M.F.: Atlas of anatomy of the hand, Edinburgh, 1976, Churchill Livingstone. Used by permission.)*

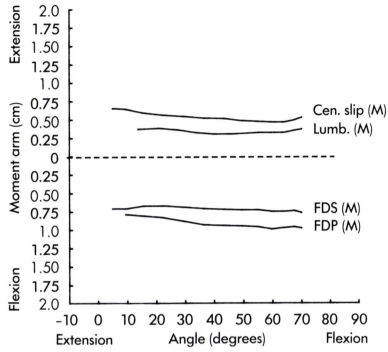

FIG 12–86.
Moment arms for flexion and extension at the proximal IP joint of the middle finger with the MCP joint in extension. (*Note:* The excursion of the extensor digitorum central slip was measured by the excursion of a fine wire running along through the tendon, in order to avoid the restraints caused by lateral attachments of the real tendon, which vary according to the position of other joints.)

ger. Brooks[53] inserts a transfer to the flexor sheath. This is at the middle line, so it causes no deviation. However, any intrinsic replacement that acts only on the MCP joint demands more total energy to produce the desired results of MCP flexion with IP extension than a replacement that is inserted to the lateral band. Suppose that it takes x units of energy to flex the MCP joint by any intrinsic muscle that is inserted into the lateral band and that it takes y units extra to hold the PIP joint extended. The total energy to achieve this posture is now $x + y$. Now, if it is to be done by a tendon transfer that does not go to the lateral band but which stops at the base of the phalanx, it will still take x units of energy to flex the MCP joint and it will still take y units of energy to extend the PIP joint, though in this case it is the extensor digitorum that provides the y units. Therefore we still use $x + y$. But think again. An extensor tendon, to extend an IP joint, must also exert the same tension proximally where it is an extensor of

the MCP joint of the same finger. Therefore the extensor digitorum unavoidably opposes the new transfer at the MCP joint. Therefore the new transfer must contract with $x + y$ to flex the MCP joint against the extensor. In fact it needs still more tension because the extensor has a larger moment arm for MCP extension than for PIP extension. So the total energy needs to be $x + 2y$ or more. The most efficient way to replace a lumbrical is to insert the replacement where it belongs—the lateral band. See also Figures 12–87 and 12–88.

INTEROSSEI

For two reasons we have determined not to attempt a detailed anatomic and mechanical account of the interossei in this book. The first reason is that this is a practical clinical manual, and most surgeons will be content to understand the group action of the intrinsic muscles rather than try to remember individual components and variations.

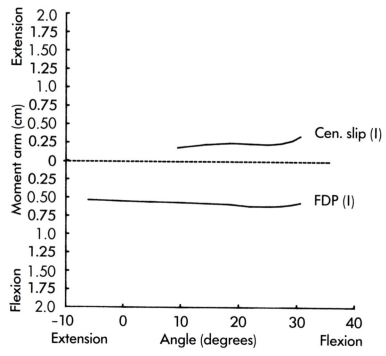

FIG 12-87.
Moment arms for flexion and extension of the FDP and of the central slip at the DI joint of the index finger with the proximal and MCP joint in extension. Central slip excursion was measured by the movement of a wire rather than of the tendon itself.

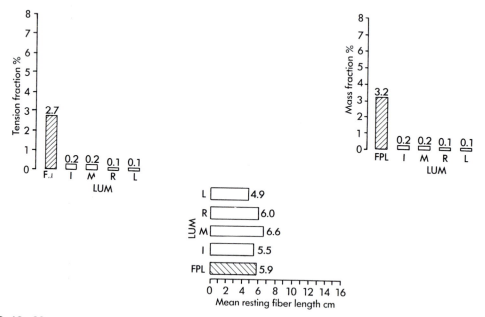

FIG 12-88.
Bar graphs of lumbrical muscles for tension fraction, mass fraction, and mean fiber length, using the FPL for comparison (values are rounded off to the nearest tenth).

The second is that good studies have been published by others, and our studies have added little to the sum of knowledge adduced by such as Landsmeer,[65] Kaplan,[63] Eyler and Markee,[62] Long et al.,[67] Salisbury,[68] and Duchenne.[61]

Factors Common to Most Interosseus Muscles

Tension.—All the interossei are short-fibered muscles that have long tendons of insertion that run through most of the muscle's length, receiving muscle fibers from both or all sides (Figs 12–89 to 12–91). For a given bulk of muscle the interossei can exert four times as much tension as the lumbricals, and they have about one fourth of the potential excursion. A glance at Table 12–2 shows the interossei to be on average as strong as the FDS with about one fourth of the potential excursion.[59]

Pennation Angles.—In all of our calculations of muscle tension we have assumed that muscle fibers are roughly parallel to their line of action. We know that there should be a correction in some muscles for pennation angles, but have not attempted it. This is partly because we have observed in living muscles that when muscles contract strongly their fibers tend to straighten out to become more nearly parallel to their line of action. However, the opposite is true in interosseus muscles, where fibers that arise from two parallel bones converge on a tendon between them. Here the origins are immobile, and the pennation angles become larger as the fibers shorten in strong contraction. We have not allowed for this in our tables, which means that the interossei may not be quite as strong as we have made them appear.[66]

Potential Excursion.—Our figures for excursion are only approximations and averages. In reality a careful dissection of a typical DI muscle shows that the muscle fibers that attach to the tendon that inserts into the lateral band are up to 50% longer than

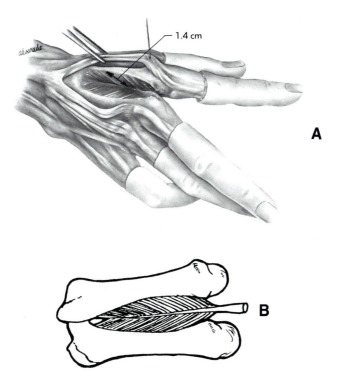

FIG 12–89.
A, the second DI muscle. The typical interosseus muscle has a long intramuscular tendon, receiving short fibers all along its length. (Contrast with the short tendon and long fibers of the lumbrical.) **B,** diagramatic simplfication of the second dorsal interosseus.

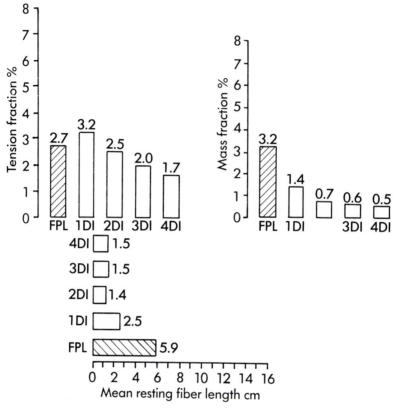

FIG 12-90.

Bar graphs of the first, second, third, and fourth dorsal interosseus muscles, showing tension fraction, mass fraction, and mean fiber length, with the FPL for comparison.

FIG 12-91.

The main tendon of the first DI is usually obscured by the two main masses of muscle fibers that converge on it from the second and first metacarpals. A little teasing apart of fibers reveals the long, strong tendon parallel to the second metacarpal.

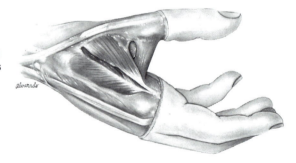

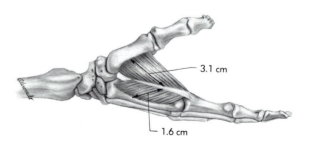

the fibers of the same muscle that insert into the tendons that go to the bone of the base of the phalanx. Some of the fibers that finally insert into bone are as short as 1.0 cm and some that insert finally into wing tendons may be almost 2.0 cm.

Moment Arms: Abduction and Adduction of the Proximal Phalanx.—These are the most consistent moment arms of the interossei. They are, as expected, a little more than half the transverse diameter of the metacarpal head and are of the order of 1 cm in the middle finger of an average hand. Note that as the MCP joint flexes and the axis moves up to the phalanx, the medial-lateral deviators become rotators. More correctly, they work around a narrower cone (see Fig 4–14). Moments for MCP flexion vary even within one muscle in one hand. The tendon fibers that insert into the lateral band, especially those from the palmar interossei, may have a flexion moment of as much as 7 or 8 mm. Tendon fibers that insert into the bone of the phalanx may vary from 2 mm to 5 mm, and all will vary also according to the position of flexion or extension of the joint. When the MCP joint is extended, the flexion moment arms may be short, but as the joint flexes the tendons bowstring a little and move forward until the moment arms for flexion increase to about 5 mm or more. Although the interossei have smaller moment arms than the lumbricals for MCP flexion, they can exert so much more tension (compare Fig 12–90 with Fig 12–88) that they are much more important MCP flexors than the lumbricals.

In hands that have ulnar palsy, all fingers may sometimes show clawing although the median-supplied lumbricals to the index and middle fingers are normal. Even when only two fingers are clawed, a light backward push on the proximal segments of the middle and index fingers usually results in flexion of the IP joints, showing that the lumbricals can only just hold the finger straight (this is latent clawing).[58]

For the IP joints the palmar interossei (PI) are more effective extensors than the lumbricals in all positions of the MCP joints. In this they are assisted to a varying extent by those DI that have some insertion into the lateral band. Eyler and Markee,[62] in their classic account of their dissections of 30 hands, noted that the DI were inserted into the bone and joint capsule in 100% of the first DI, in 60% of the second DI, in 6% of the third DI, and in 40% of the fourth DI. Some of the parts of DI that insert into the lateral band are thought to be—morphologically—PI; a full discussion of this subject is beyond the scope of this book.

Since it is not practical in any individual clinical case to try to separate the flexion moments resulting from the DI and PI and from the lumbricals and especially not the various elements of one interosseus muscle, it is perhaps better to try to obtain an overall estimate of the contributions of the intrinsic muscles as a group to the strength of grasp, in comparison, say, to the long flexors. This has been attempted by Ketchum et al.,[64] at Carville and later in Kansas, using selective blocking of nerve-muscle groups in volunteer subjects.

The moment arms of the intrinsic muscle groups at the PIP and DIP joints should be easier to measure than those at the MCP joint, since it is the lateral band that transmits the force and not a variety of insertions. However, even here there are significant variables. The lateral band at the PIP joint moves rather freely in a normal finger, from a curve that may come close to the joint axis in full flexion to a straighter, more dorsal position across the PIP joints when they are fully extended. Thus the moment arm is maximal in extension and minimal in flexion. However, the lateral band is relatively unloaded and slack when the joint is flexed, while the central slip becomes tight. In extension of the joint, in contrast, it is the lateral bands that maintain the tension and the central slip which may be slack.

In the case of paralysis of the intrinsics, if the fingers have been allowed to remain in the clawed position, there is often a gradual stretching of the dorsal expansion over the IP joint most marked in patients with insensitivity, who frequently injure their knuckles. This results both in lengthening of the hood and in a broadening of the transverse sheet of fibers that holds the lateral bands in their dorsal position. The result is a slippage of the lateral bands palmarward until they may overlie the axis of flexion-extension of the joint or even come to lie on the palmar side of it. Any tendon transfer to the lateral band in such a case would have no effect on PIP extension and might even serve as a flexor of that joint. This gradual stretching of the extensor hood, with slippage palmarward of the lateral band, is like a boutonnière deformity, but it has also been termed *hooding* from the appearance of the flexed joint.

Actions of the Interossei

Abduction and Adduction at the Metacarpophalangeal Joints.—The intrinsic muscles are the only ones to give a controlled spread and a controlled bringing together of the fingers, from side to side. When any interosseus muscle acts alone, the abduction or adduction of the finger is its most obvious and definite action, and the one for which it is best suited by its moment arm. This is an important and useful action, especially for

those who play musical instruments or who operate keyboards. However, it requires the independent action of eight muscles for four fingers and thus it is impractical to consider replacement of all of them in the case of paralysis. Bunnell[60] described a way to split four sublimis tendons to provide eight intrinsic muscle replacements, with each pair going to the same side of two adjacent fingers. There must be few surgeons who would feel justified in sacrificing the primary function of four such important muscles to serve this purpose.

First Dorsal Interosseus.—The abduction of the index finger, because of its significance in pinch, has a unique importance. The first DI is stronger than either the profundus or superficialis to the index finger. If it is to be replaced it needs a strong muscle transfer. Its attachment is entirely to the bone of the first phalanx and to the capsule of the MCP joint, and its effect is oblique between abduction and flexion, with a larger moment for abduction (Fig 12–92). To un-

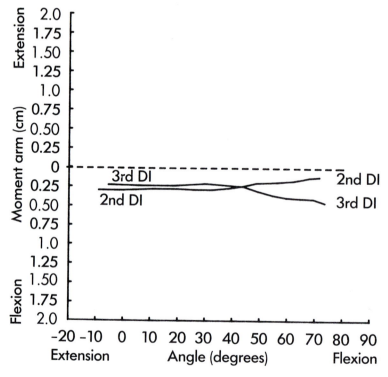

FIG 12–92.
Moment arms for flexion at the MCP joint of the middle finger of the second and third dorsal interosseus (with IP joints in extension).

derstand the first DI it is best to think of it as a pinch muscle. Therefore it must be visualized in the pinch position, i.e., with the index MCP joint flexed 45 degrees and each IP joint likewise flexed 45 degrees. In this position the first DI is an MCP flexor with a moment arm of about 0.5 cm. It is an MCP abductor with a moment arm of about 0.75 cm, and its most interesting action is that it is a rotator of the whole "bow" of the index finger around an axis that runs from the MCP joint to the point of contact between the thumb pulp and the pulp of the index fingers.

P.W.B. had noticed that in the pinch position, with the fingertip anchored by contact with the tip of the thumb, the bowed index finger could be made to rock into abduction and adduction around this axis, but he had thought that it happened because the tip of the thumb held the tip of the index finger steady. It took Agee and colleagues to demonstrate that the rocking movement is a basic function of the index MCP joint and that the axis for it is the same if the tip of the thumb is moved away (see Chapter 4).[57] The tip of the index finger cannot be actively moved mediad and laterad while the finger is curved like a bow in the position of pinch. P.W.B. has come to recognize that this oblique axis is the *real* axis of rotation and abduction-adduction of the index MCP joint. If the MCP joint remains 45 degrees flexed while the IP joints are extended, then the tip of the finger swings actively into abduction and adduction as an extension of the proximal phalanx, but the movement is *an arc* around a cone based on the same oblique axis. If the finger is then fully extended at all joints, and then abducted and adducted, the tip will still describe an arc convex to the dorsum as well as convex distally. This axis is similar in all fingers and is linked to the basic position in which a hand holds a ball with all fingers arcing into flexion and forming a composite arc from side to side while every pulp faces inward. If the hand is positioned as if holding a ball but without the ball being present, any attempt to actively abduct and adduct any finger will demonstrate that the movement involves a similar arc of motion in which the

PIP knuckles move from side to side around an arc and the fingertips move less (or, if the ball is small, the fingertips remain stationary and only rock).

The first DI stabilizes the index finger against the thumb, and it is important only in some types of pinch. For key-pinch the first dorsal interosseus is of great importance. In key-pinch the thumb pulp opposes the side of the index fingers at about the apex of the "bow" of the curved index fingers, which is the area of maximum movement around the oblique axis. Thus the first DI muscle is uniquely able to power this pinch. Without it key-pinch is possible only if the index finger is backed up by other fingers side by side.

For lateral pulp pinch, the first DI is of some value for stability. For true pulp-to-pulp pinch (or square pinch) the first DI is not really needed at all, since both index finger and thumb are stabilized by their long and short flexors, and any contraction of the first DI pulls the index finger into enough abduction-rotation to make a square pinch impossible. The important intrinsic muscle of the index finger in square pinch is the first PI.

The muscle fibers in the first DI that arise from the thumb metacarpal have the same action on the index finger as those that arise from the second metacarpal, but they also help to stabilize the thumb.

Abductor Digiti Quinti.—The ADQ, which gives bulk to the hypothenar eminence, is a very strong abductor of the little finger and is often aided in this by the fact that the EDQ tends to slip to the ulnar side of the MCP joint and serve as an extensor-abductor while the ADQ serves as a flexor-abductor. In ulnar palsy the little finger sometimes tends to rest in the abducted posture. In rheumatoid arthritis the ulnar deviation of the little finger is encouraged by the slippage of the EDQ to the ulnar side of the joint.

Since the abductions, both of the index and of the little fingers, are much appreciated actions, it would seem reasonable to restore each of them after paralysis, even if there are not enough muscles for all the fingers. However, it is no use putting in abductors if there are no adductors to oppose them. If patients have to choose between constant abduction and constant adduction, they will prefer the latter, because a hand with all fingers side-by-side and supported by one another is more useful than a spread-out hand with fingers that are always bumping into things and catching on the edges of pockets. Thus, for the abduction-adduction function of an intrinsic minus hand we recommend that the tendon transfer used for intrinsic replacement should be primarily for restoration of flexion-extension balance and should be attached to the radial side of each finger (preferably to the lateral band) except for the index finger, where it is attached to the ulnar side. This gives a parallel-fingered hand. In addition to this a separate abductor may be attached to the insertion of the first DI for the action of key-pinch if this is needed.

Stabilization and Flexion of the Metacarpophalangeal Joint.—Although abduction and adduction are the actions with the largest moment arms, most of the time the interossei contract simultaneously with one another and the adductor and abductor moments cancel one another out. In such cases the only remaining value of the interossei at the MCP joint is for flexing the joint and for impaction and stabilization of the joint itself. We are not able to give any firm figure for the moment arm for MCP flexion because of the number of different ways in which the interossei are inserted and the rather wide individual variations from person to person and from finger to finger. However, some important generalizations about the group may be made.

1. Considered as MCP flexors the muscles are powerful but with small moment arms. The tendons from some interosseus muscle fibers cross the MCP joint so far dorsally that they are not MCP flexors at all. An average moment arm for flexion of 3 mm when the joint is extended and of 6 mm in flexion may be a fair average.
2. Considered as MCP joint stabilizers,

however, the interossei need short moment arms. The shorter the better.

It is a general rule throughout the body that musculoskeletal stability is maintained by muscles and bones; ligaments are used mainly to guard against accidental or unplanned stress. This is not an absolute rule, but it is true that in the absence of external stress most joints can be controlled through their whole range of motion by muscular activity alone, even after their ligaments have been excised. In shallow-cup joints like the MCP joints this is possible only if the concave cup is kept in firm contact with the convex head all the time, especially in flexion when the subluxating effect of the long flexors is maximal. In this position the joint is held stable by the collateral ligaments and the sagittal bands of the "flexible torque tube" that surrounds the joint. However, when the ligaments are weakened, as in rheumatoid arthritis, an interosseus muscle with no moment arm for flexion or extension would still be of value as a joint stabilizer.

Intrinsic Plus.—Intrinsic-plus position of the fingers may be caused either by actual overaction or spasticity of the intrinsics or by contracture as in localized Volkmann's contracture. More commonly it occurs in rheumatoid arthritis in association with commencing loss of axis stability and early palmar subluxation of the MCP joint. This so increases the moment arm of the intrinsics for MCP flexion that it uses up the short available excursion of the intrinsics and makes it impossible for the MCP joint to extend without hyperextension of the IP joint (swan-neck deformity).

Weakness of the Intrinsic-Paralyzed Hand.—All surgeons who have wide experience in the rehabilitation of the intrinsic-paralyzed hand know that patients complain of *weakness* of grasp. In the case of high ulnar palsy there may be real weakness from the loss of a mass of profundus muscle and FCU. However, even low ulnar palsy patients say their hand is weak. We suggest that the following factors may be responsible.

1. There is an actual loss of grip strength from paralysis of the intrinsics. These muscles are more powerful than most surgeons realize and contribute significantly to grasp and very significantly to MCP flexion when the IP joints are straight (see Chapter 14).
2. The stabilizing action of interossei of the MCP joints is lost, leading to the following:
   a. The need to use the long extensor muscles as stabilizers, thus weakening the action of flexors, by having to oppose them with simultaneous contraction of long extensors.
   b. The *sense* of instability of the MCP joints in flexion *inhibits* the strong contraction of the long flexor muscles. It also inhibits the strong use of the MCP joints in full flexion. Thus "weakness" may be more pronounced in grip positions that demand full flexion (unsupported) of the MCP joint and be less pronounced in the hook position or wide grasp where the MCP joints are flexed less than 30 degrees and have more basic stability.

See also Figure 12–93.

HYPOTHENAR MUSCLES

Abductor Digiti Quinti, Flexor Digiti Quinti, and Opponens Digiti Quinti

Anatomy and Mechanics.—The ADQ, flexor digiti quinti (FDQ), and opponens digiti quinti (ODQ) form one main mass

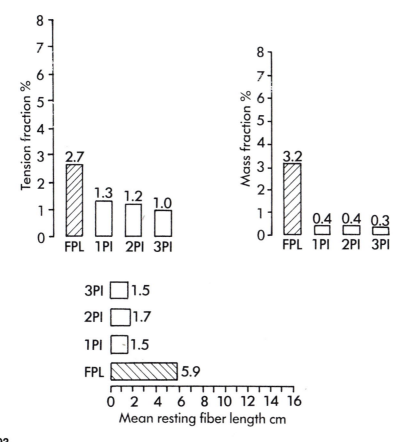

FIG 12–93.
Bar graphs for the first, second, and third PI, showing tension fraction, mass fraction, and mean fiber length, with the FPL for comparison.

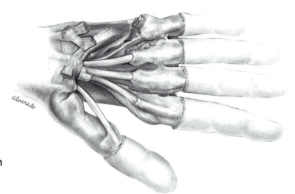

FIG 12–94.
ADQ and FDQ removed, showing stumps of origin, pisiform, and hamulus of hamate, with pisohamate ligament and intact fibers of the ODQ.

with a good deal of variation in the way the individual parts originate and are inserted. The details of the anatomy are well described by Landsmeer[73] (Figs 12–94 and 12–95).

Abductor Digiti Quinti and Flexor Digiti Quinti.—For the purpose of this mechanical review we note that most of the abductor

arises from the pisiform while most of the flexor arises from the region of the hamulus of the hamate. More important than the difference of direction of pull is the fact that the pisiform has a mobile origin and its stability depends on the FCU. This is one reason why the muscle fibers that originate on the pisiform are longer than those from the hamulus. They have to accommodate to the

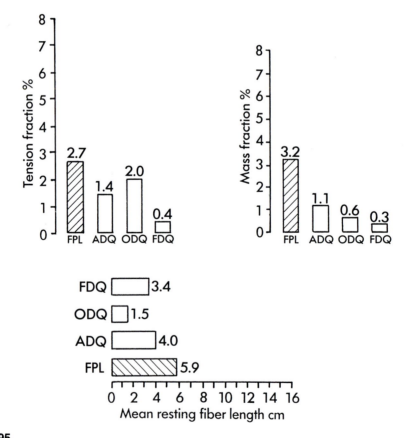

FIG 12–95.
Bar graphs of tension fraction, mass fraction, and mean fiber length of the ADQ, ODQ, and FDQ, with the FPL for comparison.

variable position of the pisiform. This bone has about a 1.4-cm range of motion in a fresh cadaver, but in life its position probably varies 0.5 cm or less. At the level of the muscle bellies the superficial or palmar portion comes mainly from the pisiform and is inserted mostly into the wing tendon. The deeper part is mainly from the hamulus and is inserted by multiple tendons into the base of the phalanx and the capsule fibers. Both parts flex the MCP joint and both parts abduct, but the named abductor has a greater moment arm for abduction and less for flexion of the MCP joint; it also contains the fibers that insert into the wing tendon and assist in extension of the IP joint. Both muscles also flex the fifth metacarpal and therefore assist in the cupping of the palm and maintenance of the metacarpal arch. In a normal hand the cupping of the palm is accompanied by a contraction of the FCU. This shows that the pisiform bone needs to be stabilized, and that the pisiform-originating muscles contribute strongly to the cupping.

Opponens Digiti Quinti.—This small muscle lies deep to the other two muscles of the hypothenar eminence and is quite distinct in that it is inserted into the shaft of the fifth metacarpal (Fig 12–96). The action is to flex and oppose the fifth metacarpal and thus to cup the palm. The fact that all of its muscle fibers are short means that though its total bulk is small, it can still exert considerable tension, many times more than

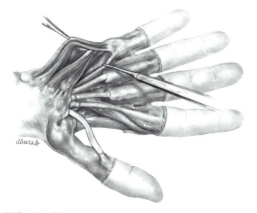

FIG 12–96.
Fibers of the ADQ and the flexor tendon pulled aside to expose the FDQ.

any lumbrical, for example. The functional significance of the hypothenar muscles must be considerable since their combined tension capability is more than that of the long flexor of the thumb. The collective action related to the cupping of the palm is probably as important as their individual actions in relation to the little finger. The hypothenar muscles are among the few muscles in the body that are actively used to absorb forces *across* their fibers. The actions of pounding on the desk or table, hitting a stapler, and the karate chop all are done on the bulk of contracted hypothenar muscles.

Surgical Significance

The hypothenar muscles are of much more significance than most surgeons accord to them. We may consider their importance under four heads. (1) MCP flexion and IP extension, (2) abduction or spreading of the hand, (3) opposition or cupping, and (4) soft tissue mass.

Metacarpophalangeal Flexion and Interphalangeal Extension.—Most surgeons recognize the loss of what is often called the intrinsic action of the small muscles to the little finger as part of a clawhand. Any operation for correction of low ulnar palsy will include correction of the clawing of the little finger. This is often accomplished by attachment of a tendon slip or tendon graft to the radial lateral band, thus ignoring the action of most of the hypothenar muscles.

In a description of one such operation Edgerton and Brand[71] commented that in the case of shared motors, the tendon to the little finger should be sutured last, and at 0.5 cm more tension than the other fingers, to compensate for metacarpal mobility. Even so, the little finger is the one most likely to lag behind the others and be incompletely corrected.

Abduction of the Hand.—Unless spreading of the fingers is important to a patient (e.g., pianists), it is probably unnecessary to have an abductor for the little finger. Given a choice for the resting position of that finger, most patients prefer it beside the others rather than spread apart. In the case of rheumatoid arthritis with ulnar drift, the abductor group of muscles becomes a deforming force. Any operation to correct ulnar drift must eliminate that ulnar pull.

Opposition or Cupping.—The ability to hold water in the hand is a basic need for any human without utensils. Millions of Asians use their hands and fingers for eating rice and fluid curries. Even though the hand may rarely be used as a cup in Western society, the hand looks and feels strange if the palm is quite flat, and it looks and behaves more strangely if the cup or palmar arch is reversed.

In low ulnar palsy cupping is lost, and in some operations to correct clawing the palmar arch becomes even further reversed. Any procedure that brings a tendon from the back of the hand, through the fourth interosseous space, and in front of the transverse intermetacarpal ligament to the little finger may correct clawing, but it thrusts the MCP joint backward, reverses the arch, and looks ugly. Tendons or grafts that approach from the palm are less deforming, and so are tendons that approach through the third interosseous space and cross to the little finger in front of the fourth metacarpal. However, almost any operation to correct clawing of the little finger has a tendency to reverse the metacarpal arch.

Various operations have been suggested to restore cupping of the palm: the tendon T and tendon Y of Bunnell,[70] and the pulley correction of Beine,[69] and attempts to combine opponens of the thumb with opponens of the little finger metacarpal. We doubt that any are really practical, the ranges of motion of the first and fifth metacarpal being so different that any common tendon action would need very careful matching of moment arms and excursions. For example, a point near the base of the first metacarpal will move as far as a point at the end of the fifth when both move into the cupped position. It may be justifiable to transfer a muscle just to control the fifth metacarpal, or perhaps the metacarpal should be stabilized at its base in a neutral position.

Soft Tissue Mass.—The mass of muscle tissue at the hypothenar eminence is a useful cushion for the hand or fist. It serves to spread pressure and diminish shear stress in grasping and protects the bones while pounding a table with the fist or in the activities of karate. Rather than trying to restore this soft contour, it is probably better to warn patients to be careful of their ulnar borders, especially as there may be sensory deficit in ulnar palsy.

Hypothenar muscles have been used for transfer for thumb opposition in median palsy. This is an excellent operation.[72, 73] The original description needs to be followed carefully to avoid a proximal bulge at the heel of the palm or loss of the motor nerve (see Figs 12–94 and 12–95).

FOREARM-MOVING MUSCLES

BRACHIORADIALIS

Anatomy and Mechanics

The BR is an elbow flexor, a weak pronator of the fully supinated forearm, and a supinator of a fully pronated forearm (its old name was supinator longus). It is of interest to the hand surgeon because it can be used as a motor for the wrist or hand in cases when most forearm muscles may be paralyzed. Its innervation is proximal to other forearm muscles supplied by the radial nerve, and its nerve root supply is higher in the cervical plexus than most muscles below the elbow, so it may survive and be useful in cases of high tetraplegia.[75]

The muscle is composed of longitudinally oriented parallel fibers that cross the elbow joint. Therefore the fibers closest to the axis of the elbow are much shorter than those more superficial. When the elbow is held at about 60 degrees short of full extension, the longest fibers may be about 20 cm in length and the shortest about 10 cm. The origin of the muscle is along the lateral supracondylar ridge of the humerus and intermuscular septum from a point 5 cm to one 10 cm proximal to the lateral epicondyle. This linear lateral origin continues distally as the origin of the ECRL (Fig 12–97).

The fibers insert on a tendon that must be fully understood if it is to be used for transfer. The apparent tendon is flat and lies

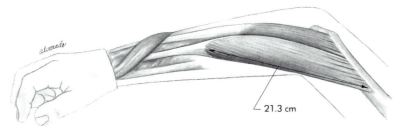

FIG 12–97.
The BR, showing origin proximal to the origin of the ECRL and showing long muscle fibers from the proximal origin and shorter fibers from the distal origin.

deep to the muscle fibers. The muscle fibers from the lowest part of the origin insert at the proximal deep end of the tendon. The highest fibers on the humerus (which are the longest) insert distally on the surface of the tendon (Fig 12–98). Having received its muscle fibers proximally, the tendon becomes narrower and thicker and oval in cross section as it goes distally until it inserts into the surface of the radius at the styloid process. There is a second tendon system for the BR that is a thin aponeurotic fascia that blends with the antebrachial fascia (Fig 12–99). This fascia serves (1) as an insertion of the muscle and (2) to bind the tendon down and prevent the muscle-tendon unit from bowstringing, which would otherwise create a webbed forearm at the elbow.

Surgical Implications

If a surgeon should think that the definitive oval-section tendon in the distal forearm were the only tendon and were to transfer it as such, the operation would fail because of

the remaining aponeurotic insertion more proximal. If the BR is to be transferred, it must be exposed in the distal three fifths of the forearm and the whole aponeurosis divided. We suggest that the flat fascia-like layer be cut a few millimeters from the edge of the muscle and folded inward. Such a rolled tendon might present a better surface for subsequent mobility in a new situation (Figs 12–100 and 12–101).

The moment arm at the elbow, measured in a dissected cadaver, is very large when the elbow is flexed and much smaller when it is extended. However, in the intact forearm the moment arm in flexion would be reduced by the restraint of the aponeurotic insertion at midforearm and by the investing fascia of the upper forearm. Our clinical measurements in young adults suggest that 7.5 cm would be about the moment arm of the upper fibers and perhaps 5.5 cm for the muscle average in elbow flexion. This might correspond with the 13-cm excursion required for total elbow range of motion. The

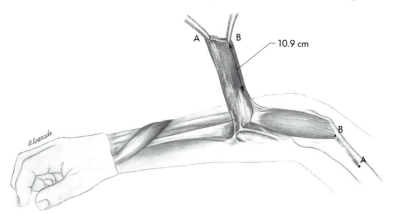

FIG 12–98.
The BR, exposing its underside. The muscle fibers from the lowest part of the linear origin *B* run on the deep aspect of the muscle to the proximal apex of the tendon. The fibers from *A,* nearly twice as long, spiral to the superficial side and insert distally on the tendon (see Fig 12–99).

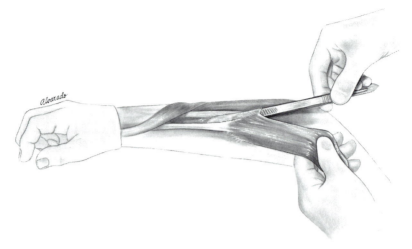

FIG 12–99.
The knife handle is under a layer of true aponeurotic tendon insertion. It must be divided if the BR is to be transferred.

muscle fibers have plenty of length for this, but if the tendon is to be transferred for wrist or finger motion, the latter would be very much weaker when the elbow was flexed. On this account we have considered moving the origin of the brachioradialis down toward the elbow when it is to be used across the wrist. This would reduce its required excursion at the elbow and therefore make more excursion available distally. However, the motor nerve and blood supply of the brachioradialis are both short at the

upper end of the muscle, and they allow very limited freedom to move the origin. This should be attempted only by experienced surgeons, but even a limited slide of the origin toward the epicondyle might give needed excursion distally (Fig 12–102).

PRONATOR TERES

Anatomy and Mechanics

The PT is a diverse and difficult muscle to analyze. Anatomically it is divided into a larger humeral and a smaller ulnar segment, between which nerves and vessels pass and may occasionally have pressure problems

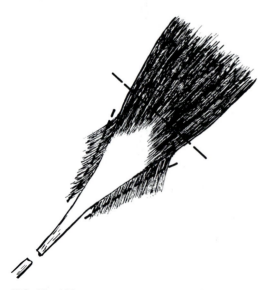

FIG 12–100.
Where the aponeurotic fascia is cut, we roll the edges in to minimize raw cut tendon edges, which encourage adhesions.

FIG 12–101.
Cross section of musculotendinous junction. **A,** cut free of fascia; **B,** edges rolled inward.

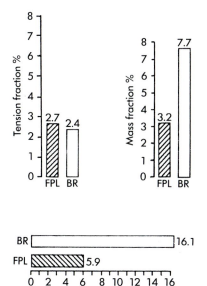

FIG 12–102.
Bar graph of BR showing tension fraction, mass fraction, and mean fiber length, with FPL for comparison.

(Fig 12–103,A). From a mechanical viewpoint we prefer to consider this muscle as three divisions that blend into one another.

Fibers From the Supracondylar Ridge With Elbow-Flexing Action.—The highest fibers of the humeral origin, arising from the supracondylar ridge, have some elbow-flexing action. These fibers are long enough to serve pronation as well as elbow flexion, but their pronation action must be somewhat weaker in full elbow flexion (Fig 12–103,B).

Muscle Mass Originating From the Epicondyle and Common Flexor.—These fibers, while not as long as those from the supracondylar ridge, are dedicated to pronation and are long enough to use a good moment arm for pronation (see Fig 12–103,B).

Deep Fibers, Including Ulnar Head and Fibers From the Intermuscular Septum.— These fibers are shorter than the superficial fibers and join the tendon of insertion from the deep side. They are short because they are close to the axis of pronation-supination and therefore do not need much excursion (see Fig 12–103,B).

We comment on these various lengths,

because if the PT is transferred for some other action, such as wrist extension, where the tendon will cross the wrist joint, then the same *additional* length of excursion will be required of all fibers, including the shortest. In fact, the PT is used regularly for wrist extension in radial palsy, and does very well. This is because the pronator winds around the radius with exactly the same obliquity after the transfer as before and therefore remains a pronator of the forearm and a flexor of the elbow even after it has become an extensor of the wrist as well. Thus all the various fiber lengths are still appropriate for their original actions. However, *every fiber* has to use a new degree of amplitude to provide wrist extension. This may be easy for the longer superficial fibers, but it must be almost impossible initially for the shortest deep fibers, which might need to double their length to allow full wrist movement. This is why in early cases of pronator transfer for radial palsy, there may be good stability but limited mobility or a link between the degree of pronation and the freedom of the wrist. This also explains why the pronator is rarely a good transfer for finger movers, because it has many short fibers.

To obtain good wrist extension, using the pronator, the tendon transfer should be attached to the ECRB at a tension a little higher than neutral with the wrist extended. As soon as the attachment has healed securely, the wrist must be progressively restored to more flexion and held at each new position by a splint at night to encourage growth in length of the short muscle fibers.

The insertion of the PT to the bone of the radius is by a flat tendon that blends with the periosteum. It may be short and difficult to handle if it is divided just where it meets the bone. A blunt elevator should be used to lift the tendon-periosteum off the bone before dividing it 1 cm or more distal to where it first meets bone. This extra length is raw and will adhere freely. It should be buried in the substance of its recipient tendon (usually the ECRB). The raw area of the radius must be covered with the belly of BR, over which the PT will move easily.

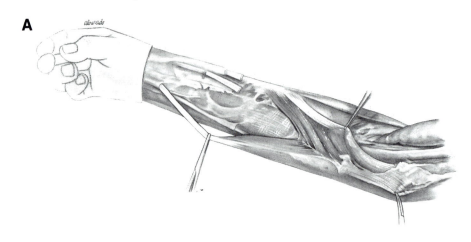

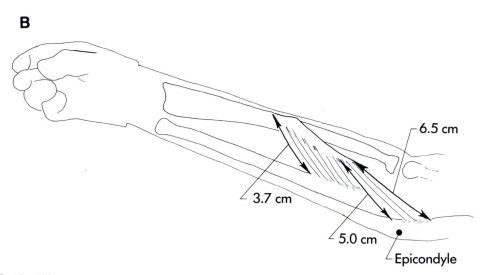

FIG 12–103.

A, PT and FCR *(below).* **B,** diagram showing differing fiber lengths of PT related to segments that are (1) elbow flexors (origin above the epicondyle, 6.5 cm), (2) superficial pronator fibers (5.0 cm), and (3) deeper fibers, closer to the axis of pronation (3.7 cm).

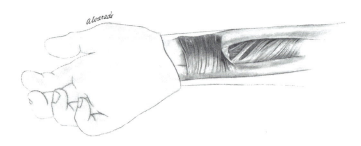

FIG 12–104.

The PQ has variable fiber length. Deep fibers close to the axis are about 1.5 cm long; the superficial fibers are about 3 cm long.

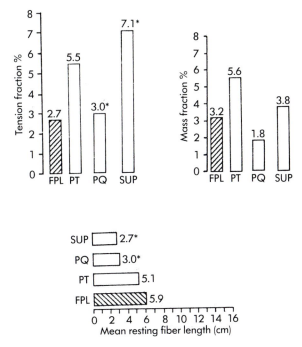

FIG 12–105.

Bar graphs of PT, PQ, and supinator showing tension fraction, mass fraction, and mean fiber length, with the FPL for comparison. (These figures are approximations.)

PRONATOR QUADRATUS

The pronator quadratus (PQ) is a difficult muscle to analyze, because every muscle fiber crosses the axis of pronation-supination at a different distance, so each muscle fiber has its own need for excursion and its own length (Fig 12–104). Thus it is difficult to calculate an average fiber length or the relative cross section. Since the PQ is never used for transfer and has no action other than pronation, we have not spent time in making calculations about its capabilities. The figures we show are approximations based on visual evaluations of numbers of dissections (Fig 12–105).

SUPINATOR

The supinator is another muscle that we have not studied in detail because of the extreme difficulty of obtaining an average fiber length when the fibers are at such varied distances from the axis and therefore have a wide variety of lengths (Fig 12–106). Also, since the supinator can never be used as a transfer, there is little need to know its exact capabilities. In our dissections we have weighed the whole muscle and have made a visual approximation of what seems to be the mean fiber length (see Fig 12–105). From these we obtain a surprisingly large figure—7.1 cm—for cross-sectional area and therefore of tension capability. We

FIG 12–106.

Supinator. Also seen is the interosseous membrane.

record this figure with little confidence that it is accurate, only because it tells us at least that the muscle is capable of very high tension, perhaps the highest of any muscle below the elbow.

REFERENCES

Static Slings

1. Brand, P.W., Beach, R.B., and Thompson, D.E.: Relative tension and potential excursion of muscles in the forearm and hand, J. Hand Surg., 6:209–219, 1981.
2. Zancolli, E.: Surgery for the quadriplegic hand with active, strong wrist extension preserved: a study of 97 cases, Clin. Orthop., 112:101, 1975.

Extensor Carpi Radialis Longus

3. Brand, P.W.: Biomechanics of tendon transfer, Orthop. Clin. North Am., 5:205–230, 1974.
4. Milford, L.: The hand, ed. 2, St. Louis, 1982, Mosby–Year Book, Inc.
5. Zancolli, E.: Structural and dynamic bases of hand surgery, ed. 2, Philadelphia, 1979, J.B. Lippincott Co., pp. 247–251.

Extensor Carpi Radialis Brevis

6. Brand, P.W.: Tendon grafting, J. Bone Joint Surg., 43B:444–453, 1961.
7. Smith, R.J.: Extensor carpi radialis brevis tendon transfer for thumb adduction: a study of power pinch, J. Hand Surg., 8:4–15, 1983.

Palmaris Longus

8. Antia, N.H.: The palmaris longus motor for lumbrical replacement, Hand, 1:139–145, 1969.
9. Brand, P.W.: Tendon grafting, J. Bone Joint Surg. 43B 3, 1961.
10. Brand, P.W.: Paralysis of the intrinsic muscles of the hand. In Rob, C., and Smith, R., editors: Operative surgery, ed. 3, Woburn, Mass., 1977, Butterworth Publishers.
11. Camitz, H.: Über die Behandlung der Oppositionslähmung, Acta Chir. Scandv., 65:7741, 1921.
12. Fritschi, E.P.: Reconstructive surgery in leprosy, Bristol, 1971, John Wright & Sons, Ltd.
13. Fritschi, E.P.: Surgical reconstruction and rehabilitation in leprosy, New Delhi, 1984, The Leprosy Mission.

14. Harvey, F.J., Chu, G., and Harvey, P.M.: Surgical availability of the plantaris tendon, J. Hand Surg., 8:243–247, 1983.
15. Littler, J.W.: Tendon transfers and arthrodeses in combined median and ulnar nerve paralysis, J. Bone Joint Surg., 31A:225–234, 1949.
16. White, W.L.: The unique, accessible and useful plantaris tendon, Plast. Reconstr. Surg., 25:133, 1960.

Flexor Carpi Ulnaris

17. Riordan, D.: Personal communication, June 1978.
18. Cranor, K.: Personal communication, June 1978.

Intrinsic Muscles of the Thumb

19. Brand, P.W.: Tendon transfers of median and ulnar nerve paralysis, Orthop. Clin. North Am., 1:447–454, 1970.
20. Bunnell, S.: Opposition of the thumb, J. Bone Joint Surg., 30 269, 1938.
21. De Vecchi, J.: Oposicion del pulgar-fisiopatologia: una nueva operacion: transplante del adductor, Bol. Soc. Cir. Uruguay 32 423–436, 1961.
22. Snow, J.W., and Fink, G.H.: Use of a transverse carpal ligament window for the pulley in tendon transfers for median nerve palsy, Plast. Reconstr. Surg., 48:238–240, 1971.

Carpometacarpal Joint

23. Agee, J.: Personal communication.
23a. Buford, W.L., Myers, L.M., and Hollister, A.M.: A modeling and simulation system for the human hand, J. Clin. Engineering, 15:445–451, 1991.
24. Kapandji, I.A.: Biomechanics of the thumb, Hand, vol., 1, Philadelphia, 1981, W.B. Saunders Co.
24a. Hollister, A., Giurintano, D.J., Buford, W.L., et al.: The axes of rotation of the thumb carpometacarpal joint, J. Orthop. Res., 10:454–560, 1992.
25. Kucyznski, K.: Carpometacarpal joint of the human thumb, J. Anat., 118:119–126, 1974.
25a. Hollister, A., Giurintano, D.J., Buford, W.L., et al.: The axes of rotation of the metacarpophalangeal and the interphalangeal joints of the thumb, J. Orthop. Res., in press.
26. Ou, C.A.: The biomechanics of the carpometacarpal joint of the thumb. Dissertation, Louisiana State University, Depart-

ment of Mechanical Engineering, December 1979.

Abductor Pollicis Brevis

27. Agee, J.: Personal communication, August 1978.

Opponens Pollicis

28. Thompson, T.C.: A modified operation for opponens paralysis, J. Bone Joint Surg., 24:632–640, 1942.

Abductor Pollicis

29. Boyes, J.H.: Intrinsic muscles of the hand. In Bunnell's surgery of the hand, ed 5. Philadelphia, 1970, J.B. Lippincott Co., pp. 457–501.

30. Smith, R.J.: Extensor carpi radialis brevis tendon transfer for thumb adduction: a study of power pinch, J. Hand Surg., 8:4–15, 1983.

Abductor Pollicis Longus

31. Edgerton, M.T., and Brand, P.W.: Restoration of abduction and adduction to the unstable thumb in median and ulnar paralysis, Plast. Reconst. Surg., 36:150, 1965.

32. Neviaser, R.J., Wilson, J.N., and Gardner, M.M.: Abductor pollicis longus transfer for replacement of first dorsal interosseous, J. Hand Surg., 5:53, 1980.

Extensor Pollicis Longus

33. Boyes, J.H.: Tendon transfers for radial palsy, Bull. Hosp. J. Dis., 21:97, 1960.

Flexor Pollicis Longus

34. Moberg, E.: Reconstructive hand surgery in tetraplegia, stroke, and cerebral palsy: some basic concepts in physiology and neurology, J. Hand Surg., 129, 1976.

Finger-Moving Muscles

35. Agee, J. and Guidera, M.: The functional significance of the juncturae tendinae in dynamic stabilization of the metacarpophalangeal joints of the fingers, Atlanta, 1980, American Society for Surgery of the Hand.

Flexor Digitorum Profundus

36. Brand, P.W., Cranor, K.C., and Ellis, J.C.: Tendon and pulleys at the metacarpophalangeal joint of a finger, J. Bone Joint Surg., 57A:779–784, 1975.

37. Long, C., Conrad, P.W., Hall, E.A., and Furler, S.L.: Intrinsic-extrinsic muscle control of the hand in power grip and precision handling: an electromyographic study, J. Bone Joint Surg., 52A:853–867, 1970.

Flexor Digitorum Superficialis

38. Biesalski, K., and Mayer, L.: Die physiologische Sehnenverpflanzung, Berlin, 1916, Julius Springer.

39. Boyes, J.H.: Tendon transfers for radial palsy, Bull. Hosp. J. Dis., 21:97, 1960.

40. Boyes, J.H. editor: Bunnell's surgery of the hand, ed 5, Philadelphia, 1970, J.B. Lippincott Co.

41. Bunnell, S.: Surgery of the intrinsic muscles of the hand other than those producing opposition of the thumb, J. Bone Joint Surg., 24:1–31, 1942.

42. Chase, R.A.: Atlas of hand surgery, Philadelphia, 1973, W.B. Saunders Co.

43. Cunningham, D.J.: Cunningham's textbook of anatomy, ed. 12, 1981, London, Oxford University Press.

44. Edgerton, M.T., and Brand, P.W.: Restoration of abduction and adduction to the unstable thumb in median and ulnar paralysis, Plast. Reconstr. Surg., 36:1950, 1965.

45. Gray, D.J., Gardner, E., and O'Rahilly, R.: The prenatal development of the skeleton and joints of the human hand, Am. J. Anat., 101:169–223, 1957.

46. Gray, H.: Gray's anatomy of the human body, ed. 29, Philadelphia 1973, Lea & Febiger.

47. Henry, A.K.: Extensile exposure, ed. 2, Edinburgh, 1970, Churchill Livingstone, p. 97.

48. Snow, J.W., and Fink, G.H.: Use of a transverse carpal ligament window for the pulley in tendon transfers for median nerve palsy, Plast. Reconstr. Surg., 48:238–240, 1971.

49. Stiles, Sir H.J., and Forrester-Brown, M.F.: Treatment of injuries of the spinal peripheral nerves, London, 1922, H. Frowde and Hodder & Stoughton, p. 166.

50. Thompson, T.C.: Modified operation for opponens paralysis, J. Bone Joint Surg., 24:632–640, 1942.

Extensor Digitorum Communis

51. Agee, J., and Guidera, M.: The functional significance of the juncturae tendinae in dynamic stabilization of the metacarpophalangeal joints of the fingers. Presented at the Annual Meeting of the American Society for Surgery of the Hand, Atlanta, 1980.

Lumbricals

52. Backhouse, K.M., and Catton, W.T.: An experimental study of the functions of the

lumbrical muscles in the human hand, J. Anat., 88:133, 1954.

53. Brooks, A.L.: A new intrinsic tendon transfer for the paralytic hand, J. Bone Joint Surg., 57A:730, 1975.

54. Bunnell, S.: Surgery of the intrinsic muscles of the hand other than those producing opposition of the thumb, J. Bone Joint Surg., 24:1–31, 1942.

55. Landsmeer, J.M.F.: The coordination of finger joint motions, J. Bone Joint Surg., 45A:1654, 1963.

56. Landsmeer, J.M.F.: Atlas of anatomy of the hand, Edinburgh, 1976, Churchill Livingstone.

Interossei

57. Agee, J., Hollister, A., and King, F.: Personal communication, 1981.

58. Andersen, J.G., and Brandsma, J.W.: Management of paralytic conditions in leprosy, All Africa Leprosy and Rehabilitation Training Centre, P.O. Box 165, Addis Ababa, Ethiopia, 1984.

59. Brand, P.W., Beach, R.B., and Thompson, D.E.: Relative tension and potential excursion of muscles in the forearm and hand, J. Hand Surg., 6:209–219, 1981.

60. Boyes, J.H., editor: Bunnell's surgery of the hand, ed. 5, Philadelphia, 1970, J.B. Lippincott Co.

61. Duchenne, G.B.: Physiology of motion, Philadelphia, 1949, J.B. Lippincott Co. (Translated by E.B. Kaplan.)

62. Eyler, C.L., and Markee, J.E.: The anatomy and function of the intrinsic musculature of the fingers, J. Bone Joint Surg., 36A:1, 1954.

63. Kaplan, E.B.: Functional and surgical anatomy of the hand, ed. 2, Philadelphia, 1966, J.B. Lippincott Co.

64. Ketchum, L.D., Brand, P.W., Thompson, D.E., and Pocock, G.S.: The determination of moments for extension of the wrist generated by muscles of the forearm, J. Hand Surg., 3 208, 1978.

65. Landsmeer, J.M.F.: Atlas of anatomy of the hand, Edinburgh, 1976, Churchill Livingstone.

66. Lieber, R.L., Babak, M.F., and Botte, M.J.: Architecture of selected wrist flexor and extensor muscles. J. Hand Surg., 15A:244–250, 1990.

67. Long, C., Conrad, P.W., Hall, E.A., and Furler, S.L.: Intrinsic-extrinsic muscle control of the hand in power grip and precision handling, J. Bone Joint Surg., 52A:853–867, 1970.

68. Salisbury, C.R.: The interosseus muscles of the hand, J. Anat., 71395, 1936.

Abductor Digiti Quinti, Flexor Digiti Quinti, and Opponens Digiti Quinti

69. Beine, A.: Combined opponens replacement of thumb and little finger: a preliminary report, Int. J. Lepr., 42:303–306, 1974.

70. Boyes, J.H., editor: Bunnell's surgery of the hand, ed. 5, Philadelphia, 1970, J.B. Lippincott Co.

71. Edgerton, M.T., and Brand, P.W.: Restoration of abduction and adduction to the unstable thumb in median and ulnar paralysis, Plast. Reconstr. Surg., 36:1950, 1965.

72. Huber, E.: Hilfsoperation bei Medianuslähmung, Dtsch. Z. Chir., 162:271–275, 1921.

73. Landsmeer, J.M.F.: Atlas of anatomy of the hand, Edinburgh, 1976, Churchill Livingstone.

74. Littler, J.W., and Cooley, S.G.E.: Opposition of the thumb and its restoration by abductor digiti quinti transfer, J. Bone Joint Surg., 45A:1389–1396, 1963.

Brachioradialis

75. Moberg, E.: Reconstructive hand surgery in tetraplegia, stroke, and cerebral palsy: some basic concepts in physiology and neurology, J. Hand Surg., 1:129, 1976.

CHAPTER 13

The Influence of Muscle Properties in Tendon Transfer

P.H. Peckham, Ph.D.

Alvin A. Freehafer, M.D.

M.W. Keith, M.D.

Many factors are important in establishing optimal results from tendon transfer surgery. The required function must be specified, the available donor muscles must be identified and evaluated, the appropriate biomechanical linkages must be chosen,[5] the transfer must be attached at the correct length, the anastomosis must be secure, and postoperative care and training must be available.[1, 15] Each of these considerations may be expected to affect the performance of the surgical procedure. This chapter focuses on one aspect of the problem encountered in performing tendon transfers, that of establishing the appropriate length during the surgical procedure. This consideration consists of determination of the available excursion (or length change) produced by the active muscle and of the relationship between muscle length and tension, or Blix curve.[6]

The length change that can be produced by a muscle is an important measure of its suitability for transfer to provide a new motor function. This measure is defined by Brand[1] as available amplitude. Bunnell,[4] Steindler,[16] and Kaplan[9] have measured the required amplitude of muscles of the upper extremity in cadavers. Required amplitude is the distance (excursion) that the tendon must move to produce a specified movement in a joint.[1] In these studies, movement of the distal portion of the cut tendon was measured as the hand was passively moved through its range of motion. This information, while extremely valuable in planning transfers, gives little information regarding the inherent capabilities of the muscle under active conditions. More recently, Brand et al.[2] have provided information regarding potential capabilities of the muscle based on anatomic features. In this chapter we present a method for establishing the available excursion for each individual muscle, providing a precise means of matching the available properties to that which is required.

The length at which the transfer is made influences the potential capabilities of the muscle. At the time of surgery, proper tension of the muscle-tendon unit is generally determined by judgment and experience. Recommendations on the tension vary from maximal tension to laxity.[10, 11] Omer and Vogel[13] used muscle stimulation to test the function of the muscle-tendon unit after reconstruction. Most surgeons probably adjust tension somewhere between, using passive range of joint motion to approximate the function of the actively contracting muscle. The relationship between length and tension, given by the Blix curve, can provide a much more precise means of choosing the length of the transfer, based on both active and passive properties. A preliminary report of our findings has been published.[8]

METHODOLOGY

The studies were performed on adult human subjects undergoing elective tendon transfer of the hand or arm. Following the administration of general anesthesia (no long-term neuromuscular blocking agents were used) and surgical preparation of the operating field, the skin was opened and the muscles to be transferred were identified. A skeletal reference point was then established by drilling a Kirschner wire into the bone near the tendon of insertion. With the arm then placed in a standardized posture of shoulder at the side, elbow at 90 degrees, and wrist neutral, a tendon marker (suture) was placed opposite to the pin and the tendon was severed distally.

Length changes were measured during passive lengthening and active contraction. For measuring length changes a calibrated scale was laid next to the tendon along the line of action of the muscle. During passive lengthening the muscle was pulled by applying force to the tendon. Frequently, the surgeon's judgment determined the maximum stretching force applied. Measurement with a calibrated force transducer showed that the retracting force was approximately 5 kg. The active unloaded shortening of the muscle was determined by delivering an electrical stimulus. An intramuscular coiled wire electrode was inserted under direct vision with a 26-gauge hypodermic needle into the muscle, and the needle was withdrawn leaving the electrode in place.[14] During insertion of the needle a low stimulus pulse rate delivered through the needle ensured that the electrode was properly positioned. The electrode was a helical coiled wire electrode wound from 45-μm type 316 stainless steel into an overall diameter of 200 μm. The electrode was insulated with polyurethane except for 10 mm at the barbed end. This electrode was utilized because its flexible nature provides for stable, reproducible contractions during movement of the muscle. The stimuli applied to the muscle were biphasic regulated stimuli with pulse amplitudes of 20 mamp and pulse frequencies of 20 Hz. The pulse width used was adjusted to activate as much of the muscle as possible with minimal or no current spread to adjacent muscles. Such contractions generally did not recruit all fibers in the muscle. In all cases the reference electrode was a 10 × 10 cm saline-saturated pad placed under the ipsilateral shoulder. Three measurements were taken for both passive lengthening and active shortening. This aspect of the study added approximately 5 minutes to the procedure for each muscle studied.

Length-tension characteristics of the muscle were then determined. The tendon suture was connected to a strain gauge, and the amount of passive and active tension generated in isometric contraction was measured at various lengths. The strain instrumentation has a sensitivity of 10 g and a natural frequency of 600 Hz. We performed all measurements to minimize experimental variability. After excursions and length-tension characteristics were determined, the tendon transfer was completed routinely. Dissection of soft tissue surrounding the muscle was performed in appropriate cases. In these instances, excursions were recorded both before and after dissection.

The stimulating electrode was also used to establish the function of the transferred muscles. Before closure of the skin, stimulation strength was increased gradually from minimum until substantial force was generated by the muscle, and the resulting movement was observed. This effectiveness of the movement was a final intraoperative assessment of transfer function.

These tests were performed on 73 muscles in 46 subjects. The vast majority of subjects (38 of 46) were spinal cord injury patients who were readmitted more than 1 year after injury for reconstructive surgery of their hand. The preserved neurologic level of these subjects was generally C6 or C7. The muscles studied were the brachioradialis, flexor carpi radialis (FCR), flexor pollicus longus, flexor digitorum superficialis, pronator teres, palmaris longus, extensor carpi ulnaris (ECU), extensor carpi radialis longus (ECRL), extensor digiti minimi (EDM), and posterior head of deltoid. These muscles represent the majority of muscles

TABLE 13–1.

Results of Tendon Excursion Measurement

Muscle	Excursion (mm)						
	Passive		Active		Total		
	Unfree	Free	Unfree	Free	Unfree	Free	
Brachioradialis	8.70	21.33	24.29	54.48	33.81	76.0	x̄
	4.91	9.09	16.49	14.77	19.51	16.84	s
	23.0	24.0	21.0	21.0	21.0	21	n
Flexor carpi radialis	13.75	25	27.43	40	40.67	65	x̄
	6.39	—	11.01	—	9.95	—	s
	8	1	7	1	6	1	n
Flexor pollicus longus	17.5	—	25.0	—	42.5	—	x̄
	10.61	—	14.14	—	24.75	—	s
	2	—	2	—	2	—	n
Flexor digitorum superficialis	15	—	27.5	—	42.5	—	x̄
	7.07	—	3.54	—	10.61	—	s
	2	—	2	—	2	—	n
Pronator teres	7.33	11.67	18.33	24.33	25.67	36.0	x̄
	4.62	5.16	2.89	2.66	4.04	5.51	s
	3	6	3	6	3	6	n
Palmaris longus	6	—	30	—	36	—	x̄
	—	—	—	—	—	—	s
	1	—	1	—	1	—	n
Extensor carpi ulnaris	10	10	25	25	35	35	x̄
	—	—	—	—	—	—	s
	1	1	1	1	1	1	n
Extensor carpi radialis longus	23.6	—	35.2	—	58.8	—	x̄
	15.08	—	15.93	—	21.32	—	s
	5	—	5	—	5	—	n
Extensor digitorum quinti	15	—	20	—	35	—	x̄
	—	—	—	—	—	—	s
	1	—	1	—	1	—	n

RESULTS

AVAILABLE EXCURSION

available for transfer in this patient population.

The results obtained for the excursion measurements are shown in Tables 13–1 and 13–2. The total excursion (sum of active plus passive) is shown in the last column, and these data are compared to the published data of the other investigators in Figures 13–1 and 13–2. The freed value represents the available excursion after perimuscular soft tissue dissection. The data show that the passive stretching accounts for only about one third of the total available ex-

cursion of the tendon, with the remaining two thirds resulting from active contraction. Soft tissue dissection[7] increased the excursion over twofold in the brachioradialis and by 50% in the FCR (one only) and pronator teres, but it had no effect in one ECU muscle.

The data obtained were not consistent from subject to subject but rather showed a

TABLE 13–2.

Tendon Excursion Measurements for Posterior Deltoid (Free)

	Passive		Active		Total	
	0	90	0	90	0	90
X̄	18.8	20.8	54.5	52.0	73.8	73.0
S	12.2	16.3	18.3	11.5	26.2	26.4

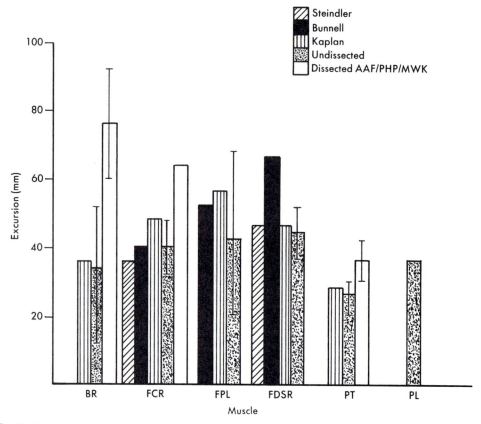

FIG 13–1.
Total available excursion of flexor muscles compared to data of Bunnell,[4] Kaplan,[9] and Steindler.[16] *BR* = brachioradialis; *FCR* = flexor carpi radialis; *FPL* = flexor pollicis longus; *FDSR* = flexor digitorum superficialis (ring); *PT* = pronator teres; *PL* = palmaris longus.

statistical variability. The data are compared to those of Bunnell,[4] Kaplan[9] and Steindler[16] in Figure 13–1 for flexor muscles and Figure 13–2 for extensor muscles. The data derived by these authors generally were determined by measurement of distal excursions in fresh or frozen material. Since each of the studies did not examine the same muscles, in this study we compare muscles that we studied with published data for these muscles. Our data compare well with those of Steindler, Bunnell, or Kaplan, whose results do not agree with one another. In each case (except the ECU for which we measured only one muscle) the values of our study are within 1 SD of other published results. Additionally, we present data for other muscles that could be used for transfers (palmaris longus, EDM, posterior head of deltoid) and are unreported to our knowledge.

LENGTH-TENSION CHARACTERISTICS

The length-tension characteristics of a muscle showed differences from subject to subject and from muscle to muscle. We present herein only those muscles studied in sufficient number to establish clear trends. The brachioradialis muscle usually was measured after soft tissue dissection. This muscle showed active properties that were generally quite flat over a 15-mm length centered approximately 5 mm longer than the reference marker (Fig 13–3). The decrease in force over this range was less than 15%. Twelve of 15 muscles followed this same trend; two muscles continually increased to peak toward the maximum length and one muscle was more peaked with less than 15% force change over a 10-mm length (−5 to +5 mm). Three muscles were measured before soft tissue dissection (Fig

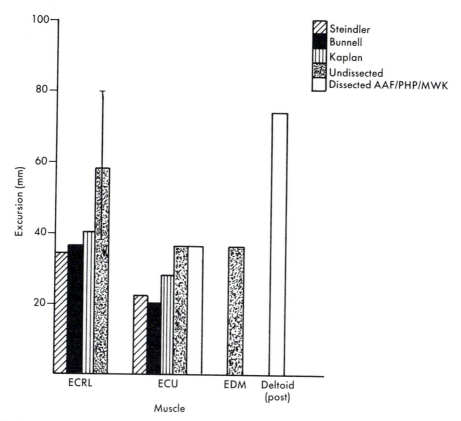

FIG 13–2.
Total available excursion of extensor muscles compared to data of Bunnell,[4] Kaplan,[9] and Steindler.[16] *ECRL* = extensor carpi radialis longus; *ECU* = extensor carpi ulnaris; *EDM* = extensor digiti minim.

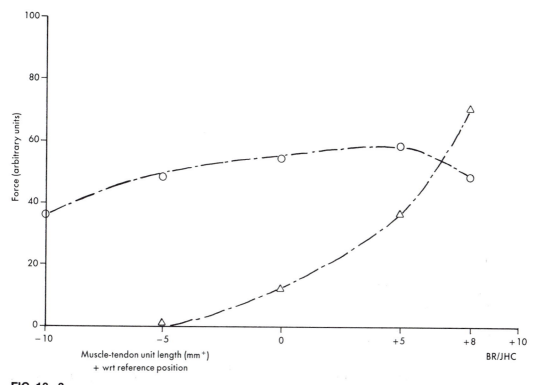

FIG 13–3.
Length-tension curve for the brachioradialis. *open triangle* = length vs. passive tension (passive lengthening); *open circle* = length vs. active tension (electrical stimulation).

13–4). In the case of undissected muscles the active force was more peaked and occurred at a shorter length than that of dissected muscles. The passive force generated by the muscle was very different from one subject to another, sometimes being quite flat, in other subjects steadily increasing with increasing length, and occasionally being quite flat, then increasing sharply. The pronator teres showed length-tension characteristics similar to those of the brachioradialis (Fig 13–5). Six of seven muscles demonstrated flat active properties over at least a 10-mm length. The midpoint of the flat portion of the active curve was more variable, by between 0 and +15 mm, with a mean of +7.1 mm and an SD of 4.7. Passive force generally increased steadily with increasing length.

The FCR showed a more peaked length-tension curve (Fig 13–6). In three of five muscles studied, there was a peak at the reference point, and one muscle peaked at 5 mm longer than the reference point. The remaining muscles showed continually in-

creasing force with increasing length. The passive force steadily increased with length in each case.

The ECRL was similar to the FCR (Fig 13–7). Each of four muscles had a peak active force curve, with three muscles peaked at 5 mm longer than the reference point and one peaked at 1 degree. The passive force steadily increased with increasing length.

The posterior head of the deltoid showed generally quite flat active force characteristics (Fig 13–8). A slight peak occurred at 15 mm beyond the reference point or longer in four of eight muscles, at 20 mm or longer in three muscles, and at 5 mm in one muscle. The passive force continually increased with increasing length.

DISCUSSION

The results of this study define the available excursion for various muscles of the upper extremity, referring to the capacity of

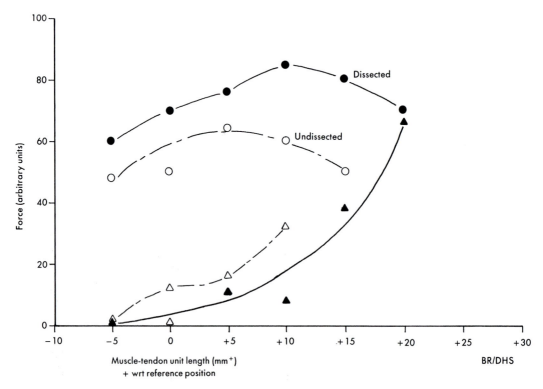

FIG 13–4.
Length-tension curve for the brachioradialis showing the effect of dissection. *open triangle* = length vs. passive tension (passive lengthening); *open circle* = length vs. active tension (electrical stimulation).

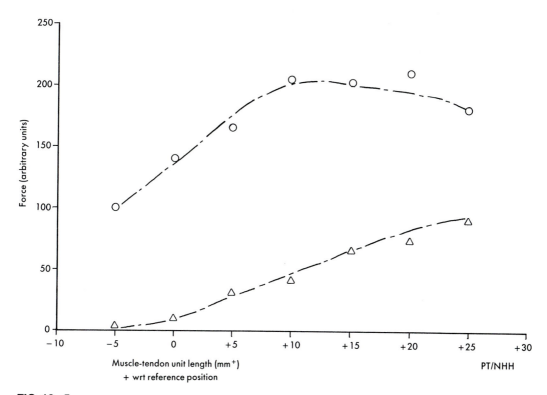

FIG 13–5.
Length-tension curve for the pronator teres. *open triangle* = length vs. passive tension (passive lengthening); *open circle* = length vs. active tension (electrical stimulation).

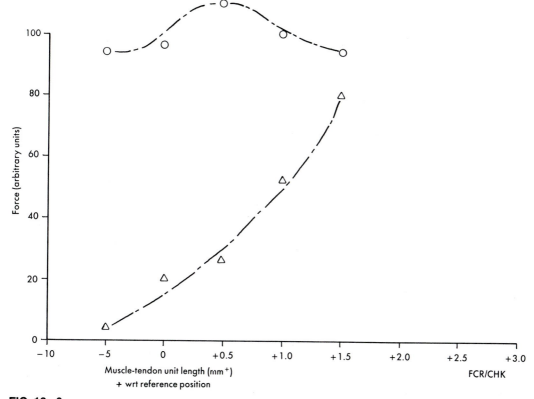

FIG 13–6.
Length-tension curve for the flexor carpi radialis. *open triangle* = length vs. passive tension (passive lengthening); *open circle* = length vs. active tension (electrical stimulation).

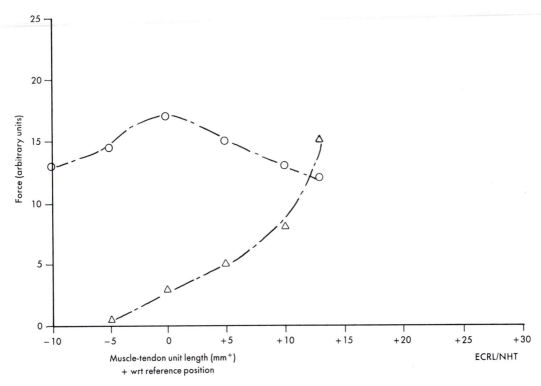

FIG 13-7.
Length-tension curve for the extensor carpi radialis longus. *open triangle* = length vs. passive tension (passive lengthening); *open circle* = length vs. active tension (electrical stimulation).

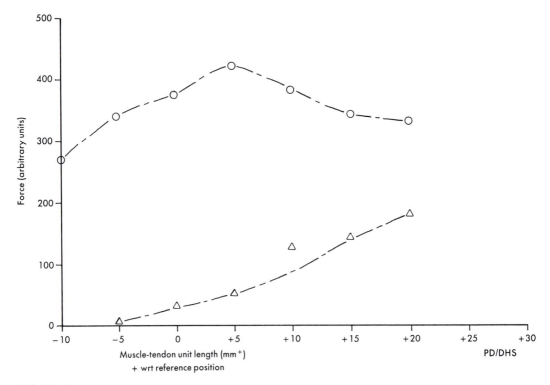

FIG 13-8.
Length-tension curve for the posterior head of the deltoid. *open triangle* length vs. passive tension (passive lengthening); *open circle* = length vs. active tension (electrical stimulation).

the muscle. The actual distance that the tendon must move to produce a certain movement is the required amplitude.[1] This latter measure is what has been reported in the literature. In performing tendon transfer, both measures must be known. The results of this study demonstrate two facts relating to performing the transfer. First, comparing our results with published data, the available excursion is generally of the same order as that required (i.e., available excursion is approximately equal to required excursion). This result suggests that muscle excursion adjusts itself to the externally imposed requirements for length change. Some evidence in the physiologic literature supports this concept, since it has been demonstrated that muscles have the capacity to adapt the number of sarcomeres with externally imposed length constraints.[17]

The available excursion measured in our studies showed considerable variability. We believe that this is an accurate reflection of the different capabilities of the muscles studied, as opposed to a systematic error. The latter possibility we believe was minimized by using a uniform retracting force for the passive lengthening and ensuring that the amount of active shortening was not a function of the level of recruitment (i.e., not a function of the number of active muscle fibers). The fact that such a variability might exist is not surprising considering the different skeletal size of the subjects and that deformities such as contractures could be present. We have not as yet tried to correct for anthropometric measurements. While the origin of this variation is not precisely identified, the important consideration is that the technique employed in these studies enables us to assess intraoperatively the available amplitude for each individual subject who is receiving the transfer. Because it has been demonstrated that this variation does exist from subject to subject, it would appear that such a measurement is important. This is especially true for tendon transfers in spinal cord injury patients, as were the majority of subjects in this study, since one generally requires that shorter-amplitude proximal muscles that have voluntary control be transferred to perform functions

in the hand which generally require longer excursions. In such a case, the full available amplitude may be used to provide the required movement of the joints.

Soft tissue dissection was performed during surgery of some muscles with the intent of increasing the available amplitude.[8] This resulted in an increase in amplitude from 0% to over 200%, depending on the muscle. In the brachioradialis, the increase in amplitude was greatest, increasing from 34 mm before to 76 mm following dissection. This dissection had little effect on the variability in the amplitude or on the percent excursion in passive lengthening and active shortening. While the persistence of this effect is not known, it appears that dissection is warranted surgically to get at least a transient increase in available excursion.

The excursion technique employed in this study also provides a means for studying muscles that otherwise may be difficult to evaluate. An example of this is the posterior head of the deltoid, which may be transferred to provide elbow extension.[12] Our study showed that this muscle had an available amplitude of 73 mm. The relationship between this amplitude and the moment arm provides the surgeon with an approximation of the anticipated range over which the elbow could be expected to perform.

The relationship between muscle force and length can be important in providing effective tendon transfers. There are two components to the total force available, the active component caused by contraction of the muscle and the passive component resulting from internal properties. The magnitude of each was described by the results of this study. The passive properties of a muscle may be important in tendon transfers for enhancing tenodesis function, but they may limit the range of motion if inserted as an antagonist to an active muscle. The magnitude of the passive force at each length determines the extent of this effect. The active properties determine the ability of the muscle to act as a force generator at each length. The active force of muscles such as the brachioradialis, pronator teres, and posterior head of the deltoid was relatively insensitive to length change over a distance of 10 mm or greater. Others, the FCR and ECRL, were more peaked in the active characteristics. To achieve maximal force after transfer, some "optimal" length of the muscle most likely should be preserved. The active force gives an approximation to the "optimal" length. For muscles with peaked characteristics, the tendon may be attached at the length at which the active force is maximal. For muscles with flatter characteristics, a wider range of lengths is possible, and the decision regarding the appropriate length may be based on other factors, such as excursion and passive properties. These factors may be particularly important in the spinal cord injury patient in whom proximal muscles, generally having limited available amplitude, are transferred distally to provide motion of the digits requiring greater amplitude.

The procedure that we use to establish the correct muscle length at surgery incorporates the results of the excursion and length-tension measurements as determined intraoperatively. Firstly, the required excursion is measured by moving all joints across which the transfer will act through the full range of motion. Secondly, the available amplitude of the motoring muscles is measured. The decision for the pairing of motor and insertion is based on providing sufficient excursion for the required function(s). Other factors (e.g., synergistic activity patterns and mechanical linkages) may be important in this decision as well. The appropriate length is determined by length-tension (Blix) characteristics. The arm and hand are placed in the posture at which maximal force is required, and the length of the motoring muscle is set at the length of peak active force. If the active characteristics are flat, approximately the midrange of the flat portion is used.

The technique of electrical stimulation of the muscle provides a measure of confidence of the function of the transfer. Following tendon transfer and suture, a low level of stimulation was delivered to the muscle to observe the movement of the joints. If appropriate movement was not obtained, then adjustments were made before final closure. This technique, combined with traditional passive ranging of the joints, provides a simple effective method for evaluating the function of the transferred unit.

The muscles evaluated in this study may not be normal. Many of the muscles receive their innervation from cord levels close to the spinal cord injury and thus may be expected to be partially denervated.[3] However, each muscle clearly was clinically graded at least "good" and had sufficient voluntary function for use in transfer. Further evidence that the muscles were not normal in function is that in some subjects, the same muscle was studied in opposite limbs with somewhat different results. This evidence, we think, demonstrates the importance of studying the muscles individually in each subject.

The effectiveness of a tendon transfer is a result of many interrelated factors, many of which may be difficult to control. Clearly, surgical judgment plays an extremely important role in integrating these factors. In this study, we have attempted to bring under more rigid control one surgical aspect that may clearly influence the result of the surgical procedure and to provide a supplementary means for checking the active function of the transfer. While this cannot and should not replace the surgeon's experience, we think that the information provided augments the information traditionally available, enabling one to perform a better surgical procedure.

ACKNOWLEDGMENTS

This study was supported in part by research grant G008005815, National Institute of Handicapped Research, Department of Education, Washington, D.C., 20202. The studies were performed at Highland View Hospital and the Veterans Administration Medical Center, Cleveland, Ohio.

REFERENCES

1. Brand, P.L.: Biomechanics of tendon transfer, Orthop. Clin. North Am. 4:205–242, 1974.
2. Brand, P.W., Beach, R.B., and Thompson,

D.E.: Relative tension and potential excursion of muscles in the forearm and hand, J. Hand Surg. 6:209–219, 1981.

3. Brandstatler, M.E., and Dinsdale, S.M.: Electrophysiological studies in the assessment of spinal cord lesions, Arch. Phys. Med. Rehabil. 57:70–74, 1976.

4. Bunnell, S.: Surgery of the hand, ed. 3, Philadelphia, 1956, J.B. Lippincott Co.

5. Cooney, W.P.: Opponensplasty: an anatomic and biomechanical study, Proceedings of the 36th annual meeting, Am. Soc. Surg. Hand., Las Vegas, 52:1981.

6. Efltman, H.: Biomechanics of muscle, J. Bone Joint Surg. 48A:363–376, 1966.

7. Freehafer, A.A., and Mast, W.A.: Transfer of the brachioradialis to improve wrist extension in high spinal cord injury, J. Bone Joint Surg. 49A:648–652, 1967.

8. Freehafer, A.A., Peckham, P.H., and Keith, M.W.: Determination of muscle-tendon unit properties during tendon transfer, J. Hand Surg. 4:331–339, 1979.

9. Kaplan, E.G.: Functional and surgical anatomy of the hand, ed. 2, Philadelphia, 1965, J.B. Lippincott Co.

10. Mayer, L.: The physiological method of tendon transplantation, Surg. Gynecol. Obstet. 22:182–297, 1916.

11. Mayer, L.: The physiological method of tendon transplantation, Surg. Gynecol. Obstet. 33:528–543, 1921.

12. Moberg, E.: Surgical treatment for absent single-hand gripped elbow extension in quadriplegia, J. Bone Joint Surg. 75A:196–206, 1975.

13. Omer, G.E., and Vogel, J.A.: Determination of physiological length of a reconstructed muscle-tendon unit through muscle stimulation, J. Bone Joint Surg. 47:304–313, 1965.

14. Peckham, P.H., Marsolais, E.G., and Mortimer J.T.: Restoration of key grip and release in the C6 tetraplegic patient through functional electrical stimulation, J. Hand Surg. 5:462–469, 1980.

15. Smith, R.J., and Hasting, H.: Principles of tendon transfer to the hand, Instr. Course Lect. 29-129–152, 1980.

16. Steindler, A.: Kinesiology of the human body, Springfield, Ill., 1955, Charles C Thomas, Publisher.

17. Tabary, J.C., et al.: Functional adaptation of sarcomere number of normal cat muscle, J. Physiol, (Paris) 72:277–291, 1976.

CHAPTER 14

An Experimental Investigation Into the Forces Internal to the Human Hand

Lynn D. Ketchum, M.D.

David E. Thompson, Ph.D.

The magnitude of forces generated by individual muscles in the human hand and forearm cannot be determined in the cadaver; nor can they be obtained in an animal whose muscles are different from the human. An in vivo method, preferably noninvasive, is the most desirable way of obtaining muscle force data.

The simplest way of determining muscle forces is to begin with a joint that is moved by just one muscle. We have used the distal interphalangeal (DIP) joint, where there is a single extensor opposed by a single flexor. The method used was as follows: all of the joints of the hand proximal to the one studied were stabilized with straps and a sandbag[4] in the manner shown in Figures 14–1 and 14–2. The subject was then asked to flex or extend the joint with a maximal effort. Flexor measurements were made with the joint in extension and extensor measurements were made with the joint in flexion. Biglund and Lippold[1] have shown that a maximal voluntary contraction transducer located at a predetermined distance from the joint axis opposes the force of the contracting muscle sufficiently to prevent motion, resulting in an isometric contraction. The force was read out directly on a strip recorder. The distance of the transducer from the joint axis multiplied by the force of the transducer is equivalent to the force of the muscle multiplied by the perpendicular distance of its tendon from the joint axis, which is the moment arm for that tendon. The moment arms of the tendons of the hand and wrist have been obtained using a direct measurement technique in six frozen cadaver specimens. These data are shown in Tables 14–1 to 14–3.

The forces of the muscle tendon unit in question may now be solved for directly, since it is the only unknown in the moment balance equation below:

$$Mf \times R = Tf \times r$$

where

MF = measured external force
from transducer
R = distance of transducer from joint axis
Tf = force of muscle-tendon unit
r = moment arm of tendon in question

Thus, at the DIP joints, the forces developed by the flexor profundi are readily extracted from measurements of external torques. At the proximal interphalangeal (PIP) joint, however, there are two flexor tendons that contribute to the total flexor moment. Having the subjects practice flexing only the PIP joint while not flexing the

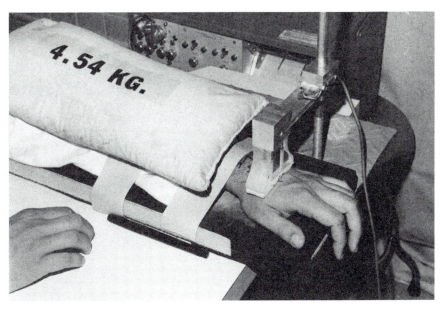

FIG 14–1.
The measurement setup for wrist extensor testing.

DIP joint will result in having only one tendon contribute to the moment at the PIP joint. The DIP joint was tested during the experiments to verify the lack of tension on the flexor side.

The isolation of forces at the metacarpophalangeal (MCP) joint was considerably more complex, since the intrinsics and extrinsics combine their flexor influences at this joint. The approach taken here was to compute the differential flexion moment as various intrinsic muscle groups were sequentially denervated using 1% lidocaine

nerve or muscle blocks. The original moment was measured with the MCP joint in flexion and the IP joints extended. It is thought that the interosseous and lumbrical muscles are contracting vigorously in this position, whereas they are relatively quiet during concurrent flexion of the MCP and IP joints.

An equation was developed for the moment balance of the MCP joint based on the assumption of equilibrium. The external moment is seen to be balanced by the moment contributions of other muscle groups.

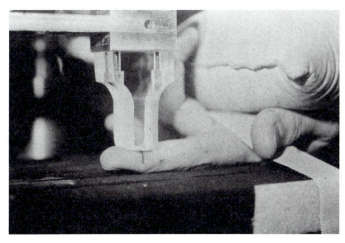

FIG 14–2.
The experimental setup for finger flexor testing at the proximal interphalangeal joint.

TABLE 14-1.

Moment Arms, Joint Diameters, and Their Ratios for the Muscles of the Fingers of the Hand, Measured in Six Fresh-Frozen Cadavers (mm)

	Index		Middle		Ring		Little	
	Mean	SD	Mean	SD	Mean	SD	Mean	SD
Profundus at DIP	6.50	0.50	7.00	0.50	6.80	1.04	6.00	0.00
Joint diameter	14.30	1.04	14.70	2.52	14.80	2.47	13.50	1.32
Ratio	0.46		0.48		0.46		0.44	
Profundus at PIP	9.80	0.45	10.70	1.20	10.40	1.14	8.50	0.61
Joint diameter	20.40	2.07	21.30	1.30	21.50	1.80	17.50	1.12
Ratio	0.48		0.50		0.48		0.49	
Sublimis	8.30	0.29	8.70	1.04	8.50	0.94	7.40	0.65
Joint diameter	20.80	2.22	21.30	1.30	21.60	1.84	17.50	1.12
Ratio	0.40		0.41		0.39		0.42	
Lumbrical at MCP	11.40	1.11	12.30	1.76	11.00	1.15	9.80	1.26
Joint diameter	28.00	2.00	32.20	2.80	28.80	3.66	25.30	2.75
Ratio	0.41		0.38		0.38		0.39	
Volar interosseus at MCP	8.50	1.32	9.00	2.29	7.80	0.50	6.80	1.04
Joint diameter	28.30	2.31	32.30	2.08	28.80	3.66	25.30	2.75
Ratio	0.30		0.28		0.27		0.27	
Extensor tendon at DIP	6.50	0.00	4.30	0.29	4.50	0.50	4.00	0.87
Joint diameter	14.30	1.04	14.70	2.52	14.80	2.47	13.50	1.32
Ratio	0.30		0.31		0.29		0.30	
Extensor tendon at PIP	5.80	0.76	6.40	0.42	6.60	0.55	5.20	0.42
Joint diameter	20.40	2.07	21.30	1.30	21.50	1.80	17.50	1.12
Ratio	0.28		0.30		0.31		0.29	
Extensor tendon at MCP	9.70	0.58	12.30	1.15	10.50	2.18	9.80	0.87
Joint diameter	28.30	2.31	32.30	2.08	27.00	1.32	25.30	2.70
Ratio	0.34		0.38		0.39		0.36	
Dorsal interosseus at MCP	4.52	0.50						
Joint diameter	25.07	1.19						
Ratio	0.18							
Lumbrical at PIP	3.58	0.77						
Joint diameter	16.23	1.92						
Ratio	0.22							

The equation at the bottom of this page was evaluated for the index, long, ring, and little fingers. By neutralizing the interossei in turn, the effects of each muscle group could be extracted. The individual moment arms for each muscle group were estimated by scaling them to the anteroposterior (AP) diameter of the MCP joint. The individual tensions in the intrinsics were obtained by dividing their moment arms into their moment contribution. The nerve and muscle blocks were performed in a series of 10 out of 40 hands tested. The first block consisted

of a 4-mL injection of 1% lidocaine into the immediate proximity of the median nerve, resulting in temporary denervation of the first and second lumbrical muscles. The moment for MCP joint flexion was taken when the subject noted hypoesthesia of the index finger. The second block was a direct block of the first volar interosseus through a 2.5-mL injection of the lidocaine into and around the first volar interosseus. A reduced moment was then measured for the MCP joint. The final block was accomplished by a 3.5-mL injection of 1% lidocaine into the

$$\text{Tf}_{1DI} \times \text{r}_{1DI} + \text{Tf}_{1VI} \times \text{r}_{1VI} + \text{T}_{1Lum} \times \text{r}_{1Lum} + \text{Tf}_{ext} \times \text{r}_{ext} = \text{MF} \times \text{R*}$$

*1DI, = first dorsal interosseous; *1Lum* = first lumbrical; *1VI* = first volar interosseous; *ext* = extrinsic flexors.

TABLE 14–2.

Moment Arms, Joint Diameters and Their Ratios for the Muscles of the Thumb, Measured in Six Fresh-Frozen Cadavers (mm)

	Thumb			Thumb	
	Mean	SD		Mean	SD
Extensor tendon at IP	6.50	0.00	Flexor pollicis longus at IP	9.80	0.76
Joint diameter	20.80	1.04	Joint diameter	20.80	1.04
Ratio	0.31		Ratio	0.47	
Extensor tendon at MP	10.70	1.15	Abductor pollicis brevis at MCP	13.20	1.04
Joint diameter	30.70	3.79	Joint diameter	30.70	3.79
Ratio	0.35		Ratio	0.43	
Extensor pollicis brevis at MCP	10.30	1.04	Adductor pollicis brevis	12.70	1.15
Joint diameter	30.70	3.79	Joint diameter	32.00	6.08
Ratio	0.34		Ratio	0.40	

belly of the first dorsal interosseus. A reduced moment was again measured.

WRIST EXTENSORS

The determination of the forces generated by the wrist movers is also difficult. Here again, we try to balance the forces internal to the hand with those external to the hand to maintain equilibrium. In Figure 14–3, the external force, *F*, is one that holds the hand down, preventing extension of the wrist.[3] The external moment is this force multiplied by the perpendicular distance, *R*, from the joint axis.

The internal forces are provided by three muscles, the wrist extensors, the extensor carpi ulnaris (ECU) and extensor carpi radialis brevis (ECRB), and longus (ECRL), each of which has a maximum force, *T*, and

an effective moment arm for extension, *r*. Each muscle contributes to the total moment that is the sum of the individual moments, *Tr*. Thus, at equilibrium, the internal and external moments are balanced:

$$FR = \underset{ECU}{Tr} + \underset{ECRB}{Tr} + \underset{ECRL}{Tr}$$

If these three extensors are paralyzed and replaced by a single tendon transfer, then the tension of the transferred tendon times the moment arm in its new position will equal the new external moment that it can support on the back of the hand.

EXTERNAL WRIST EXTENSOR MOMENT

A transducer, shown in Figure 14–1, was constructed and used to measure the torque

TABLE 14–3.

Flexion-Extension Moment Arms, Joint Diameters, and Their Ratios for the Muscles of the Wrist, Measured in Six Fresh-Frozen Cadavers (mm)

	Wrist			Wrist	
	Mean	SD		Mean	SD
Flexor carpi radialis	15.00	1.41	Extensor carpi radialis brevis	16.30	3.18
Joint diameter	41.00	1.41	Joint diameter	41.00	1.41
Ratio	0.37		Ratio	0.40	
Flexor carpi ulnaris	11.80	3.18	Extensor carpi radialis longus	12.50	2.12
Joint diameter	41.00	1.41	Joint diameter	41.00	1.41
Ratio	0.29		Ratio	0.31	
			Extensor carpi ulnaris	6.30	0.35
			Joint diameter	41.00	1.41
			Ratio	0.15	

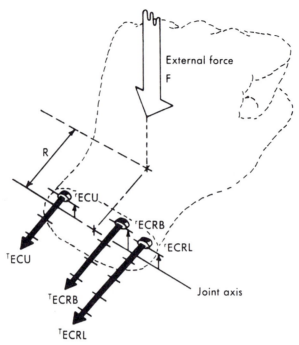

FIG 14–3.
A schematic of the moment balance between internal and external forces of the wrist during extensor testing. *R* = distance of transducer from joint axis; *ECU* = extensor carpi ulnaris; *ECRB* = extensor carpi radialis brevis; *ECRL* = extensor carpi radialis longus.

produced by the extensors and flexors of the wrist. Youm et al.[5] have shown that the flexors reciprocally relax during maximal extensor contraction. This ensures that the full power of the extensors is available for wrist extension. The transducer, when placed on the dorsum of the prone hand, will thereby indicate the maximal effort of the wrist extensors. Subjects were instructed to extend their wrist while maintaining loose, flaccid fingers. This is done to ensure that the finger extensors do not contribute to the wrist moment. The rigid transducer measures this isometric extension moment. Following a short practice session, a mean of three successive readings is recorded as "F." By measuring the forces at two different points of application, one may solve for the effective moment arm and the total moment. Using 15 subjects, selected only by their availability for testing, the mean moment produced was found to be 69.6 kg·cm.

Obviously, the preoperative extension moment of a patient with radial palsy would be zero. However, an approximation for what would be normal for that individual could be obtained through testing of the contralateral normal extremity.

INTERNAL MOMENT BALANCE

To calculate the individual contribution of each of the three wrist extensors from a total wrist extension effort, it is necessary to make three simplifying assumptions. The first is that, although the total muscle mass of any given forearm may be very different from any other, the ratio between the mass of any one muscle and another in the same arm is rather constant from arm to arm. This was verified by dissecting eight cadaver forearms and hands. By tabulating all of our measurements as ratios of muscle mass to the total muscle mass in the forearm and hand, we discovered a consistency that supports our hypothesis. The extensors ECRB and ECRL, for example, make up approximately 10% of the muscle mass below the elbow. Together they account for 76% of the wrist extensor mass. In comparison with each other, the muscle mass of the ECRB is 88% that of the ECRL.

Our second assumption is that the muscle mass is directly related to its work potential. This was stated by Elftman,[2] who also makes it clear that, since work equals force times distance, for a given mass of muscle, the tension it can generate is in inverse proportion to its maximal excursion. This assumption implies that, for muscles of similar fiber length, their maximal excursions will be roughly equivalent. This allows one to relate muscle mass to maximal tension. This is approximately true of the extensors of the wrist, and we use a common proportionality constant, K, for them all. Thus, if we use the term *mass fraction, mf*, to represent the ratio between the mass of each wrist extensor and total muscle mass, we obtain the following relationship for any given hand:

$$\frac{T}{mf} = \frac{T}{ms} = \frac{T}{mf} = K, \text{ constant}$$
$$\textbf{ECU} \quad \textbf{ECRL} \quad \textbf{ECRB}$$

A third and final assumption was invoked here: the principle of geometric similarity, which assumes that the ratio of moment arm to joint diameter is constant, even between hands. In our studies of 16 cadaver hands, it was found that, given one or two moment arms, all of the others could be deduced from a scale of proportion based on the cadaver measurements. At present we use the maximal AP diameter of the lower end of the radius near its radial border as a standard by which one may estimate all of the moment arms for any given hand. The ratios based on this diameter are as follows: ECU, 0.151; ECRB, 0.365; ECRL, 0.273. The actual measurement of the diameter is done with a skin-fold calipers firmly applied over the intact skin. A factor for the skin thickness may be deducted. Using the measurement of the AP diameter (D) of the radius, *Dap*, and referring to the ratios of the extensors shown in Table 14-2, the moment arms for wrist extensors for any subject can be determined. Combining the equations for the moment balance with the simplifying assumptions yields the following result:

$$\text{T} \quad = \text{FR}/1.146 \text{ D}$$
$$\textbf{ECU} \qquad \textbf{AP}$$
$$\text{T} \quad = 1.75 \text{ T}$$
$$\textbf{ECRL} \quad \textbf{ECU}$$
$$\text{T} \quad = 1.417 \text{ T}$$
$$\textbf{ECRB} \quad \textbf{ECU}$$

These equations allow us to make reasonable estimates for the force potential of the wrist extensor muscles. They are summarized pictorially in Figure 14-3.

DISCUSSION

The measurements of forces generated by the intrinsics of the index finger, the extrinsic flexors, and the extrinsic extensors for digits 2 through 5 are given in Tables 14-4 to 14-7. The significance of the intrinsic musculature to the overall strength of the hand is well demonstrated, particularly to the flexion of the MCP joint with extended PIP and DIP joints. For this motion, the strength of the intrinsics contributes approximately 75% to the flexion moment of the fingers.

One might question the validity of the muscle denervation technique since no electromyograms were performed. There was, however, a consistent decrease in the residual moment measured following the blocks. The forces acting to simultaneously extend the IP joint and flex the MCP joint are summarized in Table 14-4. The extrinsics were found to contribute approximately 27% to the total movement in this experiment. This attests to the force potential of the intrinsic musculature.

TABLE 14-4.

Tendon Forces Contributing to Metacarpophalangeal Joint Flexion With Simultaneous Interphalangeal Joint Extension in the Index Finger of Ten Normal Hands (kg)

	Mean	SD
First dorsal interosseous	5.44	2.52
First volar interosseous	2.77	1.17
First lumbrical	2.03	0.90
Extrinsic flexor moment	27%	5.70%

TABLE 14–5.

Muscle-Tendon Forces Generated by the Flexor Digitorum Profundus and Superficialis in 40 Normal Hands (kg)

	Mean	SD
Profundus		
Index	6.18	2.64
Middle	5.77	2.14
Ring	5.54	3.12
Little	5.27	2.61
Superficialis		
Index	6.91	2.36
Middle	7.63	2.8
Ring	6.21	2.89
Little	3.73	1.53

It should also be noted that we have used figures for the moment arm for ECU measured in the cadaver with the wrist in supination, whereas the total moment for wrist extension was measured in pronation. This

TABLE 14–6.

Ratios and Aggregate Muscle-Tendon Forces Generated by the Flexor Digitorum Profundus and Superficialis in 40 Normal Hands (kg)

	Mean	SD
Profundus:superficialis		
Index	.97	.45
Middle	.79	.32
Ring	.95	.58
Little	1.5	.73
Profundus + superficialis		
Index	13.22	3.87
Middle	13.4	4.26
Ring	11.75	5.95
Little	9.01	3.15

TABLE 14–7.

Muscle-Tendon Forces Generated by the Long Extensor Muscles in 40 Normal Hands (kg)

	Mean	SD
Index	5.98	2.86
Middle	4.47	2.01
Ring	4.38	1.66
Little	3.94	1.71

flatters the ECU, which during supination of the forearm has a much poorer moment arm for wrist extension than our figures suggest.

REFERENCES

1. Biglund, B., and Lippold, O.C.J.: Motor activity in voluntary contraction of human muscle, J. Physiol. 125:322–335, 1954.
2. Elftman, H.: Biomechanics of muscle, J. Bone Joint Surg. 48A:2, 1966.
3. Ketchum, L.D., et al.: The determination of moments for extension of the wrist generated by muscles of the forearm, J. Hand Surg. 3:205–210, 1978.
4. Ketchum, L.D., et al.: A clinical study of forces generated by the intrinsic muscles of the index finger and the extrinsic flexor and extensor muscles of the hand, J. Hand Surg. 3:571–578, 1978.
5. Youm, Y., et al.: Moment arm analysis of the prime wrist movers, presented at Fifth International Congress of Biomechanics, Jyvaskla, Finland, 1975.

CHAPTER 15

Mechanics of Tendons That Cross the Wrist

John M. Agee, M.D.

As clinical hand surgeons and therapists, we evolve our views of wrist mechanics in part from clinical and radiographic examinations of our patients. This helps mature our understanding of the kinematics of the bones and their restraining ligaments, but it is frequently inadequate to help us appreciate the tendons and their mechanics. Although we have developed intricate kinematic models of the wrist, most studies have focused on the bony anatomy, its pathology, and collapse patterns. Less well appreciated are the mechanisms by which the wrist continuously rebalances the effect of a host of muscle-tendon units. This chapter presents practical anatomic concepts that the wrist uses to balance forces by controlling the moment arms of tendons that cross its surfaces. Interestingly, the same kinematic linkages that define wrist function for load transmission simultaneously control the moment arms of the tendons.

THE WRIST BALANCES FORCES BY MOMENT ARM CONTROL

The powerful flexor carpi ulnaris (FCU) uses its patella-like pisiform and pisitriquetral joint to define its moment arm. All other wrist and digital tendons use tunnels (pulleys) in concert with the external shape that the encapsulated carpal bones present to the tendons to define their moment arms. This combination of pulleys and the changing shape that the carpus presents to its overlying tendons is crucial to understanding normal and abnormal wrist tendon mechanics.

The pulleys that control tendon moment arms are: the six extensor compartments of the wrist, the tunnel for the flexor carpi radialis (FCR) tendon, and the carpal tunnel.

CLINICAL AND ANATOMIC OBSERVATIONS: THE SIGNIFICANCE OF THE CHANGING SHAPE OF THE CARPUS

The constant sliding and gliding that occurs between the various carpal bones creates a changing three-dimensional shape that the encapsulated wrist presents to the wrist and digital tendons. One dimension, the anteroposterior (AP) diameter on its radial side, is easy to study on the normal wrist (Figs 15–1 and 15–2). Place your left thumb on the palmar side of your right wrist where the FCR tendon crosses the scaphoid and place the pulp of your left index finger at the insertion of the extensor carpi radialis brevis (ECRB) into the base of the long finger metacarpal. Note that radial deviation increases and ulnar deviation decreases the AP diameter as gauged by the distance between your thumb and finger. Although observations are most easily made with radial-ulnar deviations, this effect is maximal as observed from radial deviation in extension to ulnar deviation in flexion. This changing

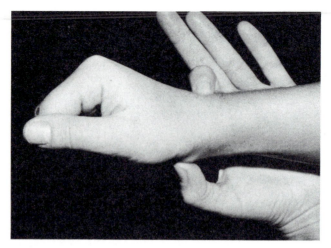

FIG 15−1.
Radial deviation increases the AP diameter of the wrist on its radial side. The thumb and index fingers mark the wide distance between the ECRB and FCR tendons that lie subcutaneous beneath their pulps. Contour lighting illustrates the palmar descent of the distal scaphoid and the FCR tendon, both of which move palmarward with combinations of radial deviation and extension.

AP diameter increases the distance between the radial wrist extensor tendons and their antagonist, the FCR. Biomechanically, this indicates an increase in the moment arm of one or both of these tendons. Anatomically, radial deviation is accompanied by the palpable palmar descent of the radial side of the carpus which, in effect, thrusts the FCR tendon in a palmar direction to optimize its moment arm for wrist flexion.

These observations of surface anatomy are enhanced by a simple anatomic experiment. On a cadaver specimen, the courses of the radial wrist extensor and flexor tendons are defined for radiographic identification by lengths of lead solder, which has been threaded along the course of each wrist extensor tendon to its insertion (Fig 15−3).

With the cadaver wrist in extension and radial deviation, a lateral radiograph documents the palmar displacement of the FCR tendon (Fig 15−4). The AP orientation of the long axis of the scaphoid, which defines the spacing of the lunate from the trapezium and trapezoid, acts to displace the radial side of the carpus and FCR tendon palmarward. In effect, wrist extension displaces the tendon away from the axis of rotation of the wrist to increase (optimize) its moment arm for the return motion of flexion. Note the lead marker for the ECRB on the dorsal side of the wrist.

As the wrist moves from radial extension to ulnar flexion, its radial side translates dorsally carrying with it the FCR in its fibro-osseous tunnel (Fig 15−5); this motion pro-

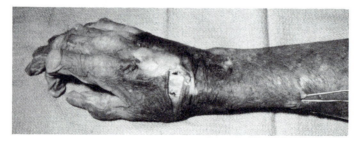

FIG 15−2.
Ulnar deviation decreases the AP diameter of the wrist. Note that the distal scaphoid and adjacent FCR tendon shift dorsally with the radial side of the carpus displacing this wrist flexor tendon closer to the wrist extensor tendons.

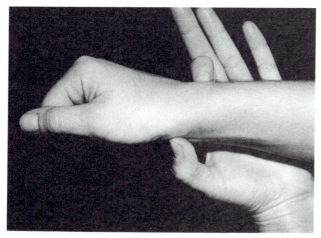

FIG 15–3.
Lead solder has been threaded along the course and through the pulleys for the radial wrist extensor tendons in an unpreserved cadaver specimen. The course of the FCR tendon is similarly instrumented on the palmar side of the specimen.

gressively decreases the moment arm of the FCR tendon for wrist flexion. At a wrist-neutral position, the scaphoid has moved precisely into position, immediately deep to the course of the radial wrist extensor tendons. With its convex dorsal radial surface accurately aligned with the adjacent margin of the radius, further flexion and ulnar deviation drapes these extensor tendons over the curved shape created by the radius and scaphoid. By this elegant maneuver, the scaphoid is repositioned to elevate the wrist

extensor tendons away from the center of the wrist, thereby preserving large moment arms for the return motion of the wrist into radial extension. The *broken line* in Figure 15–5 illustrates the direct course either wrist extensor would take from the second extensor compartment to the metacarpal bases were it not for the scaphoid that lifted its course *(solid line)* to preserve its moment arm.

Once appreciated, this concept exposes the functional limitations imposed by loss of

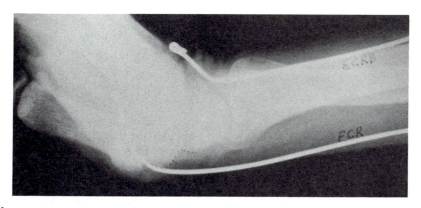

FIG 15–4.
With the wrist in extension and radial deviation, a lateral radiograph demonstrates the palmar displacement of the FCR tendon as the changing shape of the carpus separates this wrist flexor tendon from the main wrist extensor tendon, the ECRB. The *dotted line* defines the distal pole of the scaphoid. The wide separation of these two powerful wrist flexor and extensor tendons characterizes the large moment arms they use to stabilize the wrist. Note that the ECRB remains elevated from the dorsal side of the carpus as it extends from its distal radial pulley to insert on the metacarpal bases. This palpable bowstringing of tendons across an extended wrist serves to increase the moment arms and therefore the torques available to each radial wrist extensor muscle.

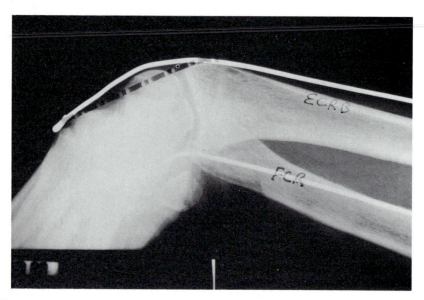

FIG 15–5.
The oblique radiograph of the flexed and ulnarly deviated wrist demonstrates the course of the ECRB tendon as it is draped over a "convex pulley" formed by the combined shape of the distal radius and the adjacent scaphoid. Were the scaphoid absent, this wrist extensor would take a straight course, illustrated by the *broken line,* from the radius to its insertion, with an associated decrease in its moment arm for wrist extension. On the palmar side, the FCR tendon is carried dorsally in its fibro-osseous tunnel as the wrist moves into ulnar deviation and flexion.

the scaphoid following proximal row carpectomy (Fig 15–6) or prosthetic replacement with currently available devices. This tendon-shift mechanism is also destroyed by partial wrist fusions that weld the scaphoid to one or several adjacent carpal bones, thereby preventing its important moment arm–defining function.

THE WRIST AS A GEARBOX: SHIFTING TENDONS WITH THE SCAPHOID

The value of the changing AP diameter of the wrist and its associated effect on the moment arms of the radial wrist flexor and extensor tendons is illustrated in Figure 15–7 (A–C). To throw a fastball, the pitcher's wind-up cocks the wrist into radial extension with the scaphoid moving into a relatively vertical position (perpendicular to a plane formed by the metacarpals). The FCR is carried palmarward, thereby securing a large moment arm for the wrist flexion necessary to start acceleration of the ball [(Fig 15–7 (A)]. As the acceleration progresses from (a) to (c), the FCR tendon exhibits a progressively smaller moment arm. In (c), the ball

receives a final push from the fingers as the FCR tendon shifts dorsally and into "high gear."

Analogous to a gearshift car accelerating from a stoplight, the acceleration of the ball is initiated with the wrist cocked back and its FCR slipped palmarward into "low gear" [see Fig 15–7 (a)]. As the wrist snaps toward ulnar flexion, the moment arm of the FCR progressively shortens, shifting its mechanical advantage from the low gear (long moment arm) necessary for the initial acceleration to the high gear (shorter moment arm) optimal for continued acceleration at higher velocities. Unlike the car, the wrist shifts the moment arms of its tendons smoothly without the inefficient jerking motion common to even skillful transmission shifting or automatic transmissions. The effect of the scaphoid on these radial wrist tendons is illustrated in Figure 15–8.

THE EXTENSOR COMPARTMENTS

The six extensor compartments control the mechanical effect of the radial

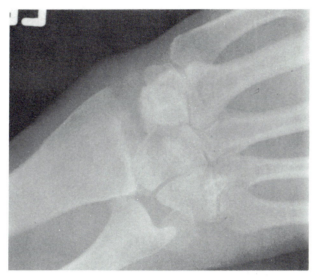

FIG 15–6.
Following proximal row carpectomy, the scaphoid is unavailable to optimize the moment arms of the tendons that normally cross its surface. Each increment of wrist flexion is therefore accompanied by an obligatory loss of its extensor moment arm as the unsupported course of the tendon allows it to fall progressively toward the center of the wrist. Similar moment arm deficiencies are exhibited by the FCR tendon.

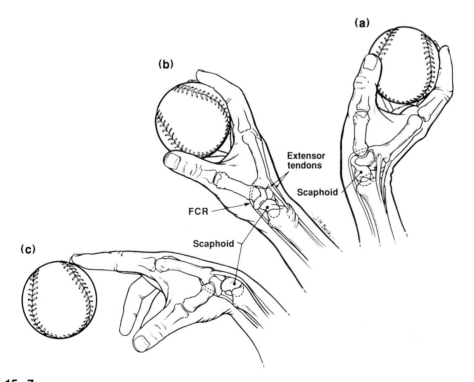

FIG 15–7.
Shifting tendons with the scaphoid. *(a)* With the wrist cocked back, the scaphoid moves palmarward into the "low gear" necessary to initiate acceleration. With acceleration, the moment arm of the FCR progressively shortens [*(b)* and *(c)*] to shift its mechanical advantage from the longest moment arm necessary for initial acceleration to the shorter moment arms or "higher gears" optimal for continued acceleration at higher velocities.

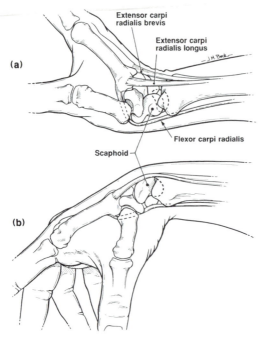

(a)

Extensor carpi
radialis brevis

Extensor carpi
radialis longus

J.H. Beck

Flexor carpi radialis

Scaphoid

(b)

FIG 15–8.
(a) The wrist extensors bowstring for maximal extensor moment arms while the vertical scaphoid carries the FCR tendon palmarward with the radial side of the carpus. *(b)* Wrist flexion creates a "convex pulley" from the combined shape of the distal radius and the repositioning of the scaphoid. This convex pulley lifts the radial wrist extensor tendons to preserve their moment arms for the return cycle into wrist extension. Simultaneously, the repositioned scaphoid moves the FCR tendon dorsally, shortening its moment arm to complement hand opening, precision digital functions, and so forth.

nerve–innervated wrist, finger, and thumb extensor muscles by defining the moment arms of their tendons with pulleys at the wrist (Figs 15–9 and 15–10). From observing and palpating the extensor tendons that cross your own wrist, note that during combined wrist and digital extension each wrist and digital tendon is restrained by the extensor retinaculum into a controlled bowstringing away from the surface of the carpal bones. Wrist flexion is accompanied by a smooth and intimate layering of the extensor tendons on the surface of the encapsulated carpal bones. In wrist flexion, the combined shape of the dorsal surface of the distal radius and proximal row of the carpus (scaphoid, lunate, and triquetrum) creates a "convex pulley" designed to lift the tendons away from the center of the wrist, thereby preserving optimal moment arms (Figs 15–5 and 15–11).

CONTROL BY THE SCAPHOID OF THE CARPAL TUNNEL TENDONS

With radial deviation, the scaphoid joins the trapezium to define the radial and dorsal radial wall of the carpal tunnel. The digital flexor tendons are forced in a palmar and ulnar direction as the distal half of the scaph-

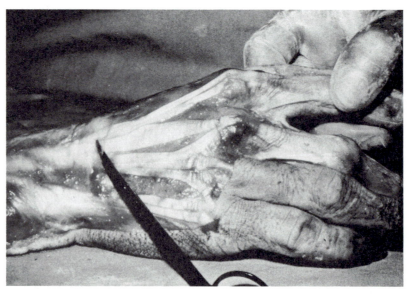

FIG 15–9.
The scissors tips define the distal end of the extensor retinaculum. Note that the extensor communis tendons diverge from their extensor compartment (pulley) to all four fingers. *(Courtesy of Daniel C. Riordan, M.D.).*

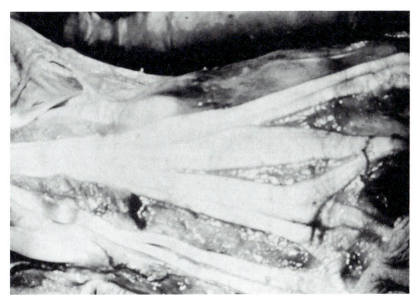

FIG 15–10.
The reflected extensor retinaculum exposes the extensor tendons precisely spaced in their pulley compartments. *(Courtesy of Daniel C. Riordan, M.D.).*

oid creates a vertical "finger" of bone around which the tendons angulate on their way to the digits (Fig 15–12). The mobile and now vertical scaphoid of the extended and radially deviated wrist forms a fulcrum for the carpal tunnel tendons on the radial side similar to that formed by the hook of the hamate on the ulnar side of the carpal tunnel. Note that in wrist extension, while the wrist extensor tendons maintain a large moment arm as they bowstring to their metacarpal insertions (see Figs 15–1 and 15–4), the repositioned scaphoid in effect pushes all nine carpal tunnel tendons toward the FCU. This wide separation of tendons transmitting force from these two powerful groups of muscles assures the large moment arms that each needs for optimal wrist stability.

CONCLUSION

Clinical assessment of wrist function requires an appreciation of wrist kinematics

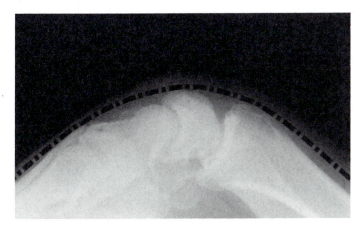

FIG 15–11.
The *broken line* depicts the course of the finger extensor tendons. With wrist flexion, the distal radius joins the dorsal prominence of the lunate to lift the finger extensor tendons dorsally to produce a relationship similar to that between the scaphoid and wrist extensor tendons.

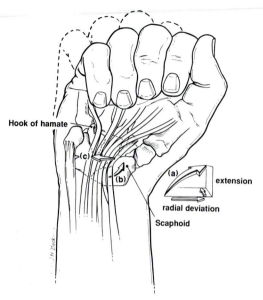

Hook of hamate

(c)

(b)

(a)

extension

radial deviation

Scaphoid

FIG 15–12.
Radial deviation in extension *(a)* positions the scaphoid as a vertical "finger" of bone to define the dorsal radial wall of the carpal tunnel as a fulcrum for the digital flexor tendons. By actively sweeping into a vertical position *(b)*, it displaces all nine digital flexor tendons toward the palmar ulnar border of the wrist *(c)*, thereby widely separating the wrist flexion effect of the digital flexors from the radial wrist extensor tendons. Wide separation of the high tension forces transmitted by these tendons is an essential element in creating the stable wrist necessary for forceful grasp by the digits.

that integrates the balance of tendon forces across the wrist with force transmission through the wrist. Of special interest and enormous importance are the kinematic mechanisms that continuously redefine the shape that the carpal bones present to the tendons that cross their surfaces as this re-shaping continuously rebalances forces at the wrist for optimal hand function. A working knowledge of these mechanisms is integral to the planning of reconstructive wrist surgery as well as to an appreciation of the limitations of currently available procedures such as limited wrist fusions and prosthetic joint replacements. In contrasting the function of a wrist with a normal scaphoid to that of one with a relatively immobile or even an absent scaphoid, I am hopeful that an audience of bright young clinicians and engineers will advance our understanding to the benefit of patients everywhere.

Index